TUNNELING PHENOMENA IN CHEMICAL PHYSICS

TUNNELING PHENOMENA IN CHEMICAL PHYSICS

By V.I. GOL'DANSKII,
Institute of Chemical Physics of the Academy of Sciences of the USSR, Moscow

L.I. TRAKHTENBERG,
Karpov Institute of Physical Chemistry, Moscow

and V.N. FLEUROV,
Institute of Chemical Physics of the Academy of Sciences of the USSR, Moscow

Translated from the Russian by
V.N. FLEUROV
Presently with the School of Physics and Astronomy, Tel Aviv University, Israel

GORDON AND BREACH SCIENCE PUBLISHERS
New York London Paris Montreux Tokyo Melbourne

Gordon and Breach Science Publishers

Post Office Box 786
Cooper Station
New York, New York 10276
United States of America

Post Office Box 197
London WC2E 9PX
England

58, rue Lhomond
75005 Paris
France

Post Office Box 161
1820 Montreux 2
Switzerland

3-14-9, Okubo
Shinjuku-ku, Tokyo
Japan

Private Bag 8
Camberwell, Victoria 3124
Australia

Library of Congress Cataloging-in-Publication Data

Gol 'danskiĭ, V. I. (Vitaliĭ Iosifovich)
[Tunnel 'nye iavleniia v khimicheskoi fizike. English]
Tunneling phenomena in chemical physics/V.I. Goldanskii, L.I. Trakhtenberg, V.N. Fleurov: translated from the Russian by V.N. Fleurov.
p. cm.
Translation of: Tunnel 'nye iavleniia v khimicheskoi fizike.
Bibliography: p.
Includes index.
ISBN 2-88124-655-9
1. Chemical reaction, Conditions and laws of. 2. Tunneling (Physics) I. Trakhtenberg, L. I. (Leonid Izraĭlevich) II. Flërov, V. N. (Viktor Nikolaevich) III. Title.
QD501.G62613 1988
541'.0421--dc19 88-2840
CIP

 Printed in Great Britain by Billing & Sons Ltd, Worcester

Contents

Preface

One of the most important manifestations of the wave properties of matter is tunneling — the ability of quantum particles to penetrate potential barriers whose height exceeds the kinetic energy of the particles, and hence to pass through the region beneath the barrier, which is absolutely forbidden to them according to classical mechanics. This fundamental property of quantum particles was realized in practice immediately after the basic principles of quantum mechanics had been formulated and a theory of nuclear decay arose. Since then, the concept of tunneling has found wide application in different fields of nuclear and elementary particle physics, in the physics and electronics of solids, and even in cosmology.

Quantum mechanics also led to a revolution in chemistry — atomic spectra and wave functions were explained and quantitatively described, notions such as chemical bonding and potential barriers were explained on a microscopic basis, and it became possible to formulate a quantitative microscopic theory of chemical reactions. The question naturally arose in the 1920s as to the possibility of non-Arrhenius type chemical reactions accompanied by underbarrier tunneling transitions of the reagents.

For a long time the development of this field came through the study of the simplest theoretical models and the search for weak deviations from the Arrhenius law in gas- and liquid-phase reactions carried out at relatively high temperatures. A detailed account of these efforts can be found in Bell's monograph.†

A new stage in the development followed from the study of low-temperature solid-phase chemical reactions. In the experimental field it was marked by the discovery of a low-temperature plateau of the chemical reaction rate constant. This was first found for formaldehyde polymerization, and later on in a number of other reactions. Theoretical work passed from the study of the simplest models of tunneling to a description of tunneling chemical reactions within the framework of the theory of radiationless electronic transitions. It is our opinion that the realization of the principal role played by intermolecular vibrations, in addition to the intramolecular vibrations considered by the conven-

† Bell, R.P. *The Tunnel Effect in Chemistry* (Chapman and Hall, London, 1980)

tional theoretical formulation of radiationless transitions, was a very important factor.

The development of the field has now reached such a level that there is the need for a new summary of the subject that considers from a unified point of view numerous theoretical and experimental results. This is how we conceived our principal task in writing this book. We also realized that, besides chemical reactions, the tunneling effect is of great importance in a number of other physical, physico-chemical and biological phenomena. That is why an appreciable portion of the book is devoted to issues such as kinetic processes in amorphous solids, quantum diffusion of atoms, surface phenomena in quantum crystals and dynamic properties of biopolymers. The joint consideration of these, at first sight, diverse issues with chemical reactions appears to be quite fruitful. The fact that all the problems deal with tunneling of heavy (of the order of the atomic mass) particles allows a unified point of view to be taken in the presentation of the material and some common properties and characteristics to be perceived in a wide range of phenomena.

Naturally, the selection of the material and the style of its presentation reflects our own interests and preferences. To a certain extent these also affect the selection of references. It was not our aim to present a comprehensive review of each issue touched upon in the book. We restricted ourselves to listing only the most important original papers containing principally new ideas or experimental results. We were able to account for results published before 1984, together with a few from that year.

The book contains ten chapters. After the presentation of general ideas on tunneling in Chapter one, the following four chapters are devoted to the low-temperature chemical reactions that may proceed owing to tunneling effects. Chapter 6 considers tunneling effects in amorphous materials. Chapter 7 presents quantum diffusion and surface phenomena in quantum crystals. Chapters 8 and 9 discuss hopping diffusion and tunnel scavenging of electrons, and stand somewhat by themselves. Chapter 10 briefly considers tunneling effects in biological systems.

The authors are deeply indebted to Drs Yu.A. Berlin, V.L. Klochikhin, A.I. Mikhailov and N.I. Chekunaev, who took the trouble to look through the book and made a number of valuable comments, and to Mrs A. Lekhnitskaya for her help in producing the figures. One of the authors, V.N. Fleurov, is presently with the school of Physics and Astronomy at the University of Tel Aviv, Israel.

CHAPTER 1
Basic Ideas in Tunneling

A quantum-mechanical problem of a particle tunneling through a potential barrier is considered in this chapter. It is emphasized that the possibility for the particle to penetrate a classically inaccessible region in which the potential-barrier height is larger than the total energy of the particle is due to the particle's wave properties. A similarity is demonstrated with the optical phenomenon of total internal reflection.

Three main types of particle tunneling are distinguished:

(i) penetration of the potential barrier by a free particle;
(ii) tunneling transition of a bound particle into the continuous spectrum;
(iii) tunneling transition of a bound particle into a bound state.

As an example, the transmission coefficient is derived for a one-dimensional rectangular barrier. To calculate the transmission coefficient for a potential barrier of arbitrary shape, use is made of the quasiclassical approximation.

Tunneling transitions of particles in the continuous spectrum are treated in many books on quantum mechanics (see e.g. [1–3]); hence in the present work most attention is paid to the qualitative features of the problem, and the transmission coefficient is derived only for illustration purposes.

Quasistationary states and tunneling from a bound to a free state have not been considered so widely in the literature, so this problem is treated here in more detail.

The problem of particle tunneling from one potential well into another (the case of greatest importance for chemical kinetics) is not properly presented in most monographs and textbooks on quantum mechanics (an exception is [4], which does discuss tunneling between two potential wells). For this reason we give most attention here to tunneling between two bound states.

A problem of tunneling in periodically shaped potential wells is also considered.

This chapter concludes with a study of tunneling transitions in disordered systems and of some basic ideas concerning Anderson localization.

1.1. TRANSMISSION AND REFLECTION COEFFICIENTS

If a particle is described quantum-mechanically then it can participate in processes that are forbidden by the laws of classical mechanics. Among these is the possibility for a quantum particle to penetrate through classically inaccessible regions where the total energy E of the particle is smaller than the potential energy U.

We begin our consideration of this phenomenon by treating a simple case: a particle propagating in the one-dimensional potential shown in Fig. 1.1, where the potential energy depends only on a single coordinate x. A classical particle with $E < U_0$ is restricted to region I $(-\infty < x < 0)$, and, on moving from left to right, it is elastically reflected by the potential wall. The particle may not penetrate region II $(0 \leqslant x < \infty)$, since if it did then this would imply a negative kinetic energy. For $E > U_0$ the particle can be at any spatial point, and, on passing from one region to the other, the particle velocity changes without changing direction.

Quantum mechanics describes the particle motion by a Schrödinger equation, which for the present case is conveniently written separately for each of the above regions:

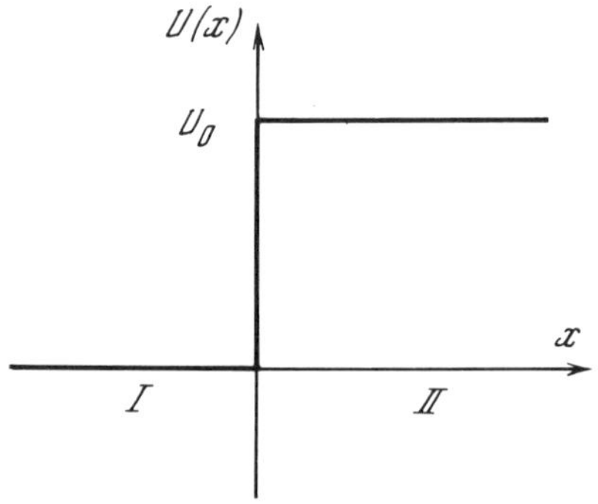

Figure 1.1 Rectangular one-dimensional wall.

$$\left.\begin{aligned} -\frac{\hbar^2}{2m}\frac{d^2\Psi(x)}{dx^2} &= E\Psi(x) && (x < 0), \\ -\frac{\hbar^2}{2m}\frac{d^2\Psi(x)}{dx^2} + U_0\Psi(x) &= E\Psi(x) && (x \geqslant 0). \end{aligned}\right\} \tag{1.1}$$

Here m is the particle mass, $\hbar$ is Planck's constant divided by 2π, and $\Psi(x)$ is the wave function of the particle. The solution of this equation in region I is

$$\Psi_I(x) = e^{ik_Ix} + Ae^{-ik_Ix}, \quad k_I = \frac{1}{\hbar}(2mE)^{1/2}. \tag{1.2}$$

The first term represents a plane wave propagating in the positive direction of the x-axis, which corresponds to the particle incident on the wall. Here the wave function normalization is chosen in such a way that the probability current density $j = (i\hbar/2m)\,[\Psi(d\Psi^*/dx) - \Psi^*(d\Psi/dx)]$ in the incident wave is equal to the particle velocity. The coefficient in the first term in (1.2) is then unity. The second term in (1.2) corresponds to the particle reflected by the wall; it represents plane wave propagating in the negative direction of the x-axis.

Solutions of (1.1) in region II can be written as

$$\Psi_{II}(x) = Be^{-\kappa x} \qquad (E < U_0), \tag{1.3}$$

$$\Psi_{II}(x) = B'e^{ik_{II}x} \qquad (E > U_0), \tag{1.3'}$$

$$\kappa = \frac{1}{\hbar}[2m(U_0 - E)]^{1/2}, \quad k_{II} = \frac{1}{\hbar}[2m(E - U_0]^{1/2}.$$

The increasing term $\sim e^{\kappa x}$ is omitted from the solution (1.3), since the wave function must be finite in the whole space. For $E > U_0$ the wave function in region II describes a particle propagating in the positive direction of the x-axis. It is obvious that there is no reflected wave in region II, so only the term corresponding to the transmitted wave is retained in the solution (1.3).

The constants A and B are determined by continuity of the wave function and its derivative at the boundary of regions I and II:

$$A = \frac{k_I - i\kappa}{k_I + i\kappa}, \quad B = \frac{2k_I}{k_I + i\kappa} \qquad (E < U_0), \tag{1.4}$$

$$A = \frac{k_I - k_{II}}{k_I + k_{II}}, \quad B' = \frac{2k_I}{k_I + k_{II}} \qquad (E > U_0), \tag{1.4'}$$

It can easily be seen that even for $E < U_0$ the wave function in region II is non zero, and hence there is a nonzero probability of finding the particle in the classically forbidden region; the corresponding probability density is

$$|\Psi_{II}(x)|^2 = \frac{4k_I^2 \exp[-2\kappa x]}{k_I^2 + \kappa^2}. \tag{1.5}$$

The probability density decays exponentially with x, and the penetration depth, at which the particle can be still found with a finite probability, is of the order of $1/\kappa = \lambda/2\pi$, where λ is the de Broglie wavelength of the particle. Thus the possibility for the particle to penetrate the classically forbidden results from its wave properties and is an essentially quantum effect.

Transmission and reflection coefficients are determined in the following way. The transmission coefficient D_0 is the ratio of the probability current density of the transmitted wave to that of the incident wave. The reflection coefficient R is the ratio of the reflected current flow to the incident one. In other words, if a stationary current of particles is incident upon a potential barrier then the ratio of the transmitted and incident currents is D_0, while the ratio of the reflected and incident currents is R.

For $E < U_0$ the reflection coefficient is unity and the transmission coefficient is zero, as in the classical case. The difference, as mentioned above, is just that in the quantum case the particle can enter region II. However, this difference results in a nontrivial consequence: a quantum particle can penetrate a finite potential barrier. This can be qualitatively understood in the following way. If this barrier width a_0 is finite then there is a finite probability density ($|\Psi_{II}(a_0)|^2$) of finding the particle to the right of the barrier (see Fig. 1.2), i.e. the particle can transverse the classically inaccessible region with a probability that decays exponentially with the barrier width (see (1.5)). A new wave

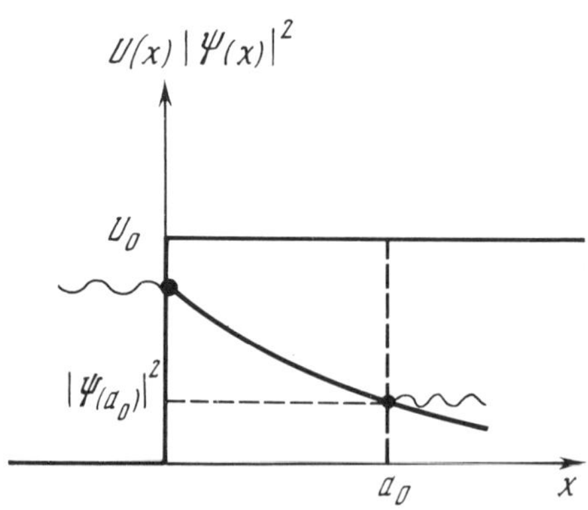

Figure 1.2 Particle penetrating through a rectangular one-dimensional potential barrier.

originates from the right-hand boundary of the potential barrier, the wave source being determined by the probability amplitude of finding the particle at the point a_0.

So there is a nonzero current density in the positive direction of the x-axis to the right of the barrier region, which results in a finite transmission coefficient. In the classical description the transmission coefficient is zero.

The crossing of a classically forbidden region plays an important part in a number of physical and chemical processes, and many theoretical and experimental studies have dealt with this phenomenon. Nevertheless, direct measurements of the transmission-coefficient dependence on the barrier width have been carried out only recently. It was observed in ref. [5] how electrons overcome a vacuum gap between two surfaces, the electron current, which is determined by the transmission coefficient, decreasing exponentially with the growing gap.

A similar phenomenon takes place in optics in the case of total internal reflection. According to geometrical optics, light does not penetrate the second medium when the angle of incidence φ at the interface of two media exceeds a certain value (the Brewster angle), which is determined by the condition $\sin \varphi_0 = n_2/n_1$ ($n_{1,2}$ are the refractive indeces of the first and second media respectively). According to wave optics, an electromagnetic wave originates in the second medium and moves along the interface (see Fig. 1.3). The amplitude of this wave

$$\mathbf{E} = \mathbf{E}_0 \exp\left\{ -\frac{2\pi}{\lambda_0} \times \left[\left(\frac{n_1}{n_2} \sin \varphi \right)^2 - 1 \right]^{1/2} \right\} \tag{1.6}$$

decreases rapidly on penetrating the second medium (cf. (1.5)). Here λ_0 is the electromagnetic wavelength, E_0 is the amplitude of the incident wave.

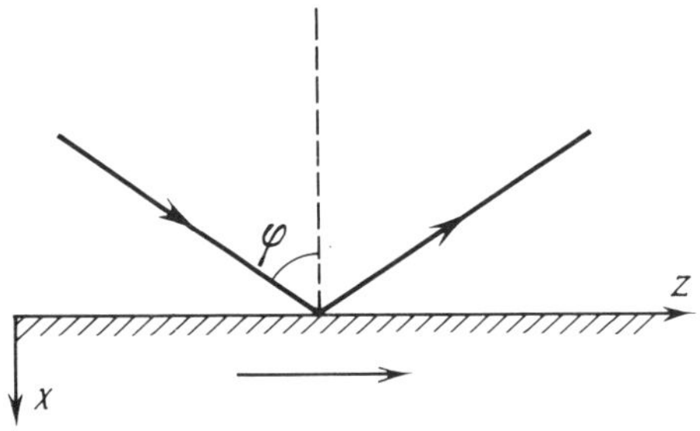

Figure 1.3 Propagation of the electromagnetic wave in the case of complete internal reflection.

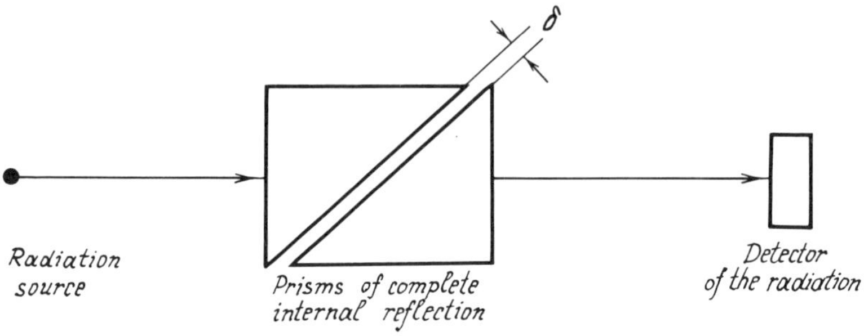

Figure 1.4 A scheme of the experiment revealing the refracted wave in the case of complete internal reflection.

One can record the wave (1.6) by means of two prisms of total internal reflection as shown in Fig. 1.4. There is a gap between the prisms, and the signal recorded by the light detector varies with the gap width δ. When the gap width increases, the signal decreases, and at $\delta \gg \lambda_0$ it becomes exponentially small according to (1.6).

There is thus an analogy between the optical and mechanical phenomena. The penetration of particles and light into regions forbidden by the laws of classical mechanics or geometrical optics is allowed when wave properties are taken into account.

It should also be emphasized that, as follows from (1.4′) at $E > U_0$, quantum mechanics results in a nonzero reflection coefficient. This is also a purely quantum effect, which does not appear from classical considerations. Below, however, we shall be interested only in the case $E < U_0$.

The transmission of a particle through a one-dimensional rectangular potential barrier of finite width is considered. When solving the Schrödinger equation, three regions are distinguished (see Fig. 1.5) and, similarly to the infinitely wide barrier case, the wave function of the particle is looked for in the form

$$\left.\begin{aligned} \Psi_I(x) &= e^{ik_I x} + A_I e^{-ik_I x} && (-\infty < x < 0), \\ \Psi_{II}(x) &= B_{II} e^{-\kappa x} + A_{II} e^{\kappa x} && (0 \leqslant x \leqslant a_0), \\ \Psi_{III}(x) &= B_{III}\, e^{ik_I x} && (a_0 < x < \infty), \end{aligned}\right\} \tag{1.7}$$

where k_I and κ are defined in (1.2) and (1.3).

By matching the wave function and its derivative at the region boundaries, the constants A_I, A_{II}, B_{II}, B_{III} can be found. B_{III} yields the transmission coefficient

$$D_0 = \frac{4k_I^2\kappa^2}{(k_I^2 + \kappa^2)\sinh^2\kappa a_0 + 4k_I^2\kappa^2} \tag{1.8}$$

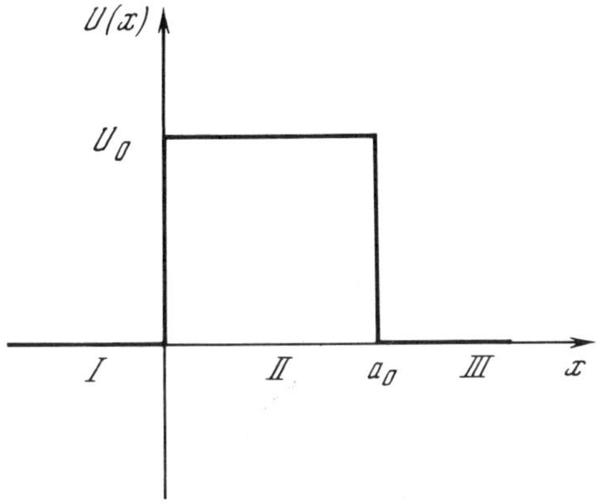

Figure 1.5 Rectangular one-dimensional barrier.

If the condition $\kappa a_0 \gg 1$ holds then $D_0 \sim e^{-2\kappa a_0}$, and the probability of particle propagation through the barrier depends exponentially on the barrier width and height.

The phenomenon described above, i.e. the propagation of the particle through the potential barrier, is called the *tunneling effect*.

1.2. THE TRANSMISSION COEFFICIENT IN THE QUASICLASSICAL APPROXIMATION

We now consider a particle tunneling through a barrier of arbitrary shape (Fig. 1.6). Let the potential energy change from one constant value ($U_{-\infty}$ at $x \to -\infty$) to another one (U_∞ at $x \to +\infty$). The transmission coefficient can be found by calculating the solutions of the corresponding Schrödinger equation, whose asymptotic form reads

$$\Psi_{-\infty}(x) = e^{ikx} + Ae^{-ikx} \qquad (x \to -\infty), \tag{2.1}$$

$$\Psi_{-\infty}(x) = Be^{ik'x} \qquad (x \to \infty), \tag{2.1'}$$

$$k = \frac{1}{\hbar}[2m(E - U_{-\infty})]^{1/2}, \quad k' = \frac{1}{\hbar}[2m(E - U_\infty)]^{1/2}, \quad E > U_{-\infty}, U_\infty.$$

Using the incident and transmitted currents j_0 and j_D respectively, the transmission coefficient is obtained as

$$D_0 = j_D / j_0. \tag{2.2}$$

The exact wave function of the system can be found only for some special functions U(x). There is, however, a technique for finding

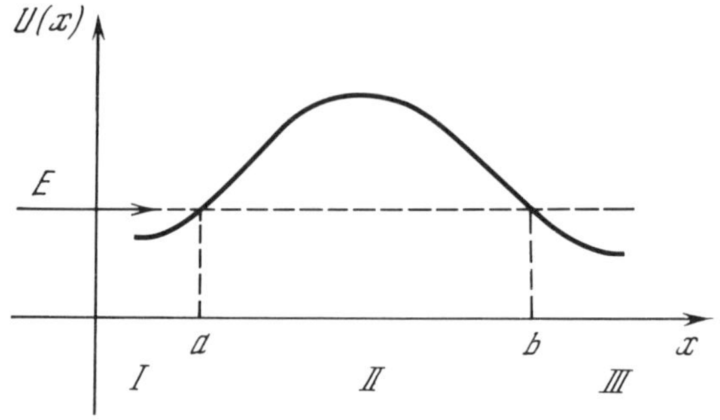

Figure 1.6 One-dimensional barrier.

approximate solutions of the Schrödinger equation when the potential is sufficiently smooth (i.e. is effectively constant over the particle wavelength); for the rectangular potential barrier the results obtained by this technique coincide with the exact ones. This is the quasiclassical (or WKB) approximation, which is widely applied in the theory of tunneling transitions. We shall not dwell upon the fundamentals of this technique or on the derivation of the formulae for the wave functions, since detailed accounts can be found in every textbook on quantum mechanics (e.g. [1–3]). Therefore only a derivation of the transmission coefficient will be carried out.

If the quasiclassical condition $(1/2\pi)\ |\partial\tilde{\lambda}/\partial x| \ll 1$, where $\tilde{\lambda} = 2\pi\hbar\ [2m\ (E - U)]^{-1/2}$, holds then the particle behavior to the left of the point $x = a$ (see Fig. 1.6) is described by two linearly independent wave functions

$$\left.\begin{aligned} \Psi_I(x) &= \frac{2C}{p^{1/2}} \cos\left(\frac{1}{\hbar}\int_x^a p\,dx - \frac{\pi}{4}\right), \\ \Psi_I'(x) &= \frac{C'}{p^{1/2}} \cos\left(\frac{1}{\hbar}\int_x^a p\,dx + \frac{\pi}{4}\right). \end{aligned}\right\} \tag{2.3}$$

To the right of the point $x = a$ the corresponding solution is

$$\left.\begin{aligned} \Psi_{II}(x) &= \frac{C}{p^{1/2}} \exp\left[-\frac{1}{\hbar}\int_a^x |p|\,dx\right], \\ \Psi_{II}'(x) &= \frac{C'}{p^{1/2}} \exp\left[\frac{1}{\hbar}\int_a^x |p|\,dx\right]. \end{aligned}\right\} \tag{2.4}$$

Similarly, there are linearly independent solutions for the regions to the right and left of the point x = b:

$$\left.\begin{aligned} \Psi_{III}(x) &= \frac{2D}{p^{1/2}} \cos\left(\frac{1}{\hbar}\int_b^x p\,dx - \frac{\pi}{4}\right), \\ \Psi'_{III}(x) &= \frac{D'}{p^{1/2}} \cos\left(\frac{1}{\hbar}\int_b^x p\,dx + \frac{\pi}{4}\right), \end{aligned}\right\} \tag{2.5}$$

and

$$\left.\begin{aligned} \Psi_{II}(x) &= \frac{D}{p^{1/2}} \exp\left[-\frac{1}{\hbar}\int_x^b |p|\,dx\right], \\ \Psi'_{II}(x) &= \frac{D'}{p^{1/2}} \exp\left[\frac{1}{\hbar}\int_x^b |p|\,dx\right]. \end{aligned}\right\} \tag{2.6}$$

The notation $p = [2m\,(E - U)]^{1/2}$ is used in (2.3)–(2.6).

If the particle propagates from left to right then the asymptotic value of the wave function in region III is given by (2.1). A linear superposition of the functions (2.5) describes a wave propagating in the positive direction of the x-axis only when the condition $D' = -2iD$ holds.

Keeping in mind the relation found for D and D′, the wave function to the left of the point x = b is represented as a superposition of solutions of (2.6):

$$\Psi(x) = \frac{D}{p^{1/2}}\left\{\exp\left[-\frac{1}{\hbar}\int_x^b |p|\,dx\right] + \frac{2}{i}\exp\left[\frac{1}{\hbar}\int_x^b |p|\,dx\right]\right\}, \tag{2.7}$$

The particle state in region II can also be described using a superposition of the solutions (2.4). This allows the constants C and C′ to be written in terms of D:

$$\left.\begin{aligned} &C = -2iD\exp[S_0], \quad C' = D\exp[-S_0], \\ &S_0 = \frac{1}{\hbar}\int_a^b |p|\,dx. \end{aligned}\right\} \tag{2.8}$$

Using the (2.8) and (2.3), the wave function in region I is represented as a superposition of the incident and reflected waves:

$$\Psi(x) = \frac{2D}{p^{1/2}}\left(e^{S_0} + \frac{1}{4}e^{-S_0}\right)\exp\left[-i\left(\frac{1}{\hbar}\int_x^a p\,dx + \frac{\pi}{4}\right)\right]$$
$$+ \frac{2D}{p^{1/2}}\left(e^{S_0} - \frac{1}{4}e^{-S_0}\right)\exp\left[i\left(\frac{1}{\hbar}\int_x^a p\,dx + \frac{3\pi}{4}\right)\right]. \quad (2.9)$$

Then the current densities for the incident and transmitted waves are calculated and substituted into (2.2), resulting in the transmission coefficient

$$D_0 = \frac{1}{\left(e^{S_0} + \frac{1}{4}e^{-S_0}\right)^2}. \quad (2.10)$$

If the condition $e^{S_0} \gg 1$ holds (which is quite often in the chemical kinetics of tunneling) then

$$D_0 \approx \exp\left[-\frac{2}{\hbar}\int_a^b |p|\,dx\right]. \quad (2.11)$$

Thus the exponential dependence of the transmission coefficient on the potential-barrier parameters obtained for the rectangular barrier holds also for arbitrarily shaped barriers.

Equation (2.11) has been derived under the assumption that the quasiclassical condition holds; however, this equation remains valid to within the preexponential factor even if the potential energy varies at one of the barrier boundaries so rapidly that the quasiclassical condition breaks down [1].

Several examples of calculations of particle transversal of a potential barrier are now discussed.

(i) Triangular barrier:

$$U(x) = U_0 + F_0 x \qquad (-U_0/F_0 \leqslant x \leqslant 0),$$
$$U(x) = U_0 - F_0 x \qquad (0 \leqslant x \leqslant U_0/F_0),$$
$$U(x) = 0 \qquad (|x| > U_0/F_0).$$

The exponential in (2.11) is easily calculated:

$$D_0 = \exp\left[-\frac{8\,(2m)^{1/2}}{3\hbar\,F_0}(U_0 - E)^{3/2}\right].$$

(ii) Parabolic barrier:

$$U(x) = U_0 - \frac{1}{2} k_0 x^2.$$

The following transmission coefficient is obtained:

$$D_0 \approx \exp\left[-\frac{2\pi}{\hbar}\left(\frac{m}{k_0}\right)^{1/2}(U_0 - E)\right]. \tag{2.12}$$

(iii) Symmetrical version of the so-called Eckart barrier:

$$U(x) = U_0/\cosh^2 k_0 x.$$

On calculating the integral in (2.11), the transmission coefficient is

$$D_0 \approx \exp\left[-\frac{2\pi}{\hbar\, k_0}(2m)\,(U_0^{1/2} - E^{1/2})\right]. \tag{2.13}$$

It should be mentioned that the Schrödinger equation for the above examples can be solved exactly, and hence the transmission coefficient can be found without applying the quasiclassical approximation (see e.g. [6]).

1.3. TUNNELING TRANSITION OF A PARTICLE FROM A BOUND TO A FREE STATE

We now consider tunneling when a particle is in the potential depicted in Fig. 1.7. A particle that is initially in a bound state penetrates the potential barrier and ends up in the continuous spectrum. The problem arises of how to calculate the probability per unit time for the particle to leave the potential well. This problem differs from that discussed

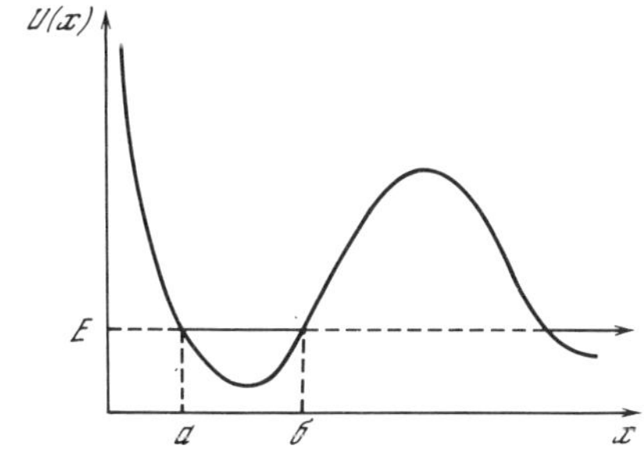

Figure 1.7 Potential energy curve corresponding to the particle transition from the bound to a free state.

above, since the question is how the particle leaves a restricted space domain rather than how it is incident from infinity onto a potential barrier. The problem of particle transmission through the potential barrier is essentially nonstationary. To consider this problem in more detail, the time-dependent Schrödinger equation for the system is written:

$$i\hbar \frac{\partial \Psi}{\partial t} = -\frac{\hbar^2}{2m} \frac{\partial^2 \Psi}{\partial x^2} + U(x)\,\Psi. \tag{3.1}$$

The equation for Ψ^* is

$$i\hbar \frac{\partial \Psi^*}{\partial t} = \frac{\hbar^2}{2m} \frac{\partial^2 \Psi^*}{\partial x^2} - U(x)\,\Psi^*(x). \tag{3.2}$$

Multiplication of (3.1) by Ψ^*, and (3.2) by Ψ, and summation of the resulting equations gives

$$i\hbar \frac{\partial |\Psi|^2}{\partial t} = \frac{\hbar^2}{2m} \left[\Psi \frac{\partial^2 \Psi^*}{\partial x^2} - \Psi^* \frac{\partial^2 \Psi}{\partial x^2} \right]. \tag{3.3}$$

Both sides of (3.3) are integrated from $-\infty$ to a point $x = x_0$ that is sufficiently far from the barrier (see Fig. 1.7), simple manipulation then give

$$\frac{\partial}{\partial t} \int_{-\infty}^{x_0} |\Psi|^2\, dx = \frac{\hbar}{2mi} \left[\Psi \frac{\partial \Psi^*}{\partial x} - \Psi^* \frac{\partial \Psi}{\partial x} \right] \Bigg|_{-\infty}^{x_0}. \tag{3.4}$$

Using the conditions that $\Psi(-\infty) = 0$ and that the asymptotics of the wave function to the right of the barrier must correspond to a wave propagating in the positive direction of the x-axis, i.e. $\Psi(x) \sim B\, e^{ikx}$ (see (2.1′)), (3.4) becomes

$$\frac{\partial}{\partial t} \int_{-\infty}^{x_0} |\Psi|^2\, dx = -\frac{\hbar k}{m} |B|^2 < 0. \tag{3.5}$$

It can therefore be seen that the probability of finding the particle within the domain $(-\infty, x_0)$ decreases with time, and hence there are no stationary solutions of the Schrödinger equation.

A nonstationary solution is looked for as an expansion

$$\Psi(x,t) = \int_0^\infty dE\; c(E)\, \Psi_E(x)\, e^{-iEt/\hbar}, \tag{3.6}$$

where $\Psi_E(x)$ are δ-normalized eigenfunctions of the operator

$$\hat{H} = -\frac{\hbar^2}{2m}\frac{\partial^2}{\partial x^2} + U(x). \tag{3.7}$$

The initial condition $\Psi(x,0) = \Psi_0(x)$ is chosen in such a way that the particle at $t = 0$ is localized in the potential well.

It is easy to find the function c(E) by multiplying both sides of (3.6) by $\Psi^*_{E'}(x)$ and integrating over x from $-\infty$ to $+\infty$. Then use of the initial condition at $t = 0$ gives

$$c(E) = \int_{-\infty}^{\infty} dx\, \Psi_0(x)\, \Psi^*_E(x). \tag{3.8}$$

The probability amplitude for finding the particle in the initial state is

$$a(t) = \int_{-\infty}^{\infty} dx\, \Psi^*_0(x)\, \Psi(x,t). \tag{3.9}$$

The probability per time unit for the particle to leave the potential well is then

$$\dot{W} = -\frac{d}{dt}\,|a(t)|^2. \tag{3.10}$$

Hence the problem is solved if the eigenfunctions of the Hamiltonian (3.7) can be found.

There is another technique for calculating the probability of a particle tunneling from a bound state to the continuous spectrum. A stationary particle flow from the potential well is assumed. The probability of finding the particle in a bound state is experimentally known to be of the form $W(t) \sim \exp[-\gamma t]$ (e.g. radiation decay law) and the equation $W(t) \sim \int_a^b dx\, |\Psi|^2$ (the integration limits a and b are defined in Fig. 1.7) suggests that a wave function be sought of the form

$$\Psi(x,t) = \Psi(x) \exp\left[-\frac{i\tilde{E}t}{\hbar} - \frac{\gamma t}{2}\right]. \tag{3.11}$$

This allows separation of variables, and substitution of (3.11) into (3.1) then gives

$$-\frac{\hbar^2}{2m}\frac{\partial^2\Psi(x)}{\partial x^2} + U(x)\,\Psi(x) = \left(\tilde{E} - \frac{i\hbar\gamma}{2}\right)\Psi(x). \tag{3.12}$$

As will be shown below, the quantity γ determines the width of the quasistationary level, and for $\gamma t \ll 1$ coincides with the probability of tunneling per unit time. In order to calculate γ and acquire a notion of the properties of quasistationary states, this problem is studied for the simplest dependence of the potential energy on coordinate,

$$U(x) = \begin{cases} \infty & (x < 0), \\ 0 & (0 \leqslant x < a), \\ U_0 & (a \leqslant x \leqslant b), \\ 0 & (x > b). \end{cases} \tag{3.13}$$

As usual, the Schrödinger equation is treated separately for each region:

$$\left.\begin{aligned} &\frac{\partial^2\Psi(x)}{\partial x^2} + \tilde{k}^2\Psi(x) = 0 \qquad (0 \leqslant x < a), \\ &\frac{\partial^2\Psi(x)}{\partial x^2} + \tilde{\kappa}^2\Psi(x) = 0 \qquad (a \leqslant x \leqslant b), \\ &\frac{\partial^2\Psi(x)}{\partial x^2} + \tilde{k}^2\Psi(x) = 0 \qquad (x > b), \end{aligned}\right\} \tag{3.14}$$

$$\tilde{k}^2 = \frac{2m}{\hbar^2}\left(\tilde{E} - \frac{i\gamma}{2}\right), \quad \tilde{\kappa}^2 = \frac{2m}{\hbar^2}\left(U_0 - \tilde{E} + \frac{i\gamma}{2}\right).$$

The solutions are

$$\left.\begin{aligned} &\Psi(x) = A_I\, e^{i\tilde{k}x} + B_I\, e^{-i\tilde{k}x} \qquad (0 \leqslant x < a), \\ &\Psi(x) = A_{II}\, e^{\tilde{\kappa}x} + B_{II}\, e^{-\tilde{\kappa}x} \qquad (a \leqslant x \leqslant b), \\ &\Psi(x) = A_{III}\, e^{i\tilde{k}x} \qquad (x > b). \end{aligned}\right\} \tag{3.15}$$

Since the potential energy at $x = 0$ becomes infinite, the wave function $\Psi(0) = 0$ and $A_I = -B_I$. Other conditions equating the wave functions and their derivatives at the region boundaries result in a transcendental equation determining the allowed values of $\tilde{E}$ and γ:

$$\frac{i\tilde{k} + \tilde{\kappa}}{i\tilde{k} - \tilde{\kappa}}\left(\frac{\tilde{\kappa}}{\tilde{k}}\tan\tilde{k}a - 1\right) e^{-\tilde{\kappa}(b-a)} + \left(\frac{\tilde{\kappa}}{\tilde{k}}\tan\tilde{k}a + 1\right) e^{\tilde{\kappa}(b-a)} = 0. \tag{3.16}$$

Under the condition $\tilde{\kappa}(b - a) \gg 1$, corresponding to a strong under-barrier decay of the wave function, the quantity γ is small, and (3.16) can be solved using an iterative technique.

Neglecting the first term in (3.6) as it is proportional to the small factor $e^{-\tilde{\kappa}(b-a)}$, the zeroth-order approximation is obtained. Then the solutions k_0 correspond to the eigevalues for the potential well derived from (3.13) at $b = \infty$.

Expanding the term in (3.16) proportional to $e^{\tilde{\kappa}(b-a)}$ in powers of $k - k_0$ and substituting $k = k_0$ into the term $\sim e^{-\tilde{\kappa}(b-a)}$, an equation for k is obtained. Using (3.14), another expression is found for k; then equating imaginary parts of this and the above expressions gives

$$\gamma = \frac{8\hbar\, \tilde{k}_0^3 \tilde{\kappa}_0}{m(\tilde{\kappa}_0^2 + \tilde{k}_0^2)(1 + \tilde{k}_0 a)} \exp[-2\tilde{k}_0(b - a)], \quad \tilde{\kappa}_0 = \tilde{\kappa}(\tilde{k}_0) \tag{3.17}$$

It is easily seen that for $U_0 \gg E$ the inequality $\tilde{\kappa}_0 a \gg 1$ holds. This inequality, together with $\hbar\tilde{k}_0/m = v$ (v is the particle velocity in the potential well), from (3.17) and (1.8) that

$$\gamma = \nu\, D_0, \quad \nu = v/2a. \tag{3.18}$$

This formula has a simple physical meaning — the probability of a tunneling transition per unit time equals the product of the transmission coefficient D_0 and the frequency ν of particle "impacts" on the wall of the potential well. Calculations using this formula can be performed when dealing with an arbitrarily shaped barrier.

It can easily be verified that γ is really the probability per unit time for the particle to leave the potential well. Substituting (3.11) into the integral (3.9) and assuming $\int dx\, \Psi_0^*(x)\, \Psi(x) \approx 1$, gives

$$a(t) = \exp[-\tfrac{1}{2}\gamma t] \exp[-i\tilde{E}t/\hbar]. \tag{3.19}$$

Next (3.19) and (3.10) result in the following probability for the tunneling transition:

$$\dot{W}(t) = \gamma \exp[-\gamma t]. \tag{3.20}$$

Then for a small time span ($\gamma t \ll 1$), $\dot{W} \approx \gamma$. The quantity $\tau = 1/\gamma$ is called the life time of the quasistationary level.

The state described by the wave function (3.11) is not a state with definite energy. There is an energy distribution of particles, i.e. there is a probability $\rho_E\, dE = |c(E)|^2\, dE$ (see (3.6)) of finding the particle in the range E to E + dE. To calculate the function ρ_E, the expressions for $\Psi_0^*(x)$ and $\Psi(x,t)$ are substituted into (3.9); then using the orthogonality of the functions $\Psi_E(x)$ gives

$$a(t) = \int_0^\infty dE\, \rho_E \exp[-iEt/\hbar]. \tag{3.21}$$

In order for the amplitude to be of the form (3.19), the energy distribution must be of the form

$$\rho_E = \frac{1}{2\pi} \frac{\hbar\gamma}{(E - \tilde{E})^2 + (\hbar\gamma/2)^2} \tag{3.22}$$

Therefore the width of the quasistationary level $\hbar\gamma$ is determined by the probability per unit time for the particle to leave the potential well.

All of the above results also hold in the three-dimensional case when the particle moves in a centrally symmetric field. The corresponding Schrödinger equation is

$$i\hbar \frac{\partial \Psi}{\partial t} = -\frac{\hbar^2}{2m} \Delta\Psi + U(|\mathbf{r}|)\,\Psi, \tag{3.23}$$

where $\mathbf{r}$ is the radius vector determining the particle position; Δ is the Laplace operator.

It is convenient to introduce spherical coordinates (r,θ,φ) and to look for a solution in the form $\Psi = \widetilde{R}(r,t)\, Y_{\ell m}(\theta,\varphi)$, where $\widetilde{R}$ and $Y_{\ell m}$ are the radial and angular parts of the wave function, ℓ, m are the quantum numbers corresponding to the eigenvalues of the angular momentum and its z-projection. Then the variables separate, and an equation for the radial part of the wave function is obtained:

$$i\hbar \frac{\partial \widetilde{R}}{\partial t} = -\frac{\hbar^2}{2m}\left[\frac{1}{r^2}\frac{\partial}{\partial r}\left(r^2 \frac{\partial \widetilde{R}}{\partial r}\right) - \frac{\ell(\ell+1)}{r^2}\widetilde{R}\right] + U(r)\,\widetilde{R}. \tag{3.24}$$

The particle behavior in the three-dimensional case is described by the function $\chi = r\widetilde{R}$ (integrals of the type (3.8), (3.9) in the three-dimensional case are $\sim \int dr\,(r\widetilde{R})(r\widetilde{R}'))$, which satisfies the equation

$$i\hbar \frac{\partial \chi}{\partial t} = -\frac{\hbar^2}{2m}\frac{\partial^2 \chi}{\partial r^2} + \left[U(r) + \frac{\ell(\ell+1)}{r^2}\right]\chi. \tag{3.25}$$

This equation at $\ell = 0$ coincides with (3.1), while at $\ell \neq 0$ the function $\widetilde{U}(r) = U(r) + \ell(\ell+1)/r^2$ takes the place of the potential energy.

It can easily be seen that the width of the quasistationary level for a particle in the potential (3.13) in the three-dimensional case is also of the form (3.17), since both the equation and all boundary conditions for $\chi(r)$ and $\Psi(r)$ coincide. (At $r = 0$ the condition $\chi(0) = 0$ follows from the finiteness of the radial part of the wave function R(r)).

As an example of a particle transition from a bound state to the continuous spectrum we briefly consider the α-decay of heavy nuclei. A detailed treatment of this question is given in [7–9], which give the original references therein.

The potential in which the α-particle moves can be approximated by (see Fig. 1.8)

$$U(r) = \begin{cases} U_0 & (0 \leqslant r \leqslant r_0), \\ \dfrac{2Ze^2}{r} & (r > r_0), \end{cases} \tag{3.26}$$

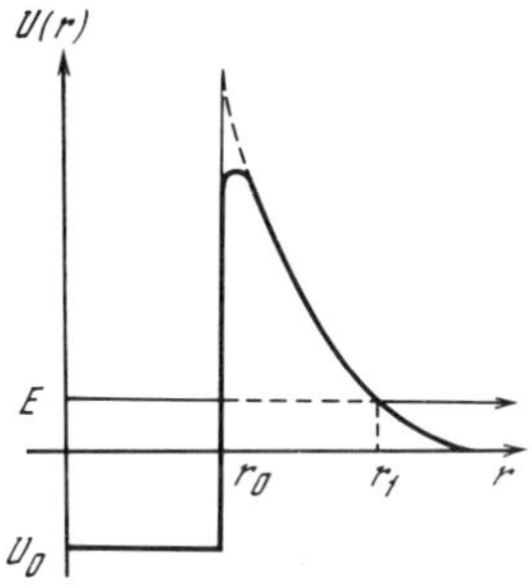

Figure 1.8 Potential energy of the α-particle.

where Z is the nuclear charge of the α-decay product. According to (3.18), the probability of α-decay per unit time is

$$\gamma = \frac{v}{2r_0} \exp\left\{ -\frac{2}{\hbar} \int_{r_0}^{r_1} dr\, [2m\,(U(r) - E)]^{1/2} \right\}, \quad U(r_1) = E. \quad (3.27)$$

Carrying out the integration and assuming $r_0/r_1 \ll 1$, the expression

$$\gamma = \frac{v}{2r_0} \exp\left\{ -\frac{4\pi\, Ze^2}{\hbar v} + \frac{4e\,(Zmr_0)^{1/2}}{\hbar} \right\} \quad (3.28)$$

is obtained.

1.4. TUNNELING TRANSITIONS OF A PARTICLE FROM A BOUND TO A BOUND STATE

We now consider the case of a particle tunneling between two potential wells (Fig. 1.9), when the potential energy is represented as the sum of two potentials U_L and U_R. Let the wave functions $\Psi_{L,R}$ and eigen-energies $E_{L,R}$ of the Schrödinger equation with the potentials U_L and U_R,

$$\left. \begin{aligned} -\frac{\hbar^2}{2m} \Delta\Psi_L + U_L \Psi_L &= E_L \Psi_L, \\ -\frac{\hbar^2}{2m} \Delta\Psi_R + U_R \Psi_R &= E_R \Psi_R, \end{aligned} \right\} \quad (4.1)$$

be known. These functions Ψ_L and Ψ_R are normalized but not orthogonal, since they are solutions of different Schrödinger equations

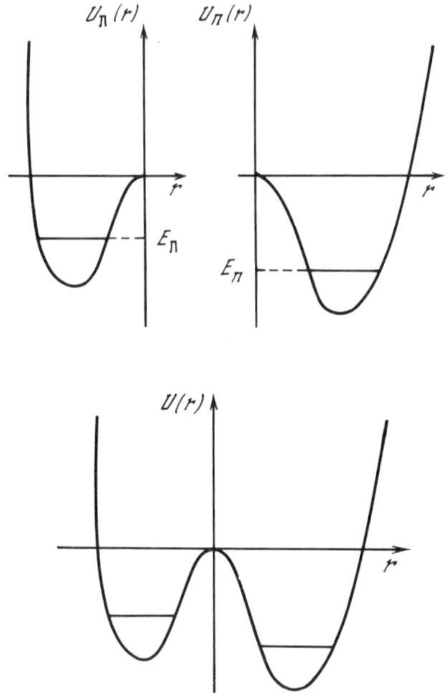

Figure 1.9 Potential energy of the tunneling particle corresponding to the particle transition from one bound state to another.

with different Hamiltonians. Generally, the potential wells are of different depth and $E_L \neq E_R$.

Of interest for us will be the probability of particle well-to-well tunneling per unit time. When solving this problem, the well-elaborated technique of conventional perturbation theory for quantum transitions is of no use, since here it is not possible to distinguish the perturbation causing the transition. In fact, initially, the potential U_R takes the role of the perturbation (if at $t = 0$ the particle is in the left-hand well), while, finally, when the particle is in the right-hand well this part is taken by the potential U_L. Apart from the fact that these perturbations are not small, they also appear to be different at the initial and final states. That is why we shall start from the Schrödinger equation with the complete Hamiltonian

$$H = -\frac{\hbar^2}{2m}\frac{\partial^2}{\partial x^2} + U_L + U_R. \tag{4.2}$$

If the resonance difference $\Delta E = E_L - E_R$ is small compared with the spacing from the level in the potential wells that is the nearest to E_L

and E_R then the wave function of the system should be sought as a linear combination

$$\Psi = a_L \Psi_L + a_R \Psi_R. \tag{4.3}$$

This expression is substituted into the Schrödinger equation with the Hamiltonian (4.2), and use is made of the fact that the wave functions Ψ_L and Ψ_R are solutions of (4.1). Then the coefficients a_L and a_R can be found from the system of equations

$$\left.\begin{array}{l} a_L(H_{LL} - E) + a_R(H_{LR} - ES) = 0, \\ a_L(H_{RL} - ES) + a_R(H_{RR} - E) = 0, \\ S = \int dx\, \Psi_L \Psi_R, \quad H_{\ell\ell'} = \int dx\, \Psi_\ell H \Psi_{\ell'}, \quad \ell,\ell' \to L,R. \end{array}\right\} \tag{4.4}$$

This system possesses nonzero solutions if its determinant equals zero. This condition yields the energy levels corresponding to the states in which the particle is spread between two wells:

$$\left.\begin{array}{l} E_{1,2} = \dfrac{1}{2}\{E_L + E_R \pm [(E_L - E_R)^2 + 4U^L_{LR} U^R_{LR}]^{1/2}\} \\ U^{L,R}_{LR} = \int dx\, \Psi_L U_{L,R} \Psi_R. \end{array}\right\} \tag{4.5}$$

In deriving (4.5), the overlap of the functions Ψ_L and Ψ_R is assumed to be small; the terms $\sim S^2$ are ommitted in (4.5) ($|U^{L,R}_{LR}| \sim S$; $|U^L_{RR}|, |U^R_{LL}| \sim S^2$, the energy is measured from the barrier top). For $E_L = E_R$ the energy levels corresponding to the particle motion simulteniously in both wells are split by the value $\Delta E_p = 2(U^L_{LR} U^R_{LR})^{1/2}$.

Making use of (4.5), the system (4.4) gives the coefficients a_L and a_R, and hence the stationary wave functions related to the energy levels E_1 and E_2:

$$\left.\begin{array}{l} \Psi_1 = [\Delta E + (\Delta E^2 + 4U^L_{LR} U^R_{LR})^{1/2}]\, \Psi_L + 2U^R_{LR} \Psi_R, \\ \Psi_2 = 2U^L_{LR} \Psi_L - [\Delta E + (\Delta E^2 + 4U^L_{LR} U^R_{LR})^{1/2}]\, \Psi_R. \end{array}\right\} \tag{4.6}$$

Taking account of the relation

$$U^L_{LR} - U^R_{LR} = \Delta E S, \tag{4.7}$$

it can be shown that the wave functions Ψ_1 and Ψ_2 are orthogonal to first order in S. The functions (4.6) can easily be normalized, but this is not necessary in the present problem.

It is easy to show that the linear combination of wave functions

$$\Psi = A_0 \Psi_1 e^{-iE_1 t} + B_0 \Psi_2 e^{-iE_2 t} \tag{4.8}$$

satisfies the time-dependent Schrödinger equation with the Hamiltonian (4.2). Therefore the wave function (4.8) describes the particle motion if it satisfies an initial condition. At $t = 0$ let the particle be in the left-hand potential and let its wave function be $\Psi(t = 0) = \Psi_L$. This condition yields the coefficients A_0 and B_0. The normalization of the wave function (4.8) at $t = 0$ coincides with that of Ψ_L (recall that $\int dx\, |\Psi_L|^2 = 1$) on the one hand; but on the other hand, the orthogonality of Ψ_1 and Ψ_2 leads to

$$\int |\Psi|^2 dx = |A_0|^2 \int |\Psi_1|^2 dx + |B_0|^2 \int |\Psi_2|^2 dx = \text{const}$$

so the normalization is conserved over the course of the time evolution.

Therefore the normalized wave function describing the behaviour of the particle is

$$\begin{aligned}\Psi = \{&[\Delta E + (\Delta E^2 + 4U_{LR}^{L}U_{LR}^{R})^{1/2})^2 \Psi_L \\ &+ 2(\Delta E + (\Delta E^2 + 4U_{LR}^{L}U_{LR}^{R})^{1/2})\, U_{LR}^{R}\Psi_R]\, e^{-iE_1t/\hbar} \\ &+ [4U_{LR}^{L}U_{LR}^{R}\Psi_L - 2(\Delta E + (\Delta E^2 + 4U_{LR}^{L}U_{LR}^{R})^{1/2})\, U_{LR}^{R}\Psi_R]\, e^{-iE_2t/\hbar}\} \\ &\times \{[\Delta E + (\Delta E^2 + 4U_{LR}^{L}U_{LR}^{R})^{1/2}]^2 + 4U_{LR}^{L}U_{LR}^{R}\}^{-1}.\end{aligned} \tag{4.9}$$

The probability amplitude for finding the particle in the left-hand well is therefore

$$a(t) = \int dx\, \Psi(x,t)\, \Psi_L(x).$$

The squared modulus of a(t) is the probability of finding the particle in the left-hand well at time t:

$$\begin{aligned}W(t) = 1 - &\frac{16[\Delta E + (\Delta E^2 + 4U_{LR}^{L}U_{LR}^{R})^{1/2}]^2\, U_{LR}^{L}U_{LR}^{R}}{\{[\Delta E + (\Delta E^2 + 4U_{LR}^{L}U_{LR}^{R})^{1/2}]^2 + 4U_{LR}^{L}U_{LR}^{R}\}^2} \\ &\times \sin^2 \frac{(\Delta E^2 + 4U_{LR}^{L}U_{LR}^{R})^{1/2}}{2\hbar}\, t.\end{aligned}$$

Only first-order terms in S are retained in (4.10).

The probability W(t) is a periodic function of time, which oscillates with period $2\pi\hbar\, (\Delta E^2 + 4U_{LR}^{L}U_{LR}^{R})^{-1/2}$ from unity to a minimum value determined by the resonance difference ΔE and the product of the matrix elements U_{LR}^{L} and U_{LR}^{R}. As follows from (4.10), the minimum probability of finding the particle in the left-hand well decreases with decreasing resonance difference.

Two limiting cases are considered.

(i) $\Delta E = 0$. Then

$$W(t) = 1 - \sin^2[(U_{LR}^{L}U_{LR}^{R})^{1/2}\, t/\hbar).$$

In this case the probability of finding the particle in the left-hand well varies from one to zero, i.e. the particle oscillates between two potential wells with period $\pi\hbar/(U^L_{LR}U^R_{LR})^{1/2}$ and spends equal time in each well. Averaging the probability of finding the particle in each well over a large (compared with the period) time span gives $\frac{1}{2}$.

(ii) $\Delta E^2 \gg 4U^L_{LR}U^R_{LR}$. Then

$$W(t) = 1 - \frac{4U^L_{LR}U^R_{LR}}{\Delta E^2}\sin^2\frac{\Delta Et}{2\hbar}.$$

It can easily be seen that the particle is mainly in the left-hand well and the averaged probability of finding the particle in this well is

$$\overline{W} = 1 - 2U^L_{LR}U^R_{LR}/\Delta E^2 \approx 1.$$

Thus, in the case of well-to-well tunneling when the transition occurs between two stationary states, the most favourable is the situation of a resonance between levels (to be more precise, thin means that the condition $|\Delta E| \leqslant 2(U^L_{LR}U^R_{LR})^{1/2}$ holds). This process is, however, reversible, i.e. the particle, on crossing over to the right-hand well, will be again found in the left-hand well with the unit probability after a time $\tau' = \pi\hbar/2(U^L_{LR}U^R_{LR})^{1/2}$. This is why the quantity τ' cannot be treated as the particle lifetime in the potential well with respect to tunneling. This kind of quantity can be introduced only for irreversible transitions.

In general (4.10) describes particle delocalization between two potential wells. The concepts of stationary and dynamic delocalization should be distinguished [10]. Stationary delocalization occurs when the levels in different potential wells coincide to within the energy difference of the corresponding stationary states. This means simply that the probabilities of finding the particle in each well are approximately equal. At the same time, in order to see the delocalization, it is necessary to make observations over a time span exceeding the period of the particle's quantum oscillations, this being a criterion of delocalization in the dynamical sense.

We will now examine how these results change if the right-hand level has a nonzero width Γ. The final width of the level might, for example, result from the fact that the right-hand level, to which the transition occurs, is not the ground state. A decay to lower levels is therefore possible. This proceeds with a high probability that exceeds, as a rule, the tunneling probability. It can therefore be assumed that

$$\Gamma \gg U^L_{LR}, U^R_{LR}. \tag{4.11}$$

Substituting the formal expression for the quasistationary energy level, $\widetilde{E}_R - \frac{1}{2}i\Gamma$, into (4.5) instead of the energy E_R, and using (4.11), gives

$$\left.\begin{aligned}&\tilde{E}_1 \approx E_L - i\frac{\Gamma}{2}\,\frac{U^L_{LR}U^R_{LR}}{\Delta\tilde{E}^2 + (\Gamma/2)^2}, \quad \tilde{E}_2 \approx \tilde{E}_R - \tfrac{1}{2}i\Gamma,\\ &\Delta\tilde{E} = E_L - \tilde{E}_R.\end{aligned}\right\} \tag{4.12}$$

Proceeding similarly to the consideration of the transitions between stationary levels, the wave function is found that describes the evolution of the system

$$\begin{aligned}\Psi = &\Big\{[(\Delta\tilde{E} + \tfrac{1}{2}i\Gamma)^2\,\Psi_L + (\Delta\tilde{E} + \tfrac{1}{2}i\Gamma)\,U^R_{LR}\Psi_R]\exp\left[-\frac{iE_Lt}{\hbar} - \tfrac{1}{2}\tilde{\gamma}t\right]\\ &+[U^L_{LR}U^R_{LR}\Psi_R - (\Delta\tilde{E} + \tfrac{1}{2}i\Gamma)\,U^R_{LR}\Psi_R]\exp\left[-\frac{i\tilde{E}_Rt}{\hbar} - \frac{\Gamma t}{2\hbar}\right]\Big\}\\ &\times[(\Delta\tilde{E} + \tfrac{1}{2}i\Gamma)^2 + U^L_{LR}U^R_{LR}]^{-1},\end{aligned}$$

$$\tilde{\gamma} = U^L_{LR}U^R_{LR}\,\frac{\Gamma/\hbar}{\Delta\tilde{E}^2 + (\tfrac{1}{2}\Gamma)^2}. \tag{4.13}$$

The probability of finding the particle in the left-hand well is given, to second order in S, by

$$\begin{aligned}W(t) = &\Big\{[\Delta\tilde{E}^2 + (\tfrac{1}{2}\Gamma)^2]^2\exp(-\tilde{\gamma}t) + (U^L_{LR}U^R_{LR})^2\exp\left(-\frac{\Gamma t}{\hbar}\right)\\ &+ 2U^L_{LR}U^R_{LR}\Big[(\Delta\tilde{E}^2 - (\tfrac{1}{2}\Gamma)^2)\cos(\Delta\tilde{E}t/\hbar)\\ &+ \Delta\tilde{E}\,\Gamma\sin\left(\frac{\Delta\tilde{E}t}{\hbar}\right)\Big]\exp\left(-\frac{\Gamma t}{2\hbar} - \tfrac{1}{2}\tilde{\gamma}t\right)\Big\}\\ &\times[(\Delta\tilde{E} + i\Gamma/2)^2 + U^L_{LR}U^R_{LR}]^{-2}.\end{aligned} \tag{4.14}$$

This probability W(t) decays exponentially with time; it includes rapidly and slowly decaying terms ($\Gamma/\hbar \gg \tilde{\gamma}$), so at $\Gamma t/\hbar \gg 1$

$$W(t) = \exp[-\tilde{\gamma}t]. \tag{4.15}$$

It is easy to see that the quantity $\tilde{\gamma}$ (see (4.13)) is the probability of particle tunneling per unit time. So this is the quantity that determines the particle lifetime in the left-hand potential well with respect to tunneling. The situation changes essentially when the finite-state level is of nonzero width. Then the process is irreversible and the nomenclature "probability per unit time" may be used, i.e. at small times ($\tilde{\gamma}t \ll 1$, but $\Gamma t/\hbar \gg 1$) the transition-probability oscillations decay rapidly and the evolution of the system is described by classical kinetics.

It is also possible to estimate the transition probability per unit time by means of quantitive considerations [11]. It is first noted that the expression

$$\rho = \frac{\Gamma/2\pi}{\Delta\tilde{E}^2 + (\frac{1}{2}\Gamma)^2}$$

is the density of states (see (3.22) and (4.13)) in the right-hand potential well at $E = E_L$; the transition takes place into a state determined by the energy conservation. The probability that the initial and final energies coincide, which is necessary for the transition, is of the order of $\rho(U^L_{LR}U^R_{LR})^{1/2}$ (these energies may coincide only to within the resonance splitting $\Delta E_r \sim (U^L_{LR}U^R_{LR})^{1/2}$). Multiplying the dimensionless probability by the frequency of resonant quantum oscillations, $\sim \Delta E_r/\hbar$, gives the tunneling probability per unit time. It can easily be verified that the corresponding expression coincides to within a numerical factor with (4.13) for $\tilde{\gamma}$.

The same problem of well-to-well particle tunneling can be treated by means of the density-matrix formulation when the subsystem under study (in the present case it is a particle tunneling between wells) interacts with a thermal bath (i.e. a molecular environment that is at equilibrium during the particle transitions).

The equations for the matrix elements of the density operator $\hat{\rho}$ are (see e.g. [12])

$$i\hbar\,\frac{\partial\rho_{ij}}{\partial t} = [H, \hat{\rho}]_{ij} - \frac{i\hbar}{\tau_{ij}}\,\rho_{ij}, \quad \gamma_{ij} = \tau_{ji}. \tag{4.16}$$

The diagonal matrix element ρ_{jj} relates to the probability of finding the particle in the state j.

Two types of relaxation times are usually introduced for diagonal and nondiagonal elements of the density matrix. The relaxation time τ_{jj} specifies the time necessary for the subsystem to reach equilibrium with the thermal bath and is due to the energy change of the subsystem studied. The constant τ_{ij} specifies the time for loss of information on the phase relation between the states i and j. τ_{ij} is influenced by any process confusing the phase; it is clear that a process contributing to τ_{jj} also contributes to τ_{ij}. It is conventional to call the constants τ_{jj} and τ_{ij} the times of longitudinal and transversal relaxation respectively.

Let the subsystem studied be specified by two relaxation times: the lifetime of the state into which the particle is transformed, $\tau_{tr} = \hbar/\Gamma$, and the relaxation time of the nondiagonal matrix elements, τ_{LR}. Then the wavefunctions Ψ and Ψ_R, up to the S, enable us to write the equations for the matrix elements of the density operator as†

† This type of equation was first introduced for nuclear spins [13], where τ_{jj} and τ_{ij} were defined as the relaxation times of the longitudinal and transversal components of the magnetic momentum.

$$
\begin{aligned}
i\hbar \frac{\partial \rho_{LL}}{\partial t} &= U^{R}_{LR}(\rho_{RL} - \rho_{LR}), \\
i\hbar \frac{\partial \rho_{RR}}{\partial t} &= U^{L}_{LR}(\rho_{LR} - \rho_{RL}) - i\Gamma\,\rho_{RR}, \\
i\hbar \frac{\partial \rho_{LR}}{\partial t} &= (\Delta\widetilde{E} - i\hbar/\tau_{LR})\,\rho_{LR} - U^{L}_{LR}\rho_{LL} + U^{R}_{LR}\rho_{RR}, \\
\rho_{LR} &= \rho^{*}_{RL}, \quad \rho_{\ell'\ell''} = \int dx\, \Psi_{\ell'}\,\hat{\rho}\,\Psi_{\ell''}, \quad \ell',\ell'' \to L,R.
\end{aligned}
\tag{4.17}
$$

In fact, the longitudinal and transversal relaxation times are, as a rule, much shorter than the lifetime of the particle with respect to tunneling. First, this results in $\rho_{LL} \gg \rho_{RR}$. Secondly, the relaxation of the nondiagonal matrix elements proceeds rapidly, and by the time $t \gg \tau_{LR}$ they are stationary. Hence for $t \gg \tau_{LR}$ it can be assumed in (4.17) that $\rho_{RR} = 0$, $\dot{\rho}_{LR} = \dot{\rho}_{RL} = 0$ and the following equation is obtained for ρ_{LL}:

$$
\rho_{LL} = \exp[-\gamma' t], \qquad \gamma' = \frac{2}{\tau_{LR}} \frac{U^{L}_{LR}U^{R}_{LR}}{\Delta E^{2} + (\hbar/\tau_{LR})^{2}}. \tag{4.18}
$$

It can be verified that if the only way of relaxation is the decay of the final state to which the particle transforms then (4.18) coincides with the probability of finding the particle in the left-hand well (4.15) obtained from the Schrödinger equation. In this case $\tau_{LR} = 2\tau_{R} = 2/\Gamma$; and substitution of this into γ' gives $\gamma' = \tilde{\gamma}$ (see (4.13)).

Another case that is often realized experimentally is $\Delta\widetilde{E} = 0, \Gamma = 0$. These conditions correspond to transitions between the ground states of symmetrical potential wells. Since $\Gamma = 0$, the diagonal element ρ_{RR} of the density matrix cannot be assumed small, whereas the nondiagonal matrix elements rapidly achieve their stationary values, as before, owing to the large transverse relaxation. This fact, as well as (4.17), which allows ρ_{LR} and ρ_{RL} to be expressed in terms of ρ_{LL} and ρ_{RR}, leads to equations determining the diagonal matrix elements of the density matrix

$$
\left.
\begin{aligned}
\frac{\partial \rho_{LL}}{\partial \tau} &= -\frac{2\tau_{LR}}{\hbar^{2}}\,\overline{U}^{2}\rho_{LL} + \frac{2\tau_{LR}}{\hbar^{2}}\,\overline{U}^{2}\rho_{RR}, \\
\frac{\partial \rho_{RR}}{\partial t} &= -\frac{2\tau_{LR}}{\hbar^{2}}\,\overline{U}^{2}\rho_{RR} + \frac{2\tau_{LR}}{\hbar^{2}}\,\overline{U}^{2}\rho_{LL}.
\end{aligned}
\right\}
\tag{4.19}
$$

In deriving (4.19), use was made of the equation $U^{L}_{LR} = U^{R}_{LR}$, which holds at $\Delta E = 0$ (see (4.7)), and the notation $\overline{U} = U^{L}_{LR} = U^{R}_{LR}$ has been introduced.

Thus if the transversal relaxation time is much shorter than the lifetime of the particle with respect to tunneling then the probabilities ρ_{LL} and ρ_{RR} satisfy the classical kinetic equations even for $\Gamma = 0$, the tunneling probability per unit time being

$$\gamma'' = 2\tau_{LR}\,\overline{U}^2/\hbar^2. \tag{4.20}$$

On solving (4.19), the probability of finding the particle in the left-hand potential well is obtained as

$$\rho_{LL} = \tfrac{1}{2}(1 + \exp[-2\gamma'' t]). \tag{4.21}$$

As mentioned above, the solutions (4.18) and (4.21) are valid if $\gamma', \gamma'' \ll 1/\tau_{LR}$ and $t \gg \tau_{LR}$ when the probability oscillations have decayed. However (4.17) can be solved without additional assumptions. This can be done by means of the Laplace transform, which converts the system (4.17) of differential equations into a system of linear algebraic equations. After solving this system, the inverse transform can be applied to give the solution of (4.17).

The solutions (4.18) and (4.21) contain terms decaying exponentially with a characteristic time equal to the tunneling lifetime of the particle, as well as oscillating terms whose decay is specified by the relaxation time τ_{LR}.

Therefore the following situations are possible when the particle tunnels from one potential well to another.

(i) If the quantum oscillation period $\tau' = \frac{1}{2}\pi\hbar\,(U^L_{LR}U^R_{LR})^{-1/2}$ is smaller than the longitudinal and transversal relaxation time then the process is coherent; the tunneling probability oscillates with time, i.e., it is described by quantum kinetics.

(ii) When the particle lifetime with respect to tunneling is much longer than the longitudinal and transversal relaxation, the coherency breaks down, and the probability of finding the particle in a certain state (initial or final) can be calculated by means of the quasi-classical kinetic equations.

(iii) If the quantum oscillation period and the relaxation time are of the same order then the tunneling probability is found by means of (4.17), whose solution leads to the diagonal elements of the density matrix and hence to the desired probability of finding the particle in one or another state.

All three of the above cases might be realized in the same system. The relaxation and tunneling times are known to depend on the temperature in various ways. With decreasing temperature the thermal molecular motion is frozen, resulting in an abrupt fall of the transversal relaxation rate; the tunneling probability per unit time also decreases at low temperatures, but only as far as a certain limit. Therefore at low temperatures the tunneling rate might exceed the relaxation rate, and coherent tunneling proceeds. On increasing the temperature, the relaxation rate might increase more rapidly then the rate of particle tunneling. Thus by varying the temperature it is possible to pass from low temperatures to the quasiclassical kinetics characteristic of the high-temperature region.

It should be pointed out that coherent effects are observed at low temperatures in experiments on methyl-group tunneling.

1.5. TUNNELING IN A PERIODIC POTENTIAL

The problem of how a particle behaves in a periodic potential is the principal problem in solid-state physics. It forms the background of the band theory of the solid state, crystal-lattice vibrations, defect diffusion in quantum crystals etc. We shall not dwell upon all these questions here, but shall rather concern ourselves only with specific features of tunneling in periodic potentials.

If the relaxation rate exceeds the probability of tunneling per unit time then, while staying in a potential well, the particle has time to equilibrate with the environment and to localize. It is thus possible to say that the process starts anew after each transition, i.e. information about the preceeding transitions is not retained. Therefore under such conditions, the periodic structure introduces no new features into tunneling as compared with transitions between two symmetrical potential wells.

The situation is essentially different if the relaxation rate is much smaller than the tunneling rate. Tunneling then becomes coherent, i.e. the process is not restricted by tunneling between two wells, but coherent motion of the particle along the whole periodic potential takes place. This implies a delocalization of the particle and its conversion into a quasiparticle.

We now examine weak relaxation when the particle behaviour is governed by the Schrödinger equation with a periodic potential. A one-dimensional potential with period $\tilde{a}$ is considered for the sake of simplicity (Fig. 1.10a). It can be represented as a sum of the potential wells as shown in Fig. 1.10b. A solution of the Schrödinger equation with the Hamiltonian

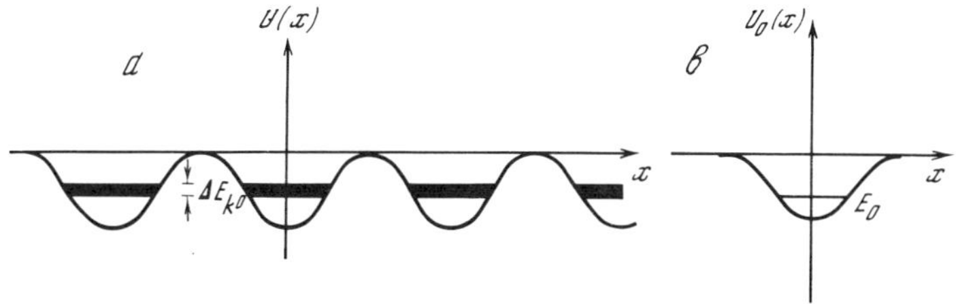

Figure 1.10 One-dimensional periodic potential.

$$H = -\frac{\hbar^2}{2m}\frac{d^2}{dx^2} + \sum_{\ell'} U_0(x - \ell'\tilde{a}) \tag{5.1}$$

is sought as a linear combination of the wave functions for the ground states of the individual potential wells:

$$\left.\begin{aligned} &\Psi(x) = \sum_{\ell'} b_{\ell'}\Psi_0^{\ell'}(x), \\ &\left[-\frac{\hbar^2}{2m}\frac{d^2}{dx^2} + U_0(x - \ell'\tilde{a})\right]\Psi_0^{\ell'}(x) = E_0\Psi_0^{\ell'}(x). \end{aligned}\right\} \tag{5.2}$$

The spacing between the excited, E_n, and ground states is assumed to be much greater than the matrix elements:

$$U_{\ell'\ell''}^{\ell_0} = \int_{-\infty}^{\infty} dx\, \Psi_0^{\ell'}(x)\, U_0(x - \ell_0\tilde{a})\, \Psi_0^{\ell''}(x), \quad E_n - E_0 \gg |U_{\ell'\ell''}^{\ell_0}|, \quad \ell' \neq \ell''. \tag{5.3}$$

Substituting the wave function (5.2) into the Schrödinger equation with the Hamiltonian (5.1) gives

$$\sum_{\ell'} b_{\ell'}\left\{E_0 + \sum_{\ell_0 \neq \ell'} U(x - \ell_0\tilde{a}) - E\right\}\Psi_0^{\ell'} = 0. \tag{5.4}$$

Now multiplying (5.4) by the wave function $\Psi_0^{\ell''}$ and integrating over x from $-\infty$ to $+\infty$, the equation determining the coefficients $b_{\ell'}$ is found:

$$\left.\begin{aligned} &\sum_{\ell'} b_{\ell'}\left\{(E_0 - E)\, S^{\ell''\ell'} + \sum_{\ell_0 \neq \ell'} U_{\ell''\ell'}^{\ell_0}\right\} = 0, \\ &S^{\ell''\ell'} = \int_{-\infty}^{\infty} dx\, \Psi_0^{\ell''}(x)\, \Psi_0^{\ell'}(x). \end{aligned}\right\} \tag{5.5}$$

The wave function (5.2) satisfies Bloch's theorem (see e.g. [14]), which in our notation reads

$$\Psi(x + \ell'\tilde{a}) = \exp[ik^0\ell'\tilde{a}]\,\Psi(x), \tag{5.6}$$

where k^0 is the wavenumber, which lies in the range $-\pi/\tilde{a} < k^0 \leqslant \pi/a$ (in the three dimensional case it is a quasiparticle wavevector). It follows from (5.6) that the coefficients $b_{\ell'}$ are proportional to $\exp[ik^0\ell'\tilde{a}]$.

If only the overlaps of the wave function relating to the neighboring potential wells are taken into account (the matrix elements $U^{\ell_0}_{\ell'\ell''}$, decrease rapidly with increasing differences $\ell' - \ell''$, $\ell' - \ell_0$) then the energy is given by

$$E(k^0) = E_0 + 2\tilde{A}\cos(k^0\tilde{a}), \quad \tilde{A} = U^{\ell'}_{\ell',\ell'\pm 1} = U^{\ell'\pm 1}_{\ell',\ell'\pm 1}. \tag{5.7}$$

Therefore the possibility of tunneling transitions splits the levels of separate potential wells, thus leading to an energy band corresponding to states of the particle propagating along the whole potential, the bandwidth being $\Delta E_{k^0} = 4|\tilde{A}|$. It is possible to construct in a similar fashion the energy band of the excited states. If the matrix elements (5.3) are not small then the energy bands of the ground and excited states may overlap.

Knowing the bandwidth, the time of the particle transition between the neighboring potential wells can be estimated by means of the uncertainty principle:

$$\tau \approx \pi\hbar/2|\tilde{A}|. \tag{5.8}$$

The above approach is used when obtaining the energy bands of tightly bound electrons, defectons, vacancions and some other quasiparticles. Calculations in real crystals are often much more involved (for example the interactions of the tunneling particle are sometimes taken into account); the principal features of tunneling obtained under the above simplified considerations are nevertheless valid.

1.6. ANDERSON LOCALIZATION

The mechanism discussed above of how the wave function of a particle tunneling between two equivalent or nonequivalent wells is formed may, in a sense, be treated as a simple example of Anderson localization — the phenomenon that plays a fundamental role in the physics of disordered solids. This can be explained by a simple consideration. A system of two potential wells is considered, and the particle is put in one of them at an instant $t = 0$. The problem is then to find the probability that the particle is in the same well after an infinite time. As we know, the answer depends on the relation between the well-to-well energy difference and the overlap resonance integral I. If

$$|\epsilon_1 - \epsilon_2| \gg I \tag{6.1}$$

then, with an overwhelming probability, the particle, even after infinite time, remains localized in the same well in which it was initially put. In this case the eigenfunctions are

$$\Psi_I = c_1\varphi_1 + c_2\varphi_2, \quad \Psi_{II} = c_2\varphi_1 - c_1\varphi_2 \tag{6.2}$$

where φ_1 and φ_2 are the eigenfunctions of the wells 1 and 2 when tunneling is neglected. If the inequality (6.1) holds, then

$$|c_1|^2 \gg |c_2|^2$$

also holds, which corresponds to the particle localization in the well 1. In the opposite case

$$|\epsilon_1 - \epsilon_2| \ll I \tag{6.3}$$

the particle is delocalized between the two wells, i.e. it may be found in each well with equal probabilities:

$$c_1 \approx 1/\sqrt{2}, \quad c_2 \approx 1/\sqrt{2}. \tag{6.4}$$

It is easy to see that a similar situation occurs in a system with a larger number of wells. For example, in the case of three wells the particle remains localized in a well if the condition (6.1) is valid for all three possible tunneling transitions between wells 1 and 2, 2 and 3, 1 and 3 respectively. If the condition (6.3) holds for one of these transitions (e.g. for the transition $1 \rightarrow 2$) then, depending on the initial conditions, the particle is either localized in well 3 or is partially delocalized and spread between wells 1 and 2. Then the system is specified by a set of wave functions: $\Psi_I = \varphi_1$, $\Psi_{II} = \varphi_2$ and $\Psi_{III} = \varphi_3$ in the first case, and $\Psi_{III} = \varphi_3$, $\Psi_{I,II} = (\varphi_1 \pm \varphi_2)/\sqrt{2}$ in the second case. But if the condition (6.3) holds for all three transitions then the particle is delocalized among all three wells and is specified by the wave functions

$$\Psi_{I,II,III} = \frac{1}{\sqrt{3}}(\varphi_1 \pm \varphi_2 \pm \varphi_3).$$

Similar considerations are possible for systems with any number of wells. If this number is finite then the answer is easily formulated in a general form. The particle, put in a well i, can be found after a large enough time with equal probability in all other wells j of the system for which the condition

$$|\epsilon_i - \epsilon_j| \ll I_{ij} \tag{6.5}$$

holds. Then the wave function takes the form

$$\Psi = \frac{1}{N_\ell^{1/2}} \sum_j (\pm \varphi_j), \tag{6.6}$$

where the sum over j is only for those wells for which the condition (6.5) holds, and N_ℓ is the number of these wells.

From this point of view, systems containing an infinite number of wells are of special interest. If the condition (6.5) is valid for all wells then the particle is completely delocalized and the probability of finding it in any well is negligibly small. Moreover, arranging these wells in a regular lattice leads to a conventional band description of the crystalline solids, the bandwidth being determined by the overlap integral I and the functions (6.6) becoming Bloch functions. If the condition (6.5) holds for only some of the wells, a special analysis is necessary to answer the question of whether the particle remains localized in a finite space domain or is delocalized.

Anderson was the first to formulate and (approximately) solve this problem in 1958 [15]. He considered the following model. Let a particle (an electron is, as a rule, implied) move over the sites of a regular lattice, their energies being a random function of the site number. This means that the electron energy ϵ_i at site i does not depend on its energy at any other site. Anderson also made a simplifying (but not essential) assumption that these energies are uniformly distributed within a range of width W. This means that the energy distribution function has the form

$$\rho(\epsilon) = \begin{cases} 1/W & (|\epsilon| \leqslant \tfrac{1}{2}W), \\ 0 & (|\epsilon| > \tfrac{1}{2}W). \end{cases} \tag{6.7}$$

Such a system is described by means of the Hamiltonian

$$H = \sum_j \epsilon_j a_j^+ a_j + \sum_{j,m\neq 0} I(m)\, a_j^+ a_{j+m}, \tag{6.8}$$

where a_j^+, a_j are the electron creation and annihilation operators at site j; I(m) is the tunneling integral for transitions from site j to site j + m. This model is usually called the Anderson model. A large number of approximate analytical studies (e.g. [16,17]) and computer simulations have been made of this problem. Although no mathematically rigorous solution has yet been found, the general qualitative pattern is on the whole clear enough.

The principal question when studying the Anderson model is whether an electron with a given energy ϵ_0 propagates over any large distance or whether it is localized in a finite spatial domain? Usually the physics of crystalline solids deals with the electrons moving freely over a lattice, and a disorder (lattice defects) results only in their scattering. This is why raising the question of electron localization due to disorder seemed, at first sight, to be surprising. The answer appeared even more surprising Anderson's principal assertion [15] is that there are energy regions in which electrons are spatially localized (Fig. 1.11). These are sparated from the regions with delocalized electrons by boundary energies, which are called mobility edges. One of the most interesting consequences is as follows: a disordered metal with the Fermi surface lying below the mobility edge, i.e. in the region of localized electrons, does not conduct an electric current and hence is an insulator. It is assumed, for example, that the forbidden energy gap in amorphous chalcogenide semiconductors is caused by Anderson localization (see e.g. [18]).

In the Anderson model with the uniform distribution (6.7) the answer to the question of whether or not, electrons are localized depends on the parameters I(m) and W. If attention is restricted to the nearest-neighbor approximation for I(m) then the model contains only one dimensionless parameter W/I specifying the system disorder. At high-enough values of this parameter nearly all the states turn out to be localized, whereas delocalized states appear only when the parameter W/I is smaller than a critical value $(W/I)_c$. If the energies ϵ_i are uniformly distributed (see (6.7)) then electrons with all energies delocalize. However, if the distribution is nonuniform, as for example in Fig. 1.11, then mobility edges and regions of localized and delocalized electrons arise.

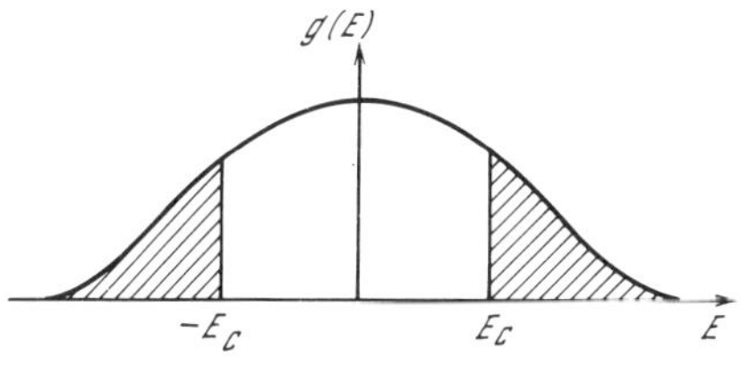

Figure 1.11 Density of states of a disordered conductor. E_c and $-E_c$ are the mobility edges. The regions of the localized states are shaded.

The considerations given at the beginning of this section help in evaluating the critical value $(W/I)_c$. This approach was first proposed by Thouless [19] (see also [20,21]). Without claiming to be complete and mathematically rigorous, it gives a clear qualitative description of localization and even allows satisfactory quantitative results to be obtained. Anderson [15] has suggested the following definition of localization. Let an electron be initially ($t = 0$) at site i; then its wave function coincides with the site wave function

$$\Psi(t)|_{t=0} = \varphi_i. \tag{6.9}$$

The electron is localized in the vicinity of site i if a finite limit

$$\lim_{t \to \infty} |\Psi(t)|^2 > 0 \tag{6.10}$$

exists. In this case the probability of finding the electron at site i is certainly nonzero even after a long time. On the other hand, accounting for the normalization of the wave function $\Psi(t)$, (6.10) results in a rapid decrease of the probability of finding the electron in any site j with increasing distance between sites i and j. This corresponds to the localization of the electron in a finite spatial domain near site i. If, however,

$$\lim_{t \to \infty} |\Psi(t)|^2 = 0, \tag{6.11}$$

then the electron has left site i and spread practically over all lattice sites, and the probability of finding it in any site is of the order of N^{-1}, where N is the total number of lattice sites.

We now consider an electron with energy ϵ_0 and distinguish a narrow energy range $(-2I + \epsilon_0, 2I + \epsilon_0)$. The electron wave functions of the neighboring sites i and j are assumed to be hybridized if both energies ϵ_i and ϵ_j lie in this range, i.e. an equality similar to (6.5) holds for them. Otherwise, the inequality (6.3) is assumed and the wave functions of the sites are not hybridized. Such a segregation procedure allows application of percolation theory (see e.g. [21]).

In the lattice individual clusters can be distinguished that unite sites with energies lying in the range $(-2I + \epsilon_0, 2I + \epsilon_0)$. The electron wave function in such a cluster has a form similar to (6.6), where the sum is now over all the cluster sites and N_ℓ is now the number of sites. In order to find out whether or not the electron is localized, it is necessary to know whether the cluster is finite or infinite. In the percolation theory this problem is known as the site problem and it has been analysed in detail numerically for many different lattices. The answer depends on the concentration $x(\epsilon_0)$ of the sites with $\epsilon_0 \in (-2I + \epsilon_0, 2I + \epsilon_0)$.

The critical value x_c is known for each lattice ($x_c \approx 0.1$ for three-dimensional lattices; see e.g. [21]), so if the inequality

$$x(\epsilon_0) \geqslant x_c \tag{6.12}$$

holds, an infinite cluster arises. In the localization problem this means that the electron is delocalized. However, if the concentration $x(\epsilon_0)$ is small, i.e. $x(\epsilon_0) < x_c$, then only finite clusters exist, and an electron with energy ϵ_0 is localized, i.e. it is in a dashed region in Fig. 1.11.

In the simplest version of the Anderson model, when the energy distribution is assumed to be uniform and is specified by the function (6.7), the concentration $x(\epsilon_0) = 4I/W$ does not depend on energy. In this case the following criterion is obtained for the mobility edge:

$$(W/I)_c = 4/x_c \tag{6.13}$$

In carrying out this analysis, rather crude assumptions have been made. First, the intermediate region, $|\epsilon_i - \epsilon_j| \sim 2I$, in which the hybridization of the wave functions is also intermediate, has been neglected. Secondly, only nearest neighbors have been taken into account, and the direct hybridization of the states corresponding to distant sites with close energies has been neglected. Thirdly, an approximation of "pair interaction's" has been used, i.e. hybridization via virtual electron transition to a third site with another energy has not been taken into account. All these contributions are, in principle, small, but it is not *a priori*, clear to what extent they can effect the estimate (6.13). And in any case, the above considerations cannot claim to be a proof of the existence of the mobility edge. They do give, however, a clear physical picture of localization, and comparison of the results following from (6.13) with those of numerical simulations gives quite satisfactory agreement.

References

1. Landau, L.D. and Lifshitz, E.M. *Quantum Mechanics. Nonrelativistic Theory.* Pergamon, Oxford.
2. Migdal, A.B. (1975) *Qualitative Methods in Quantum Theory.* (In Russian.) Nauka, Moscow.
3. Messiah, A. (1961–1962) *Quantum Mechanics*, Vols. 1 and 2. North-Holland, Amsterdam.
4. Nikitin, E.E. (1970) *Theory of Elementary Atomic–Molecular Processes in Gases.* (In Russian.) Khimia, Moscow.
5. Binning, G., Rohrer, H., Gerber, Ch. and Weibel, E. (1982) *Phys. Rev. Lett.*, **49**, 57.
6. Galitskii, V.M., Karnakov, B.M. and Kogan, V.I. (1981) *Problems on Quantum Mechanics.* (In Russian.) Nauka, Moscow.

7. Blokhintsev, D.I. (1976) *Fundamentals of Quantum Mechanics* (In Russian.) Nauka, Moscow.
8. Goldanskii, V.I. and Leikin, E. (1958) *Conversions of Atomic Nuclei.* (In Russian.) Publishing House of the Academy of Sciences of the USSR, Moscow.
9. Baz, A.I., Zeldovich, Ya.B. and Perelomov, A.M. (1971) *Scattering, Reactions, and Decays in Nonrelativistic Quantum Mechanics.* (In Russian.) Nauka, Moscow.
10. Blumenfeld, L.A. Goldanskii, V.I., Podgoretskii, M.I. and Chernavskii, D.S. (1967) *Zh. Strukt. Khim.*, **8**, 854.
11. Klochikhin, V.L., Pshezhetskii, S.Ya. and Trakhtenberg, L.I. (1980) *Teor. Eksp. Khim.*, **16**, 361.
12. Pantell, R.H. and Puthoff, H.E. (1969) *Fundamentals of Quantum Electronics.* Wiley New York.
13. Bloch, F. (1946) *Phys. Rev.*, **70**, 460.
14. Ziman, J.M. (1972) *Principles of the Theory of Solids.* Cambridge University Press.
15. Anderson, P.W. (1958) *Phys. Rev.*, **109**, 1492.
16. Lifshits, I.M., Gredeskul, S.A. and Pastur, L.A. (1982) *Introduction to the Theory of Disordered Systems.* (In Russian.) Nauka, Moscow.
17. Sadovskii, M.V. (1981) *Usp. Fiz. Nauk*, **133**, 223.
18. Thouless, D.F. (1974) *Phys. Rep.*, **13C**, 94.
19. Efros, A.L. (1978) *Usp. Fiz. Nauk*, **126**, 41.
20. Shklovskii, B.I. and Efros A.L. (1979) *Electronic Properties of Doped Semiconductors.* (In Russian.) Nauka, Moscow.

CHAPTER 2
Tunneling in Chemical Kinetics. History

This chapter considers the development of the concepts of tunneling transitions in chemical reactions. The presentation starts from general ideas on tunneling, and a criterion is discussed for the temperature below which tunneling is exponentially stronger than the Arrhenius overbarrier transitions. Some problems of electron tunneling are considered briefly.

Experimental studies of radiation-induced formaldehyde polymerization, which have resulted in the discovery of the low-temperature plateau of the chemical reaction rate, are discussed in detail. The necessity for the concept of molecular tunneling is pointed out, and a comparison of its parameters with those specifying other low-temperature processes is made.

Later experimental studies (after the discovery in 1972 of the low-temperature quantum limit) of heavy-particle tunneling in chemical reactions are then discussed. Their interpretations are given and different mechanisms for the temperature dependence of the tunneling rate constant are studied. Special attention is paid to the theory of radiationless transitions, and the conclusion is made that this theory needs to be generalized in order to account for intermolecular motions.

A simplified quasiclassical approach is suggested for calculating the tunneling rate constant within a generalized theory of radiationless transitions, and it is emphasized that this model adequately describes the experimental data on tunneling of heavy particles.

2.1. STARTING POINTS

The Arrhenius law (1889), describing the dependence of chemical reaction rate on temperature T, is one of the most fundamental laws of chemical kinetics. This law is based on the Maxwell–Boltzmann energy (velocity) distribution of reacting particles and on the assumption that the reacting particles are able to overcome a certain potential barrier of height E, called the activation energy.

Mathematically, the Arrhenius law is a proportionality between the reaction rate constant and the exponential factor $\exp[-\tilde{E}/k_B T]$, where k_B is the Boltzmann constant. This implies that spontaneous chemical reactions cease completely when temperature tends to absolute zero.

Quantum mechanics has brought three very important innovations into chemical kinetics: (1) the activation barrier is explained as a consequence of Pauli's exclusion principle and exchange repulsion of electrons; (2) the continuous Maxwell–Boltzmann distributions for electronic, vibrational and rotational energies are replaced by discrete ones, the very important concept of zero-point vibrations that do not vanish as $T \rightarrow 0$ having arisen; (3) the activation barrier has been shown to be surmounted not only via classical "corpuscular" overbarrier transitions, but also via tunneling "wave" penetration through it.

The prehistory of tunneling concepts in chemical kinetics has its origins in the work of Hund [1], where delocalization of particles between two identical potential wells was considered. After that, several papers were published, directly devoted to the analysis of the possible role of tunneling in chemical reactions [2–4].

During the following decades, Bell [5–9] was the leader in developing tunneling ideas in chemical reaction dynamics. He was the first to consider a number of problems related to particle penetration through one-dimensional barriers of various shapes — in particular, through the parabolic barrier and Eckart's barrier,

$$U(x) = E_0/\cosh^2(2x/d).$$

Bell's books [8,9] analyse in great detail some possible tunneling contributions to the transfer reactions of hydrogen atoms and ions in liquid and gaseous phases, as well as the kinetic isotope effects observed in the deuteration of hydrogen-containing compounds. Tunneling in chemical reactions nevertheless remained for a long time a rather exotic phenomenon, since the low-temperature range was very poorly studied.

General considerations indicate that a particle with mass m and velocity v (energy $E = \frac{1}{2} mv^2$) exhibits wave properties if its de Broglie wavelength $\lambdabar = \hbar/(2m\tilde{E})^{1/2}$ is comparable to a classical scale — in the present case with the potential barrier width d. In addition, in order that the classical overbarrier transition not be so fast as to hinder tunneling, the barrier height must be much greater than T.

So, for underbarrier tunneling to dominate the Arrhenius-type transitions, the following inequality must be satisfied:

$$\lambda\!\!\!^{-}/d > k_B T/\widetilde{E}, \tag{1.1}$$

i.e.

$$T < \frac{\hbar}{k_B \sqrt{2}\, d} \left(\frac{\widetilde{E}}{m} \right)^{1/2}, \tag{1.1'}$$

i.e. the temperature must be lower than a certain "tunneling temperature" T_t. To define this temperature more precisely, it is necessary to compare the rates of the overbarrier and tunneling transitions averaged over all particle energies (from 0 to ∞). In the simplest Gamov representation the tunneling rate is

$$W = \nu_0 \exp\left[-\frac{2}{\hbar} \int_a^b |p|\, dx \right], \tag{1.2'}$$

where ν_0 is the frequency of particle oscillations in the potential well, p is the momentum, and a and b are the coordinates of the barrier boundaries.

The tunneling temperature was derived by Goldanskii [10,11]† as an upper limit at which tunneling transitions exponentially dominate over Arrhenius-type transitions:

$$T_t = \frac{\sqrt{2}\, \hbar}{\pi k_B d} \left(\frac{\widetilde{E}}{m} \right)^{1/2}. \tag{1.3}$$

This temperature is shown to be the same both for Maxwell–Boltzmann distributions for any symmetric barriers with parabolic tops [10] and for a discrete level distribution (including the zero-point level) with a symmetric barrier formed by two overlapping Morse curves [11].

The great significance of the work reported in [10,11] is the existence of a low-temperature limit of the chemical reaction rate for any shape of activation barrier, since even for the Eckart barrier, which has infinite width at E = 0, the tunneling probability for a quantum-mechanical energy distribution remains finite owing to zero-point vibrations. It should be noted that references [10,11], as well as all early works on tunneling in chemical reactions, ignore the difficult case when no level-to-level resonance in both the initial and final potential wells

† In [10,11] the tunneling temperature is given as $T_t = (\hbar/\pi k_B d \sqrt{2}) \sqrt{2\widetilde{E}/m}$ since there d is the barrier *half*-width, rather than the total width.

exists, though this can suppress tunneling completely, since no phonons exist at T → 0. In fact, [10] deals with tunneling of a plane wave, while [11] considers two mirror-symmetric potential wells — level-to-level resonance taking place in both cases.

Khristov [12] extended the results of [10] to estimate correction factors that appear in the tunneling temperature at higher temperatures when under-barrier and over-barrier transitions contribute equally.

Because in 1959 *Doklady Akademii Nauk SSSR* was not routinaly translated into English, references [10,11] were for a long time not discussed in the Western scientific literature. Recently, however, interesting comments [13,14] have been published in which the main conclusions of [10,11] on the separation of the activation and tunneling regions and on the low-temperature limit of tunneling probability have been discussed. LeRoy, Murai and Williams [13] pointed out the decisive role of zero-point vibration (first taken into account in [11]) for the appearance of such a low-temperature limit under real conditions. Buhks and Jortner [14] derived the tunneling temperature, equivalent to that of [10,11], within the theory of radiationless nonadiabatic multiphonon processes. They showed that the general character of the results in [10,11] follows from the equivalence of the Gamov exponential tunneling factor and the Franck–Condon factor, which characterizes the overlapping of the vibrational (nuclear) wave-functions in the initial and final states.

In chemical reactions, the characteristic values are E = 1 eV, d = 1 Å; (1.3) then yields the tunneling temperature T_t = 320 K and T_t = 240 K for hydrogen and deuterium atoms respectively. For heavier atoms and various molecules this temperature is lower; hence detection and investigation of the reactions in which tunneling dominates the activation requires studies in the cryochemical region. This was actually done in the 1950s in connection with the problem of high concentrations of free radicals. Furthermore, during that period techniques were developed to observe very slow solid-phase reactions, such as radio-spectroscopy and tagged atoms.

Higher tunneling temperatures are characteristic of such peculiar hydrogen isotopes as "new atoms" in which protons are substituted by other lighter positive elementary particles, for example by muons (muonium, Mu) or by positrons (positronium, Ps). Equation (1.3) gives T_t = 960 K for muonium and T_t = 9600 K for positronium, and therefore in the "new-atom" reactions in the liquid phase tunneling may give a considerable or even the dominant contribution at room or higher temperatures.

A possible role of tunneling in "new-atom" chemistry was first discussed in [15–18]. In [15,16] the reaction rates of Ps and H atoms with Fe^3 and Fe^2 in aqueous solutions were compared. Positronium was found to react about 3000 times faster than H in both cases. The conclusion was then made [15] that the mechanism of reactions of Ps and H with Fe^3 is similar to those with Fe^2, the latter reactions involving complex formation. An alternative explanation of Ps and H oxidation by Fe^3 ions by means of long-range electron tunneling was therefore rejected. However, the fact that the reaction-rate ratio for Ps and H is two decimal orders higher than $(m_H/m_{P_s})^{1/2} \approx 30$ does not prove the dominance of tunneling for Ps.

This example shows very clearly the possibility of a semiclassical isotope kinetic effect, which is due to different zero-point energies of bonds formed by isotopes with different masses.

At the wavenumber $\omega_0/2\pi c \approx 3000\ \text{cm}^{-1}$ (ω_0 is the vibration frequency, c is the velocity of light, which is characteristic of Z–H bonds, the zero-point energy for Ps, Mu, H, D and T is close (for Z = C) to 5, 0.5, 0.18, 0.13 and 0.1 eV respectively. Therefore the isotope kinetic effect for Arrhenius-type reactions with the participation of hydrogen, deuterium and tritium compounds is characterized by a factor exp[550/T] for H-to-D alteration and exp[670/T] for H-to-T alteration; these factors are due simply to differences in the activation energies, which are 0.05 eV for H–D and 0.08 for H–T. This fact is often ignored, so that strong variations of the low-temperature reaction of ordinary hydrogen, deuterium and tritium compounds and the enhancement of these variations at low temperatures are treated as an unambiguous evidence of tunneling. However, only a low-temperature plateau of the reaction rate can be considered as a convincing proof. The plateau implies tunneling, although tunneling does not necessary yield a plateau. Referring to the considerations below, we can say that the low-temperature limit of the reaction rate is undoubtedly of quantum-mechanical origin.

In connection with consequences of the semiclassical isotope kinetic effect, which arises in the classical transitions owing to different zero-point frequencies, it should be noted that in muonium and especially in positronium the zero-point energies can be even higher than the potential barrier. Besides, the chemical bonds in Ps and H are of very different character because Ps does not contain a heavy nucleus. Therefore, we shall not discuss the kinetics of "new atoms" below.

An important problem of electron tunneling should nevertheless be considered, at least briefly. Substituting E = 1 eV, d = 1 Å, and the electron mass into (1.3) gives T_t = 13600 K, so that any electron transfer seems to occur via tunneling. Indeed, electron tunneling in chemical systems generally includes certain nuclear displacements, so that violations of the Franck–Condon principle should be taken into account. This fact was first mentioned by Libby [9] and then analyzed in [20–23], which are devoted to electron tunneling in redox interactions of ions in solutions and in electrode reactions. Since electron tunneling and nuclear displacement are coupled phenomena, there exists a temperature range in which the Gamov tunneling probability approaches a plateau ($T < T_{t_e}$), although the barrier is as narrow as a few tens of ångströms, while the Franck–Condon probability of nuclear displacement decreases further with temperature and approaches its plateau at much lower temperatures ($T < T_{t_N}$). (The subscripts N had e stand for nuclei and electrons respectively.) For example, in [24] (see also [25]) a process is studied in which an electron is rapidly transferred ($K \sim 10^8\ s^{-1}$) from the outer-sphere high-spin ion of bivalent Fe^2 to the inner-sphere low-spin trivalent iron of the complex anion $[Fe^{III}(CN)_6]^{3-}$ in the course of the transformation of unstable Turnbull's blue $Fe^{2+}[Fe^{III}(CN)_6]^{3-}$ into stable Prussian blue $Fe^{3+}[Fe^{II}(CN)_6]^{4-}$. In this case the tunneling parameters are $E_e = 4$ eV, $d_e = 10$ Å, and therefore $T_{t_e} = 2700$ K.

However, the corresponding displacement of the iron nuclei over a distance of the order of the zero-point vibration amplitude (d_N = 0.1 Å) accompanied by surmounting of the barrier $E_N = 0.025$–0.1 eV results in a lower plateau temperature $T_{t_N} = 50$–100 K, which has been observed experimentally [24]. It should be recalled that, as mentioned in [14], the expression for T_t derived in [10,11] can be used to describe the low-temperature plateau for both the Gamov tunneling probability and the Franck–Condon factor.

The above phenomenon of coupled long-range underbarrier (tunneling) electron transfer and weak (of the order of the vibration amplitude) nuclear displacements (without chemical conversions) is called electron–nuclear tunneling and is depicted in Fig. 2.1. The electron transfer is schematically represented by a vertical arrow between the curves e(D) and e(A). The vertical axis in both plots of the figure shows the potential energy of the system.

According to the Franck–Condon principle, this process should proceed in the vertical direction, i.e. without a variation of the donor-to-acceptor distance when an electron is transferred between them. (After this the donor takes the role of the acceptor and vice versa in the case of the reversed transfer.) In fact, the equilibrium distance between nuclei of the reducer donating the electron decreases by an amount d_N. As a rule, $d_e \gg d_N \approx \Delta$, where $\Delta = (\hbar/M\omega)^{1/2}$ is the amplitude of the atomic vibrations in the molecule (M is the reduced mass of the vibrat-

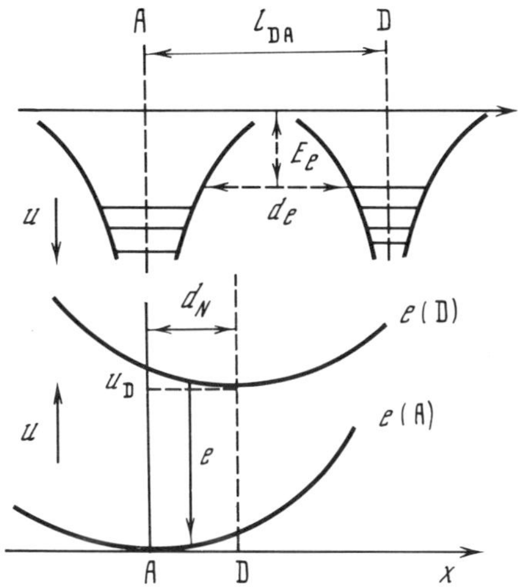

Figure 2.1 Electron-nuclear tunneling in the electron transition from the reducer-donor (D) to the oxidizer-acceptor (A). Above — one of the external (valent) electrons tunnels over the distance d_e between two potential wells formed by Coulomb attraction of the electron to the nuclei D and A (the electron levels in the D and A wells do not generally coincide); below — the potential curves of the interatomic interaction in the reducer molecule before (curve e(D)) and after (curve e(A)) departure of the electron.

ing atoms and ω is the vibration frequency). So the displacement of the nuclei results in the appearance of a factor $F_v \ll 1$, called the Franck–Condon factor, in the expression for the rate constant of the redox process.

Electron–nuclear tunneling was discovered by Chance and De Vault in 1966, and the results were published in [26,27]. They were the first to observe the low-temperature (120–124 K) plateau in the rate of electron chemical transfer in the process of the cytochrome-C oxidation by chlorophyll in the photosynthesis of *Chromatium D* bacterium, the rate constant at the plateau being $K = 300\ s^{-1}$. The foundation was thereby laid for a rapid development of electron-tunneling studies in biological processes, biopolymers and membranes (see e.g. [28,29]). The electron transfer in such systems is similar to the abovementioned fast conversion of Turnbull's blue into Prussian blue, which has been studied by Mössbauer emission spectroscopy [24,25].

Another class of redox tunneling processes is relevant to the long-range electron transfer from traps to scavengers or that between ions with variable valence in frozen irradiated vitreous alkaline water–alcohol solutions. This class of reactions, in which rate plateaus are also observed, is of essential importance in radiation chemistry. Studies of these processes started from the observation of a plateau in the rate of trapped-electron scavenging by 0^- ions: $e_{tr}^- + 0^-$ products in 10 M NaOH solution [30,31]. This direction in the study of electron tunneling is reviewed in detail in [32].

There is another reason to mention [30,31], since so-called polychronic (nonexponential) kinetics [33,34] was first considered there for tunneling reactions in which the spread of the reactants with respect to two parameters (height and width of the potential barrier) is to be taken into account. In polychronic kinetics the reagent concentrations are known to decrease linearly with the logarithm of the time. A thorough analysis of such kinetic curves using the algorithm given in [30,31] allows data to be obtained on the spatial and energy distributions of electron traps in irradiated solids and, in general, of any reactant subjected to tunneling.

After De Vault and Chance's discovery [26,27] it became necessary to consider the role of electronic–vibrational interactions and Franck–Condon factors in redox conversions of complicated biomolecules. Concepts of radiationless transitions were used [35–39]. References [40–44] describe some versions of electron tunneling in biopolymers in terms of this theory as generalized electron–nuclear tunneling.

Thus the low-temperature limit of the electron transfer rate in redox processes is the first convincing evidence of electron–nuclear tunneling, and has led to extensive studies of this effect in biology and chemistry.

However, this kind of electron transfer can hardly be referred to as a chemical reaction in the genuine sense of the word; more naturally it is related to tunneling phenomena in solid-state physics and electronics. According to the definition of a chemical reaction, it is necessary to have a rearrangement of atoms and change of the lengths and/or angles of the valence bonds, and it is by no means sufficient just to have small atomic displacements at distances of the order of the zero-point vibration amplitude.

2.2. DISCOVERY OF THE QUANTUM LOW-TEMPERATURE LIMIT OF THE CHEMICAL REACTION RATE

The low-temperature quantum limit of the chemical reaction rate was first observed in spontaneous chain growth during the radiation-induced polymerization of solid formaldehyde (the studies were performed in 1970–1973 [45–50]). The results obtained are described in [49,50] (see also the reviews [51–56]).

In the experiments the polymer chain lengths and growth rate were measured at 4.2–140 K by high-sensitivity and rather fast ($\tau_c \approx 0.3$–1 s) calorimeters of adiabatic and diathermal types [44–50].

The most important result is rather high radiation-chemical polymerization yield, i.e. the number of converted molecules per 100 eV of absorbed energy of penetrating radiation.

Since for reactions other than those of chain type the yield is close to unity, it is possible to assume $g \approx 1$ for the initiation of a polymer chain and $G = g\tilde{\nu} = \tilde{\nu}$ for polymer formation, i.e. the yield is numerically approximately equal to the kinetic length of the chain. Measurements [45–48] have shown that $\tilde{\nu} \approx 10^7$ at 140 K, 10^5 at 77 K and 10^3 at 4.2 K. Thus it was demonstrated for the first time that spontaneous chemical transformations occur even at liquid-helium temperatures. Later the existence of such transformations was convincingly proved by the observation of postpolymerization of formaldehyde at 5 K after the γ-source was switched off [51]. Kinetic measurements [46–50] give the time constant of a polymer chain growth τ and the average time necessary to add a link to the chain, $\tau_0 = \tau/\tilde{\nu}$. At about 140 K the temperature dependence is found to be satisfactorily described by the Arrhenius law with $E = 0.1$ eV, but at lower temperatures growth is slower and approaches a plateau at $T \leqslant 12$ K (see Fig. 2.2). The limit $(\tau_0)_{T\to 0} = 10^{-2}$ s, while an extrapolation according to the Arrhenius law with $E = 0.1$ eV gives $(\tau_0)_{10K} = 10^{30}$ years and $(\tau_0)_{4K} = 10^{100}$ years.

Formaldehyde polymerization is certainly a typical chemical reaction in the proper sense of the word. In the course of the polymerization the stacks of plane triangular CH_2O molecules with double C=O bonds are transformed into long chains of tetrahedral $C\diagup^{O}\diagdown C\diagup^{O}$ bonds, the substitution of the van der Waals radii of the C and O atoms by shorter radii of the valence C—O bonds (see Fig. 2.3) resulting in a noticeable (by about 40%) increase of the substance density.

To explain the low-temperature limit of the formaldehyde-chain growth rate, the authors of [49,50] introduced the concept of the molecular tunneling as illustrated in Fig. 2.4. Here the dissociation energy of the C—O bond is $D \approx 4$ eV, the sublimation energy of the crystalline formaldehyde is $\Delta H_s \approx 0.3$–0.4 eV, the polymerization energy $Q \approx 0.37$ eV and the activation energy $\tilde{E} \approx 0.1$ eV.

According to this initial simple concept, the addition of a new link to the polymer chain involves the penetration of a monomer CH_2O

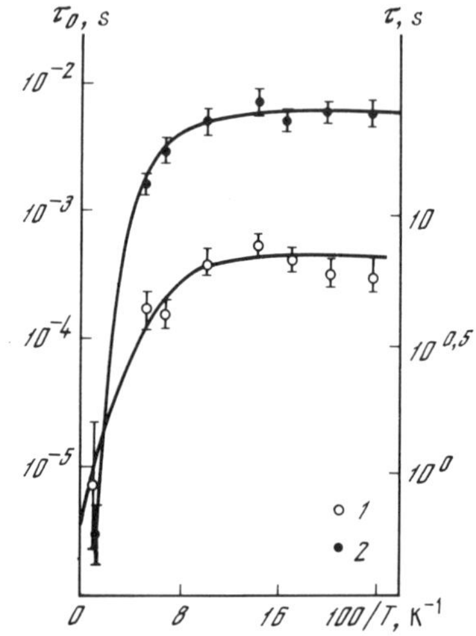

Figure 2.2 Temperature dependence (in the Arrhenius coordinates) of the growth time of the polymer chain, τ (1), and of the average time of the new link addition to this chain, τ_0 (2).

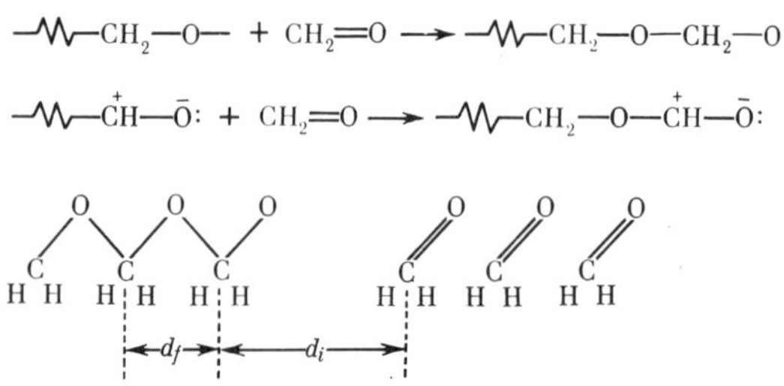

Figure 2.3 Possible elementary processes and a simplified geometrical scheme of the formaldehyde polimerization.

molecule through the activation barrier. Then the systems is prepared for the next stage of growth via the quantum diffusion of the cavity that appears near the newly formed end group of the polymer chain owing to contraction of the substance.

To evaluate the molecular tunneling rate, the authors [49,50] used the concept of radiationless relaxation of electronically excited molecular states, applying the formulation proposed by Siebrand [57,58]. Thus chemical reactions in solids are considered as radiationless transitions in which the initial reactants play the role of an electronically

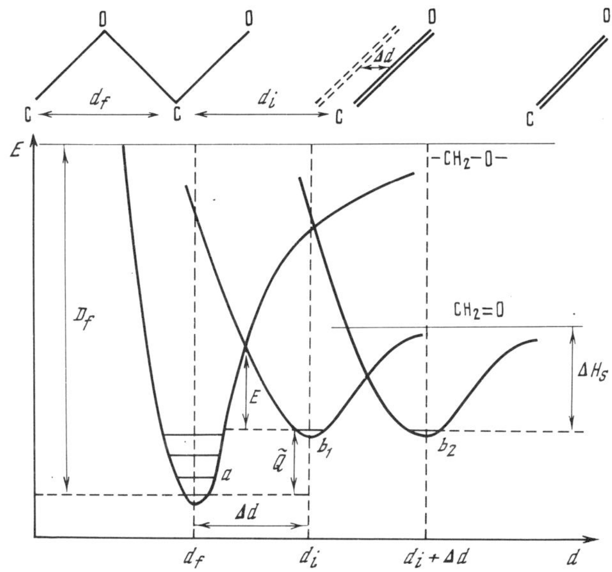

Figure 2.4 Molecular tunneling at an example of a simplified pattern of the formaldehyde molecules displacing in the course of its solid phase polymerization. Above — a scheme of the arrangement of molecules in the original monomer (double bonds C=0 are removed both from each other and from the end of the growing polymer chain) and of bonds —C—O—C—O— in the polymer chain (the distance between the neighbouring C atoms is here $d_f = d_i - \Delta d$); below — the vertical axis shows the potential energy, the horizontal axis shows the distance between the neighbouring C atoms.

excited state, while the products play the role of the ground state of the system.

Now the probability of a radiationless transition is defined as the product of the squared matrix element of the electronic transition $L_e^2 = |\langle \Psi_{ef}|L|\Psi_{ef}\rangle|^2$ and the Franck-Condon factor $F_v = |\langle \Psi_{vf}|\Psi_{vi}\rangle|^2$, i.e.

$$W_{fi} = (2\pi/\hbar)\, L_e^2 F_v \rho_f = \omega_{eff}\, F_v, \tag{2.1}$$

where Ψ_e and Ψ_v are the electronic and vibrational wave functions, L_e is the operator of the transition-inducing interaction (in particular, the nonadiabaticity operator), and $\rho_f = (\hbar\omega_f)^{-1}$ is the density of vibrational levels in the ground state. The wave function $\Psi_e(x) \sim \exp[-x/\alpha]$ (for $x > \alpha$), where $\alpha = (\hbar/(2m\tilde{E})^{1/2}$ is the damping parameter, i.e. the scale of the electron density localization, which is approximately equal to the length of the valence bond at $E \approx 1$ eV. The wave function

$$\Psi_v(X) \sim \exp[-(X/\Delta)^2] \quad (X > \Delta), \tag{2.2}$$

where $\Delta = (\hbar/M\omega)^{1/2}$ is the nuclear vibration amplitude, ω is the vibration frequency and M is the nuclear mass. The tunneling length (or, more generally, the configurational displacement length) is a very important parameter in the problem of electron and molecular tunneling; it will be denoted by d_e for electrons and d_N for heavy particles.

$d_e \gg \alpha \gg d_N = \Delta$ for electron–nuclear tunneling in redox processes [40,44]. In this case the long-range electron transfer from the donor to the acceptor is not followed by a chemical reaction, and the nuclei of the donor and acceptor shift over $\Delta \approx 0.1$ Å just as in the case of intramolecular radiationless transitions.

For molecular tunneling in chemical reactions $d_N \gg \Delta$. For large electron ($d_e \gg \alpha$) and nucleus ($d_N \gg \Delta$) displacements the matrix elements in the transition probability are determined by the asymptotic behaviour of the electronic and vibrational wave functions. Then the matrix element of the electronic transition takes the form of the ordinary Gamov tunneling factor $L_e^2 \sim \exp[-\xi d_e(m\tilde{E})^{1/2}/\hbar]$, the factor $\xi \approx 1$ depending on the barrier shape. The vibrational (nuclear) Franck–Condon factor depends exponentially on the squared nuclear displacement: $F_v \sim \exp[-(d_N(M\omega/\hbar)^{1/2})^2]$, where $\omega = \omega_i \ll \omega_f$.

Parameters of various processes, which can be interpreted in terms of radiationless transitions, are listed in Table 2.1.

A treatment of the experimental results on formaldehyde polymerization carried out in [49,50] in terms of the theory from [57,58] gives the activation barrier for the polymer chain growth, $d_N = 0.4$–0.5 Å, in good agreement with the expected increase of the polymer density as compared with that of the monomer. It is interesting that the product $d(M_{CH_2O}E)^{1/2}$ determining the tunneling probability is close to a similar product for spontaneous nuclear decay in which the

Table 2.1 General features of processes interpreted in terms of the theory of radiationless transitions

Process	d_e	$\omega_{ef}/\omega_0{}^a$	d_N	F_v	$\omega_{exp}\,(s^{-1})$
Electronic–vibrational relaxation		10^{-3}		10^{-3}	10^{14}–10^{8}
Electron tunneling in reduction–oxidation reactions					
without taking account of displacements of the nuclei	(10 × 30)	10^{-8}	0	1	10^{6}–10^{-2}
taking account of displacements of the nuclei	10	10^{-5}		10^{-3}	
Molecular tunneling		10^{-3}		10^{-4}	10^{10}–10^{-4}

[a] $\omega_0 = 10^{14}\ s^{-1}$ is the typical value of the preexponential factor for chemical reactions.

barrier width approaches 10^{-12} cm and the height is of the order of 10^6 eV. An alternative model of formaldehyde polymerization [59,60] was later proposed in which τ was considered as a characteristic of the molecular tunneling transformation of a primary radiation-induced active center into a secondary one this immediately leading to a barrierless growth of long polymer chains.

Nowadays, there are about ten examples of low-temperature reaction-rate limits, which were observed in many laboratories in the USSR, the USA, Canada and Japan. These works are reviewed below. In comments [61,61] on the last book by Bell [9] the chemical conversions observed at ultralow temperatures (say, at 4.2 K), the discovery and investigation of the low-temperature reaction-rate limits, and their description in terms of radiationless transition theory are mentioned as the principal features of contemporary studies on heavy-particle tunneling.

Before proceeding to discussion of the theoretical and experimental results obtained since 1973 [49,50], it is worth noting once more that tunneling does not necessary result in a low-temperature reaction-rate limit — but such a limit, if it really exists, is the best proof of tunneling, or, more generally of the quantum nature of activation-barrier penetration. It should be emphasized that the barrier can substantially decrease and narrow (or even disappear), as predicted by a modified theory of radiationless transitions described below [63–66]. This theory considers two-mode Franck–Condon factors, reflecting the presence not only of *intra*molecular high-frequency vibrations, but also of *inter*molecular low-frequency vibrations, as well other types of intermolecular motion, and accounting for their interaction. There is the possibility of considerable narrowing and lowering of the activation barrier, going so far as its complete disappearance (Fig. 2.5). Molecular tunneling may therefore be, in a broad sense, described as a process that proceeds both by means of the penetration of heavy particles through the activation barrier and removal of this barrier through intermolecular motion [63–66].

Figure 2.5 shows three versions of different arrangements of the reactions. The intermolecular vibrations change the distance between neighboring groups from Δd_{min} (left), through equilibrium Δd_0 (central) to Δd_{max} (right). It is clear that the activation barrier narrows and decreases in height on moving to the left, and the reaction probability increases on moving to the right; i.e. when the reactants are separated from each other the activation barrier widens and increases in height and the reaction probability decreases.

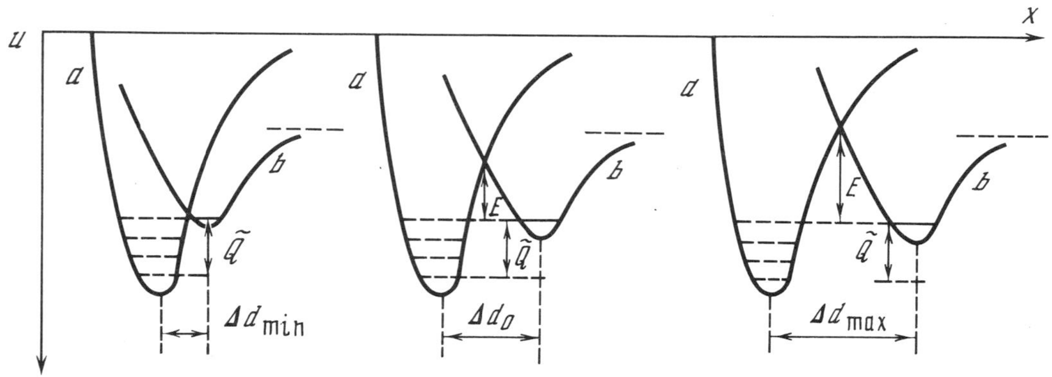

Figure 2.5 A generalized pattern illustrating the origin of the quantum low temperature limit of the chemical reaction rate.

2.3. OTHER EXAMPLES OF QUANTUM LOW-TEMPERATURE REACTION-RATE LIMITS (EXPERIMENTAL DATA OBTAINED IN 1975–1983)

After studies of radiation-induced solid-phase polymerization [45–50], which had led to the discovery of quantum low-temperature reaction-rate limits [49–50], such limits were detected and studied in a number of chemical reactions (see the reviews [51–56]). All experimental data on low-temperature kinetic plateaus are presented in Fig. 2.6 and are discussed briefly below.

Frauenfelder *et al.* [67–70] thoroughly investigated kinetics of the ligand rebinding (CO, O_2) with the heme iron of myoglobin (Mb), of isolated α- and β-chains in hemoglobin, and of protoheme. These studies were carried out over a wide temperature range (2–320 K) and time scales of 10^{-6}–10^3 s. The bonds were ruptured by a short laser pulse ($\approx$ 1 μs), the consequent rebinding then being measured by spectroscopic techniques to attain a plateau at temperatures below 10 K.

To interpret the data obtained, the iron atom in the initial compounds, bonded by six coordinated bonds, is assumed to be in the heme plane. On abstraction of O_2 or CO, the atom leaves the plane and takes a new equilibrium position. It is not until exothermal ligand rebinding that the Fe atom returns to the heme plane, i.e. it shifts over the distance d by either surmounting the barrier height $\tilde{E}$ or by tunneling through it. At low temperatures tunneling prevails, thus resulting in a low-temperature plateau of the Fe-ligand rebinding rate.

Frauenfelder and collaboraters [67–70] observed a typical polychronic Fe–O_2 and Fe–CO rebinting. An analysis of the kinetic curves using the methods developed in [30,31,33,34] allowed them to calculate the height distribution of the potential barriers $g(\tilde{E})$ and to assume that the barrier width is related to the height as $d(\tilde{E}) = d_0(\tilde{E}/E_p)^\delta$, where $E_p \approx 1$ kcal/mol, $\delta = 1.5$–0.4. From the Gamov tunneling-rate function, the barrier width was found to be $d_0 = 0.5 \pm 0.1$ Å. This simple pattern of molecular tunneling seems to be directly verified by studies of the isotope effects in the kinetics of binding to myoglobin iron at 20 K and 10 K of $^{12}C^{16}O$, $^{13}C^{16}O$ and $^{12}C^{18}O$ molecules [71]. A detailed analysis of the kinetic curves under conditions of polychronic kinetics led to the proposal of a rather complicated system of several (up to four) potential barriers, which hinder the backward motion of ligands liberated from the heme and dispersed from the protein molecule into the solvent. The role of diffusion in every stage of Fe–ligand rebinding was later studied in detail [72].

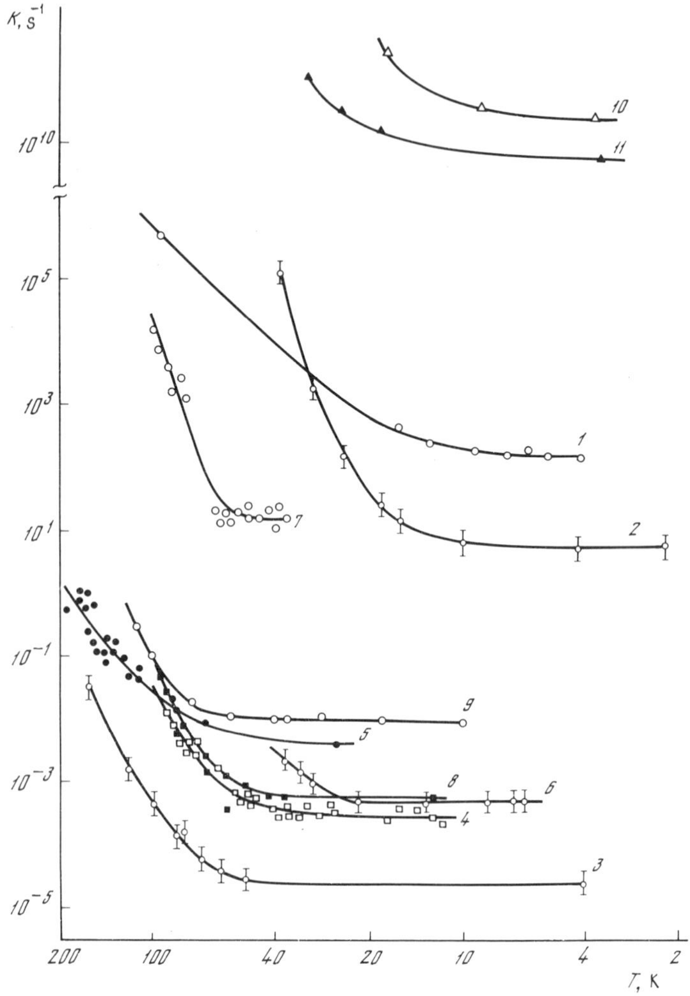

Figure 2.6 Examples of the low temperature limit of the rate of chemical reactions observed in various laboratories in USSR, USA, Canada and Japan.

The plot is presented in the Arrhenius coordinates: the temperature values are represented in the absciss axis as the 1/T values, the rate constants in the ordinate axis are plotted in the logarithmic scale. 1 — growth of the formaldehyde polymerization chains (USSR, 1973,/46–50); 2 — Fe–CO rebinding after the laser flash photolysis in hemeglobine (USA, 1975,/69/); 3 — isomerization of the radical pairs in — irradiated dimethylglyoxime (Japan, 1977,/82/); 4,5 — hydrogen abstraction by methyl radicals in the frozen glassy methanol (4) and ethanol (5) (Canada; USA, 1977,/81,13/); 6 — H atoms transfer in the sterically hindered isomerization of the aryl radicals (USA, 1978,/84/); 7 — formation of the C–O bonds in the cyclopentadienil biradicals (USA, 1979,/75/); 8 — chain hydrobromination of ethylen (USSR, 1978,/76–78/); 9 — chlorination of butyl chloride by the molecular chlorine (USSR, 1980,/79,80/); 10,11 — photoinitiated conversion of the ordinary and deuterated rodopsin into prelumirodopsin (USA, 1977,/95/).

However, direct experiments performed recently using the EXAFS technique show that the distance between the Fe atom and CO ligand changes by only d = (0.05 ± 0.03)Å in the photolytic rupture of bonds in carboxymyoglobin [73]. The kinetics of the geminate recombination of CO with myoglobin has been studied at room temperature on a time scale of the order of nanoseconds [74] and compared with earlier results obtained at low temperatures [67–72]. From this it appears that, according to the above generalized approach to molecular tunneling including intermolecular motion as the principle feature [63–66] (see Fig. 2.5), the low-temperature plateau of Fe–ligand rebinding [69,71] is due not only to small Fe and CO displacements but also to the accompanying increase in size of the residual (now five-coordinated) heme resulting from lengthening of the bonds between the central Fe atom and the four proximate (pyrrole) nitrogen atoms.

Another example of a low-temperature plateau, interpreted as a consequence of a heavy multi-atomic group tunneling was observed by Buchwalter and Closs [75] in the photolysis of 2,3-diazabicyclo [2.2.1]hept-2-ene 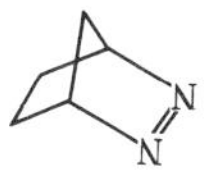

molecules in a vitreous cyclohexane matrix, whose primary product is the biradical cyclopentane-1,3-diyl  (CPDI). On switching off the light source, the ESR signal of this radical decreases owing to its conversion into bicyclo[2.1.0]pentane 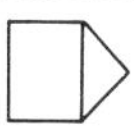 (BCP). The gas–liquid chromatography identification of the final product allows one to prefer this version of the CPDI conversion as compared to that resulting in [2.1.0]-1-bicyclo[2.1.0]pent-1(5)ene  (CP).

CPDI → BCP kinetics turns out to be non-exponential, the rate constant as calculated from the monomolecular law decreasing four times during an hour at 5.5 K. In deriving the Arrhenius and tunneling parameters, the initial slopes of the kinetic curves are used.

These experiments were performed at temperatures of 1.3–50 K. The Arrhenius dependence of the CPDI → BCP conversion rate is observed above 20 K, the activation energy being E ≈ 0.1 eV and the pre-exponential factor 10^8 s^{-1}. At lower temperatures the rate approaches a plateau. As assumed in [75] this plateau is due to the methylene group CH_2 tunneling (M = 14 dalton) during the conversion of the plane CPDI radical to a BCP molecule bent along the line of the bridging C—C bond and to the simple parabolic shape of the potential barrier. They thus obtained the barrier width d = 0.64 Å. The conversion of CPDI into cyclopenten would have corresponded to a hydrogen-atom shift with an unrealistic barrier width d = $0.64(14)^{1/2}$ Å = 2.4 Å. Another piece of evidence against hydrogen tunneling is the relatively weak kinetic isotope effect at the low-temperature plateau when deuterating the initial compounds — within a time sufficient for the effective CPDI → BCP conversion only about 20% of the deuterated byradical is converted.

The utility of the kinetic isotope effect in revealing the reaction mechanisms of the low-temperature plateau is clearly shown in the study of ethylene hydrobromination at 30–90 K [54,76–78]. When this reaction is initiated by irradiation with ^{60}CO γ-rays at low temperatures, long chains are observed: $\widetilde{\nu} = 10^5$ at 90 K, 2×10^3 at 70 K, and 8000 at 40 K [54,76], just as in formaldehyde polymerization. The duration of an elementary act of chain growth (formation of a new C_2H_5Br molecule) approaches a plateau at T ≈ 60 K, with $\tau_0 = 0.1$ s [77].

The scheme of chain growth in ethylene hydrobromination after γ-ray initiated HBr → H + Br decay is

$$\dot{B}r + C_2H_4 \longrightarrow C_2H_4Br + 0.35\ eV, \tag{3.1}$$

$$HBr + \dot{C}_2H_4Br \longrightarrow \dot{B}r + C_2H_5Br + 0.45\ eV \tag{3.2}$$

(the chain does not grow via H atoms at low temperatures, since the second stage $HBr + \dot{C}_2H_5 \longrightarrow \dot{H} + C_2H_5Br - 0.9$ eV is endothermic).

The question arises as to whether the observed reaction rate is limited owing to hydrogen tunneling in (3.2) or due to reactant molecules re-approaching one another in (3.1) and (3.2).

To solve this problem, the elementary acts of the formation of chains in ethylene hydrobromination have been compared for the systems $DBr + C_2H_4$, $HBr + C_2D_4$ and $DBr + C_2D_4$ [78].

The simplest Gamov description of hydrogen tunneling gives $\tau_0^{-1} \simeq \nu_0 \exp[-\beta_0 d_H (M_H E_H)^{1/2}/\hbar]$, where $\beta_0 \approx 1$ and depends on the barrier shape. Since in experiments $\tau_0^{-1} \approx 10\ s^{-1}$ and ν_0 is usually 10^8–$10^{13}\ s^{-1}$, deuteration would have resulted in decreasing the tunneling rate by 3–5 orders of magnitude. On the other hand, the Franck–Condon factors determining the rates of (3.1) and (3.2) are very weakly affected by the deuteration of HBr and C_2H_4; the expected decrease of the reaction rates is only a factor of 2–4. This pattern is observed in the work reported in [78], this being a convincing argument for taking into account not only pure tunneling of heavy particles through the activation barrier, but also the reduction of the barrier by intermolecular motion [63–66].

Furthermore, the transport of hydrogen atoms from HBr to C_2H_5Br does not limit the ethylene hydrobromination rate in the Arrhenius region (T = 90 K, E = 0.17 eV), which clearly follows from the absence of the strong semiclassical isotope kinetic effect in this region caused by the difference between the frequencies of zero-point vibrations of Z–H and Z–D bonds.

To conclude the list of examples of the low-temperature kinetic plateaus due to the heavy atoms and multiatomic groups transfer in chemical conversions, we should mention the photochlorination of vitreous matrices of butyl chloride, methylcyclohexane and ethanol [79,80] (the methylcyclohexane matrix has been studied in the greatest detail [80]. Cl_2 photolysis is carried out using a pulsed nitrogen laser with an intensity of 5×10^{15} photons cm^{-2} s^{-1}. These intensities do not result in any significant heating of the reaction zone, which is only 1–2 K above the thermostate temperature of 20 K. The rate of the reaction of chlorine matrix radicals was determined from the abundance of both chlorine and radicals. It was shown by photometric and ESR techniques that in the range 10–60 K the chlorination rate constant does not depend on temperature, but above 60 K it grows substantially with temperature [79,81]. This phenomenon can be interpreted in terms of tunneling with due account taken of the effect of the approach of reactants upon intermolecular vibrations.

Among the processes exhibiting low-temperature plateaus of special importance are those including hydrogen transitions since in this case tunneling is very probable owing to the low hydrogen mass. Hudson, Shiotany and Williams [81] were the first to observe the low-temperature limit of the chemical reaction rate for hydrogen transfer. The ESR technique was used to study hydrogen abstraction from the methanol molecule by methyl radicals at temperatures of 15–89 K. Similarly to [67–70,75], non-exponential kinetics was observed, so the rate constants were determined from the initial slopes of the kinetic curves. At 40 K a low-temperature plateau was observed, while at higher temperatures the rate increased noticeably with the temperature. It was shown that the temperature dependence of the rate constant cannot be explained within the framework of one-dimensional hydrogen tunneling for reasonable parameters of the potential barrier [81].

LeRoy, Murai and Williams [13] found that the rate constant for hydrogen abstraction from the ethanol molecule by the methyl radical behaves in a similar way at temperatures of 13–99 K [13], the low-temperature limit being achieved at 40–50 K. This paper also studied the reaction of the methyl radicals with methanol within the temperature range 13–99 K, the results being very close to those of [81].

The model of solid-phase low-temperature reactions proposed in [13] interprets the experimental data on the hydrogen abstraction from molecules by methyl radicals. This model will be discussed in the next section.

Another example of the low-temperature limit of the hydrogen-transfer rate constant was observed by Toriyama, Nunome and Iwasaki [82], who studied radical pair transformation in irradiated dimethylglyoxime at 4.2–140 K by ESR. The same reaction, but at higher temperatures, was earlier studied by Lebedev and Yakimchenko [83], who found a noticeable deviation from the Arrhenius law and a considerable isotope effect (H/D). The experimental results obtained there were compared with calculations performed on the basis of one-dimensional tunneling through an asymmetric Eckart barrier of height 15 kcal/mol and half-width 0.4 Å. The preexponential frequency factor is taken to be 7.6×10^{12} s^{-1}. Experiments and calculations were shown to be in good agreement at $T \approx 50$ K only if at $T = 4.2$ K they differ by an order of magnitude, i.e. no agreement between the theory and experiment is possible over the whole temperature range.

Intramolecular hydrogen transport has also been shown to have a low-temperature limit of the rate constant [84]. Brunton, Gray, Griller, Barclay and Ingold [84,85] studied the isomerization of aryl radicals, using ESR, over a wide temperature range from 28 K to room temperature. The temperature dependence of the isomerization rate constant was found to differ strongly from the Arrhenius one, the rate being almost constant at temperatures below 100 K, i.e. there is a low-temperature plateau. In addition, the low-temperature isomerization of deuterated aryl radicals was studied. Measurements performed at 123–233 K show a considerable deviation from the Arrhenius law and a strong kinetic isotope effect. The experimental data have been treated in terms of one-dimensional tunneling through potential barriers of different shapes, but no agreement between experiment and theory has been obtained for any barrier shape.

As well as the reactions with low-temperature rate limits, the tunneling mechanism is believed to be involved in hydrogen abstraction from acetonitrile by the methyl radical. In this reaction a decrease of the CH_3 radical abundance is followed by an increase of the CH_2CN radical abundance, the total radical abundance being constant [86–88,13]. Measurements [13] indicate that the temperature dependence of the reaction rate constant at 69–112 K is characterized by an apparent activation energy of 3–10 kJ/mol, which is much smaller than that for the reaction in the gaseous phase ($\approx$ 42 kJ/mol). This implies hydrogen tunneling from the acetonitrile molecule to the radical. Finally, Sprague [88] has directly shown that this reaction proceeds via tunneling, since the kinetic isotope effect (H/D) at 77 K is found to be 20 times larger than without tunneling.

An indication of tunneling is also observed in the hydrogen abstraction from methyl isocyanide by methyl radicals. The activation energy of this reaction at 77–125 K is 6–20 kJ/mol, but it does not proceed when H is replaced by D [13].

Tunneling effects are thus of importance in the reactions of atomic hydrogen with molecules. These have been studied over a wide temperature range for a number of compounds. Dubinskaya and Butyagin [89] have studied the temperature dependence of the hydrogen-abstraction rate constant from polyisobutylene and from polyethylenoxide at 100–150 K and have obtained rather low activation energies: 2 kcal/mol for polyisobutylene and 1.2 kcal/mol for polyethylenoxide. As mentioned in [89,90], this points to the process having a tunneling character. Similar reactions with the participation of hexadecane and malonic acid are examined in [91]. Their rate constants are shown to change weakly at temperatures below 150 K, the activation energy being 1–2 kcal/mol, while at 200–250 K the activation energy for the hydrogen reaction in malonic acid is 6 kcal/mol. All this indicates that below 150 K the reactions of atomic hydrogen proceed via tunneling.

The reactions of atomic hydrogen with molecules have been investigated at extremely low temperatures (10–30 K). Iwasaki *et al.* have studied the reaction of thermalized hydrogen atoms with ethane molecules in xenon and methane matrices using ESR [92,93]. The radiolysis or photolysis of HI molecules at 4.2 K produces hydrogen atoms, which are immediately captured by the inert matrix. At higher temperatures the hydrogen atoms become mobile and react with ethane.

Interesting tunneling effects were revealed by Wang and Willard [94], who investigated reactions of hydrogen and deuterium atoms stabilized in alcan glasses at 5–29 K. For the hydrogen reactions the activation energy is positive at 18–29 K, zero at 10–18 K and negative at temperatures below 10 K. For deuterium the activation energy is constant at 18–29 K, and then decreases with temperature, not going through any minimum.

In 1977, the low-temperature plateau of the rate of photoinitiated rhodopsin conversion into prelumirhodopsin was discovered by means of the picosecond laser spectroscopy [95]. This reaction remains as yet the only example with a plateau observed in both the normal and deuterated compounds, since, in contrast with the radiation-induced conversions of dimethylglyoxime [82,83], the kinetic isotope effect at the rhodopsin plateau is rather weak at temperatures below 20 K: $K_H/K_D = 7$. A rather large rate is characteristic of the rhodopsin → prelumirhodopsin conversion, which is $K_H \approx 2.8 \times 10^{10}\ s^{-1}$ at the low-temperature plateau.

Different opinions concerning the mechanism that controls the rate of rhodopsin → prelumirhodopsin conversion at low temperatures have been put forward. Peters *et al.* [95] believe that in this case one or two hydrogen atoms tunnel to nitrogen over a distance of 0.5–0.9 Å. References [96–98] also suggest a tunneling mechanism for hydrogen, but the tunneling is assumed to occur in opsin. On the other hand, Green *et al.* [99], Hurleg *et al.* [100] and Doukas *et al.* [101] believe that the low-temperature plateau is determined by the cis–trans isomerization of the chromaphor–retinal bonded to opsin. A final conclusion in favour of one or another version cannot yet be made. Keeping in mind the rather large reaction rate, the possibility cannot be excluded that there is participation of excited rhodopsin molecules in the initial stage of photoreception.

2.4. TECHNIQUES FOR CALCULATING TUNNELING RATE CONSTANTS IN SOLID-PHASE REACTIONS

Early papers discussing the role of tunneling in chemical reactions made use of equation (1.2) for the transition probability per unit time, which reproduced qualitatively correctly the dependence of W on the parameters of the potential barrier through which the particle tunneled. The temperature dependence of the rate constant was assumed to be determined by the thermal occupation of the higher energy levels by the particle (see [9]); the potential barrier looks lower and narrower for the transitions from higher levels, which increases the probability.

The most complete treatment of the experimental data on hydrogen abstraction by methyl radicals was carried out, using this mechanism, by LeRoy, Murai and Williams [13]. The r.m.s. technique determines two presumably independent parameters of the Eckart and Gauss potential barriers, i.e. their width and height. Temperature dependences were studied for five solid-phase tunneling abstractions of hydrogen by methyl radicals. For these, the parameters giving the best fit of theory and experiment are as follows: the barrier width varies for different reagents from 1.7 to 2.3 Å and the height from 11 to 15 kcal/mol. But, in fact, the height and width of the potential barriers in chemical reactions cannot be treated as independent. According to the linear model for the reactions AH + B → A + BH, the potential energy of hydrogen atoms transfering from A to B is determined by just one parameter — the A-to-B spacing [102,103]. Each spacing unambiguously determines both the width and height of the potential barrier. So the barrier width, 1.7–2.3 Å, obtained in [13] corresponds to rather larger A-to-B spacings, 3.9–4.5 Å, and for these the interaction between molecules is small and the barrier height must significantly exceed 11–15 kcal/mol. LEPS calculations (London, Eiring, Poliany and Sato; see e.g. [104]) result in higher potential barriers that are 3–5 times as high.

A higher level occupation by the tunneling particle has also been studied by Jortner and Ulstrup [105] using the theory of radiationless transitions. They managed to follow the crossover from low-temperature tunneling to the high-temperature limit described by an activation formula. A radiationless transition of the particle between two parabolic terms was investigated, the motion in each term being specified by the frequency ω. According to this model, the low-temperature limit of the chemical reaction rate is reached at $T \approx \hbar\omega/k_B$, ω being an intermolecular frequency; this implies a rather high temperature of the order of several hundreds of degrees, in contrary with the experimental value, which is only several tens of degrees.

Therefore the temperature dependence caused by the thermal occupation of the higher energy levels by the tunneling particle, though producing a low-temperature rate limit, does not describe the laws experimentally observed in solid-phase tunneling reactions if reasonable parameters are used. Moreover, it can easily be verified that this mechanism is capable of producing a temperature-dependent rate constant for the tunneling reactions only at temperatures where overbarrier transitions prevail. This is demonstrated here for the parabolic potential barrier [106] by using the quasiclassical probability for particle transition from the nth level:

$$W_n = \nu_n \exp\left[-\frac{\pi}{\hbar}\left(\frac{M}{2\widetilde{E}_0}\right)^{1/2} x_0 \widetilde{E}_n\right], \tag{4.1}$$

where $\widetilde{E}_0$, $\widetilde{E}_n$ are the potential-barrier heights for the ground and nth levels respectively, and x_0 is the barrier width at the ground level. The occupation ratio of the nth level, p_n, is related to that of the ground level, p_0, by $p_n/p_0 = \exp[-(E_0 - E_n)/k_BT]$. The occupation of the higher levels need only be taken into account when $p_nW_n \geqslant p_0W_0$. Using (4.1), the condition

$$T \geqslant \frac{\hbar}{\pi k_B x_0}\left(\frac{2\widetilde{E}_0}{M}\right)^{1/2} \tag{4.2}$$

follows from this inequality. Comparing (1.3) and (4.2), the thermal occupation of the higher levels by the tunneling particle may be neglected at temperatures where tunneling prevails.

The contribution of tunneling from higher vibrational levels is studied in [107] using the example of hydrogen abstraction from acetonitrile by methyl radicals. The LEPS technique was used to calculate the potential energy of the abstracted hydrogen [104]. It was shown that, according to the above mechanism of higher-level occupation, the rate constant might have started increasing only above 300 K, while, in fact, a two order-of-magnitude increase of the tunneling rate constant is already revealed within the temperature range 69–112 K [13,86,87].

The fact that the higher-vibrational-level occupation at low temperatures does not significantly affect the tunneling rate is easily understood. An increase of the tunneling probability on transfering to a higher level is compensated by the small occupation of this level at the temperatures considered. It is because of the competition of these factors that the equilibrium occupation of the higher energy levels by the tunneling particle starts to play a role only at high enough temperatures. Therefore three temperature ranges can be distinguished:

(i) a wide temperature range where tunneling from the ground state prevails ($T < T_t$);

(ii) a wide temperature range where overbarrier transitions prevail ($T > T_t$);

(iii) a narrow range at $T \approx T_t$ including temperatures corresponding to the maximum contributions of the higher vibrational levels to the rate constant.

Here T_t is defined by (1.3).

According to our considerations, then, the thermal occupation of the higher levels by the tunneling particle may be disregarded, and only the ground-level transitions will henceforth be taken into account.

The next important step in the study of solid-phase tunneling reactions was the introduction of the environmental reorganization caused by the transition. Marcus [21,108] was the first to consider this mechanism for electron transfer, the environment being treated as classical. A further development in this direction was due to Levich and Dogonadze, who made use of the theory of radiationless transitions (see [22,109]) and the polaron concept [111] in studying some specific features of electron transfer in polar media (the latter being approximated as harmonic). It is of note that this approach explains the peculiar temperature dependence of the electron-transfer rate constant, which is independent of temperature (quantum medium) at low temperatures, but on heating has an activation-like behavior. At high temperatures the results obtained by means of the theory of radiationless transitions coincide with the Marcus formula [21,108] to within the preexponential factor.

When calculating the tunneling rate constant the above papers use the adiabatic approximation, which ceases to be valid if the tunneling time exceeds the period of molecular vibrations in the medium. Ivanov and Kozhushner [112,113] elaborated a theory of the electron tunnel transitions, taking the finite tunneling time into account.

Ovchinnikov and Ovchinnikova [114] studied electron transfer, taking account of the whole continuous spectrum of dielectric losses. This theory assumes only a linear response of the dielectric environment to the field variation caused by the electron transfer, rather than using an oscillatory model for the medium. The dependence of the electron transfer rate on the temperature and the heat of reaction was studied and the role of high-frequency molecular vibrations in solvate complexes was also discussed.

A detailed presentation of electron-transfer theory and an application of the theory of radiationless transitions in this field can be found in [115–117]. The results obtained in this way have been applied to calculations of the rate constants for heavy-particle tunneling. Franck–Condon factors have been evaluated, with due account taken of variations of the equilibrium coordinates and the phonon frequencies caused by the transition [14,50,105,116,118]. Transitions with rearrangements of the nuclear subsystems of molecules [119] have also been considered. Aleksandrov and Smedarchina [120] studied the effect of the condense environment on chemical reaction rates in terms of the general theory of quasistationary fluctuations. Aleksandrov and Goldanskii [121] demonstrated that the reaction rate in a cell depends on the viscosity of the environment [121].

These investigations show that the temperature dependence of the predicted rate constants in solid-phase chemical reactions, in which heavy particles tunnel, is in many respects determined by the energy of the environmental reorganization. The latter is large only for charge transfer in a polar host or in the case of significant conformational reorganization of the molecules participating in the reaction. As for hydrogen abstraction by methyl radicals, for which the tunneling mechanism of hydrogen transfer can be assumed to be proved, the energy of the environmental reorganization for this is small and cannot account for the observed temperature dependence of the rate constant. The same holds for the radical pair transformation in dimethylglyoxime.

This is the reason to look for some other mechanisms explaining the temperature dependence of the rate constant of tunneling radical reactions. The role of the phonon subsystem in the transitions should be examined. Since the energy levels in the initial and final potential wells do not, as a rule, coincide, the particle is transfered only when one or more phonons are absorbed and/or emitted. This takes place simultaneously with tunneling if the intermolecular (phonon) subsystem interacts with the transfering particle. To be more exact, the coupling of the inter- and intramolecular subsystems should be dealt with, since the potential wells between which the transfer proceeds are usually formed as a result of particle interaction within the molecule.

The coupling may, on one hand, be caused by the influence (during transition) of the intramolecular subsystem on the intermolecular one, i.e. the environmental changes due to the transition [21–23,108,109] and, on the other hand, by the phonons affecting the intramolecular subsystem when the energies of the levels occupied by the tunneling particle depend on the phonon coordinates. This mechanism is taken into account when dealing with the tunneling transitions that give rise to anomalous properties of amorphous solids at low temperatures [122] (see also Chapter 6).

Another mechanism of phonon influence on the system in which the transition occurs is possible. Intermolecular vibrations cause oscillations of the distances between reactants; i.e. the potential wells between which the heavy particle is transfered are periodically displaced with respect to each other (see Fig. 2.5). This results in a dependence of the potential-barrier parameters on the phonon coordinates or, in other words, in the appearance of a fluctuating barrier [123] determining the coupling of two subsystems.

References [63–66] show that this latter mechanism results in a temperature dependence of the tunneling rate constant. The higher the temperature, the larger the thermal vibration amplitude and the smaller the effective tunneling distance; hence the tunneling probability is higher.

To confirm that this model produces an exponential dependence of the rate constant on temperature, the following evaluation can be made [124]. The potential wells of the initial and final states vibrate classically and approach each other with probability proportional to $\exp[-\gamma_R(R - R_0)^2/2k_BT]$, where γ_R is the force constant and R is the well-to-well spacing, whose equilibrium value is R_0. The tunneling probability per unit time is given by (1.2), which is conveniently written

$$W(R) = \nu_0 \exp[-J(R)], \quad J(R) = \frac{2}{\hbar}\int_a^b dr\{2M[U(R,r) - E(R)]\}^{1/2}. \quad (4.3)$$

Here E(R) is the energy level and U(R,r) is the potential energy of the tunneling particle; r is the intramolecular coordinate.

The function J(R) is now expanded in powers of the reactant displacement from equilibrium, with terms up to quadratic being retained:

$$J(R) = J(R_0) + J'(R_0)(R - R_0) + \frac{1}{2} J''(R_0)(R - R_0)^2. \tag{4.4}$$

Then (4.3) is averaged over the reactant displacement from equilibrium to give

$$\overline{W} = \nu_0 \left[1 + J''(R_0) \frac{k_B T}{\gamma_R}\right]^{-1} \exp\left\{- J(R_0) + \frac{[J'(R_0)]^2 k_B T}{2[\gamma_R + J''(R_0) k_B T]}\right\}. \tag{4.5}$$

It follows from (4.5) that thermal fluctuations of the potential-barrier parameters result in an exponential dependence of the tunneling rate constant on temperature. It should be emphasized that for a particle that is transfered between two parabolic wells the approximate relation (4.4) turns out to be exact.

Retaining only two terms in the expansion (4.4), i.e. assuming $J''(R_0) = 0$, the exponent depends linearly on temperature:

$$\overline{W} = \nu_0 \exp\left\{- J(R_0) + [J'(R_0)]^2 \frac{k_B T}{2\gamma_R}\right\}. \tag{4.6}$$

This formula was first obtained in [63]. If, however, the quadratic term is retained in (4.4) then at high temperatures the rate constant has the usual activation form. In fact (4.5) is more conveniently written as

$$\overline{W} = \nu_0 \exp\left\{- J(R_0) + \frac{[J'(R_0)]^2}{2J''(R_0)} - \frac{\gamma_R [J'(R_0)]^2}{2J''(R_0) [\gamma_R + J''(R_0) k_B T]}\right\}, \tag{4.7}$$

from which at $T > \gamma_R / J''(R_0) k_B$ there immediately follows an Arrhenius-type activation dependence. This expression (4.7) for the rate constant was first obtained in [65]. Calculations of the function J(R) for realistic potential barriers [66] show that, considering the rate constants of the tunneling reactions [13,81,82], only the linear term in the expansion (4.4) need be retained.

It is possible to use the estimate (4.5) for the tunneling rate constant only if the intermolecular vibrations are treated as classical; i.e. (4.5) is appropriate only if the temperature is close to the Debye temperature. At low temperatures the solid-state lattice is a quantum-dynamical system and should be treated on equal footing with the intramolecular subsystem. A consistent calculation of rate constants for solid-phase tunneling reactions shows that their values and temperature dependences are in many respects determined by the phonon spectrum of the solid. And the low-temperature limit corresponds to the temperature at which the intermolecular vibrations are frozen [66].

At present the model of solid-phase reactions incorporating the effect of intermolecular vibrations on the potential barrier has become generally accepted. Apart from the references given above, there are papers by Ovchinnikov and Benderskii and their collaborators [125–127] where special attention is paid to studies of the influence of phase transitions on chemical reactions [127]. McKinnon and Hurd [128] and Siebrand, Wildman and Zgierski [129] have carried out comparisons of the temperature dependence due to higher-level occupation [13] with that proposed in [63], and come down in favor of the latter one. In particular, the predicted linear temperature dependence of the logarithm of the rate constant [63] has been experimentally verified. A physically similar model of the elementary processes of proton transfer was studied by Krishtalik [130]. Ivanov and Kozhushner [131] considered nonadiabatic effects when accounting for the influence of intermolecular vibrations on the tunneling transfer rate of an atom in a solid.

The approach of reactants in gas-phase reactions involving tunneling has been studied by George and Miller [102], Marcus and Coltrin [132] and Altcorn and Schatz [133], who showed that the most favorable path at low temperatures does not coincide with the reaction coordinate. Instead, there is penetration of the barrier, with the path shape depending on the energy of the relative motion of the reactants. This description of reactant approach was also used by Ovchinnikova [64] in studying tunneling reactions in a solid.

Chemical reactions due to the tunneling transitions of particles are not the only manifestations of the tunneling effect in chemical kinetics. ESR spectra are known to depend on molecular motions if their frequencies are comparable to the corresponding splittings in the spectra. If the frequency is smaller than the splitting then the molecules and their fractions can be treated as motionless.

The frequency of the classical motion has an activation-type temperature-dependence, so decreasing the temperature should halt all molecular motions in a solid. However, in some cases this cannot be done owing to tunneling arising in the system with a frequency that remains finite even at zero temperature. Moreover, quantum motions can result in ESR spectra that differ from those of the motionless or classically moving particles. Thus, for example, it is pointed out in [134] that there is a possibility of using radiospectroscopy to observe dynamic delocalization of particles in chemical reactions.

Thus the interpretation of ESR spectra — one of the most powerful techniques in chemical physics — appears in many cases to be impossible without incorporation of tunneling rotations. The quantum motion of nuclei is most pronounced in the ESR spectra of radicals containing methyl groups. Quantum rotation of one methyl group and influence on the ESR spectrum were studied theoretically by Freed [135]; this theory is based on a model in which a rigid symmetrical methyl group rotates in a potential field generated by the remaining part of the radical and by the environment.

Freed's theory has been generalized in various respects: in particular, investigations have been made of the rotation of a nonrigid asymmetrical methyl group, the anisotropy of the hyperfine interaction, and methyl-group relaxation in ESR spectra. Of great interest is a theory of quantum-correlated rotations of two methyl groups and its influence on the ESR spectrum [136].

Unfortunately there is as yet no detailed review of the principal results on methyl-group rotations and their role in the interpretation of spectra and relaxation in magnetic resonance that reflects different points of view concerning the relation between quantum and classical rotations.

References

1. Hund, F. (1927) *Z. Phys.*, **43**, 805.
2. Bourgin, D.F. (1929) *Proc. Natl Acad. Sci.* USA, **15**, 357.
3. Roginsky, S.Z. and Rozenkevich, L. (1930) *Z. Phys. Chem.*, **10B**, 47.
4. Wigner, E. (1932) *Z. Phys. Chem.*, **19B**, 1903.
5. Bell, R.P. (1933) *Proc. R. Soc. Lond.*, **A139**, 466.
6. Bell, R.P. (1935) *Proc. R. Soc. Lond.*, **A148**, 241.
7. Bell, R.P. (1937) *Proc. R. Soc. Lond.*, **A158**, 128.
8. Bell, R.P. (1973) *The Proton in Chemistry*. 2nd edn. Cornell University Press, Ithaca, N.Y.
9. Bell, R.P. (1980) *The Tunnel Effect in Chemistry*. Chapman and Hall, London.
10. Goldanskii, V.I. (1959) *Dokl. AN SSSR*, **124**, 1261.
11. Goldanskii, V.I. (1959) *Dokl. AN SSSR*, **127**, 1037.
12. Khristov, S.G. (1961) *Dokl. AN SSSR*, **136**, 663.
13. Le Roy, R.J., Murai, H. and Williams, F. (1980) *J. Am. Chem. Soc.*, **102**, 2325.
14. Buhks, E. and Jortner, J. (1980) *J. Phys. Chem.*, **84**, 3370.
15. Goldanskii, V.I., Firsov, V.G. and Shantarovich, V.P. (1964) *Dockl. AN SSSR*, **155**, 636.
16. Goldanskii, V.I., Firsov, V.G. and Shantarovich, V.P. (1965) *Kinetika i Kataliz*, **6**, 564.
17. Goldanskii, V.I. and Firsov, V.G. (1971) *Annu. Rev. Phys. Chem.*, **22**, 209.
18. Firsov, V.G. and Goldanskii, V.I. (1975) In: *Inorganic Chemistry*. Vol. 8 Radiochemistry ed. A.G. Maddock), pp. 1–48. Butterworths, London.
19. Libby, W.F. (1952) *J. Phys. Chem.*, **56**, 863.
20. Marcus, R.A. Zwolinski, B.J. and Eyring, H. (1954) *J. Phys. Chem.*, **58**, 432.

21. Marcus, R.A. (1956) *J. Chem. Phys.*, **24**, 966.
22. Levich, V.G. and Dogonadze R.R. (1959) *Dokl. AN SSSR*, **124**, 123.
23. Levich, V.G. (1965) *Adv. Electrochem.*, **4**, 249.
24. Alekseev, V.P., Goldanskii, V.I., Prusakov, V.E. Nefed'ev, A.V., Stukan, R.S. (1972) *Pisma ZhETF*, **16**, 65.
25. Goldanskii, V.I. (1983) *Pure Appl. Chem.*, **55**, 11.
26. De Vault, D. and Chance, B. (1966) *Biophys. J.*, **6**, 825.
27. Chance B., De Vault D., Legallais V. *et al.* (1967) In: *Fast Reactions and Primary Processes in Chemical Kinetics* (ed. S. Claesson), p. 437. Almqvist and Wiksel, Stockholm.
28. Tunneling in Biological Systems (ed. B. Chance). (1979) *Academic Press*, New York.
29. De Vault, D. (1980) *Q. Rev. Biophys.*, **13**, 387.
30. Zamaraev, K.I., Khairutdinov, R.F., Mikhailov, A.I. and Goldanskii, V.I. (1971) *Dokl. AN SSSR*, **199**, 640.
31. Goldanskii, V.I., Zamaraev, K.I., Mikhailov, A.I. and Khairutdinov, R.F. (1973) *Problei.. f the Kinetics of Elementary Chemical Reactions.* (In Russian.), p. 68. Nauka, Moscow.
32. Zamaraev, K.I. and Khairutdinov, R.F. (1980) *Sov. Sci. Rev.*, **2**, 357.
33. Primack, W. (1955) *Phys. Rev.*, **100**, 1677.
34. Primack, W. (1960) *J. Appl. Phys.*, **31**, 1524.
35. Radiationless Processes in Molecules and Crystals (ed. F.K. Fong). (1976) Springer-Verlag, Berlin.
36. Robinson, G.W. (1974) In: *Excited States* (ed. E.C. Lim). Academic Press., New York.
37. Freed K.F. (1972) In: *Stereo- and Theoretical Chemistry*, p. 98. Springer Verlag, Berlin.
38. Jortner J. (1971) In: *Proc. 2nd Int. Symp. on Organic Chemistry*, p. 128. Butterworth, London.
39. Henry, B.H. and Kasha, M. (1968) *Annu. Rev. Phys. Chem.*, **19**, 161.
40. Grigorov, L.N. and Chernavskii, D.S. (1972) *Biofizika*, **17**, 195.
41. Blumenfeld, L.A. and Chernavskii, D.S. (1973) *J. Theor. Biol.*, **39**, 1.
42. Hopfield, J.J. (1974) *Proc. Natl. Acad. Sci.* USA, **71**, 3640.
43. Hopfield, J.J. (1977) *Biophys. J.*, **18**, 311.
44. Jortner, J. (1976) *J. Chem. Phys.*, **64**, 4860.
45. Kiryukhin, D.P., Kaplan, A.M., Barkalov, I.M. and Goldanskii, V.I. (1970) *Vysokomol. Soedin.*, **126**, 491.
46. Kiryukhin, D.P., Kaplan, A.M., Barkalov, I.M. and Goldanskii, V.I. (1971) *Dokl. AN SSSR*, **199**, 857.
47. Kiryukhin, D.P., Kaplan, A.M., Barkalov, I.M. and Goldanskii, V.I. (1972) *Vysokomol. Soedin.*, **14A**, 2115.
48. Kiryukhin, D.P., Kaplán, A.M., Barkalov, I.M. and Goldanskii, V.I. (1972) *Dokl. AN SSSR*, **206**, 147.
49. Goldanskii, V.I., Frank-Kamenetskii, M.D. and Barkalov, I.M. *Dokl. AN SSSR*, **211**, 133.
50. Goldanskii, V.I., Frank-Kamenetskii, M.D. and Barkalov, I.M. (1973) *Science*, **182**, 1344.
51. Kiryukhin, D.P., Barkalov, I.M. and Goldanskii, V.I. (1974) *Vysokomol. Soedin.*, **165**, 565.
52. Goldanskii, V.I. (1975) *Usp. Khim.*, **44**, 2121.
53. Goldanskii, V.I. (1976) *Annu. Rev. Phys. Chem.*, **27**, 85.
54. Goldanskii V.I. (1979) In: *Tunneling in Biological Systems* (ed. B. Chance). Academic Press, New York.
55. Goldanskii, V.I. (1978–1979) *Chem. Scripta*, **13**, 1.
56. Goldanskii, V.I. (1979) *Nature*, **279**, 109.
57. Siebrand, W. (1967) *J. Chem. Phys.*, **46**, 440.

58. Siebrand, W. (1967) *J. Chem. Phys.*, **47**, 2411.
59. Gerasimov, G.N., Bazilevskii, M.V., Tikhomirov, V.A. and Abkin, A.D. (1979) *Dokl. AN SSSR*, **244**, 1379.
60. Bazilevskii, M.V., Gerasimov, G.N., Kitrosskii, L.N. *et al.* (1981) *Kinetika i Kataliz*, **22**, 1134.
61. Marcus, R.A. (1981) *Nature*, **293**, 683.
62. Le Roy, R.J. (1982) *Phys. Today*, **35**, 85.
63. Klochikhin, V.L., Pshezhetskii, S.Ya. and Trakhtenberg, L.I. (1978) *Dokl. AN SSSR*, **239**, 879.
64. Ovchinnikova, M.Ya. (1979) *Chem. Phys.*, **36**, 85.
65. Benderskii, V.A., Goldanskii, V.I. and Ovchinnikov, A.A. (1980) *Chem. Phys. Lett.*, **73**, 492.
66. Trakhtenberg, L.I., Klochikhin, V.L. and Pshezhetskii, S.Ya. (1981) *Chem. Phys.*, **59**, 191; (1982) *Ibid.*, **69**, 121.
67. Austin, R.H., Beeson, K.W., Eisenstein, L., Frauenfelder, H. and Gunsalus, I.C. (1975) *Biochemistry*, **14**, 5355.
68. Alberding, N., Austin, R.H., Chan, S.S., Eisenstein, L., Frauenfelder, H., Gunsalus, I.C. and Nordlund, T.M. (1976) *J. Chem. Phys.*, **65**, 4701.
69. Alberding, N., Austin, R.H., Beeson, K.W., Chan, S.S., Eisenstein, L., Frauenfelder, H. and Nordlung, T.M. (1976) *Science*, **192**, 1002.
70. Alberding, N., Chan, S.S., Eisenstein, L., Frauenfelder, H., Good, H., Gunsalus, I.C., Nordlund, T.M., Perutz, M.F., Reynolds, A.H. and Sorensen, L.B. (1978) *Biochemistry*, **17**, 43.
71. Alben, J.O., Beece, D., Bowne, S.F., Eisenstein, L., Frauenfelder, H., Good, D., Marden, M.C., Moh, P.P., Reinisch, L., Reynolds, A.H. and Yue, K.T. (1980) *Phys. Rev. Lett.*, **44**, 1157.
72. Debrunner P. and Frauenfelder, H. (1982) In: *Experiences in Biochemical Perception*, p. 327. Academic Press, New York.
73. Chance, B., Fischetti, R. and Power, L. (1983) *Biochemistry*, **22**, 3820.
74. Henry, E.R., Sommer, J.H. Hofrichter, J. and Eaton, W.A. (1983) *J. Mol. Biol.*, **166**, 443.
75. Buchwalter, S.L. and Closs, G.H. (1979) *J. Am. Chem. Soc.*, **101**, 4688.
76. Kiryukhin, D.P., Barkalov, I.M. and Goldanskii, V.I. (1978) *Dokl. AN SSSR*, **238**, 388.
77. Kiryukhin, D.P., Barkalov, I.M. and Goldanskii, V.I. (1979) *J. Chem. Phys.*, **76**, 1013.
78. Barkalov, I.M., Goldanskii, V.I., Kiryukhin, D.P. and Zanin, A.M. (1980) *Chem. Phys. Lett.*, **73**, 273.
79. Misochko, E.Ya., Philippov, P.G., Benderskii, V.A., Ovchinnikov, A.A., Barkalov, I.M. and Kiryukhin, D.P. (1980) *Dokl. AN SSSR*, **253**, 163.
80. Philippov, P.G., Misochko, E.Ya., Kulikov, A.V., Benderskii, V.A. and Ovchinnikov, A.A. (1981) *Dokl. AN SSSR*, **256**, 1173.
81. Hudson, R.L., Shiotany, M. and Williams, F. (1977) *Chem. Phys. Lett.*, **48**, 193.
82. Toriyama, K., Nunome, K. and Iwasaki, M. (1977) *J. Am. Chem. Soc.*, **99**, 5823.
83. Yakimchenko, O.E. and Lebedev Ya.S. (1971) *Int. J. Rad. Phys. Chem.*, **3**, 17.
84. Brunton, G., Gray, J.A., Griller, D., Barclay, L.R. and Ingold, K.V. (1978) *J. Am. Chem. Soc.*, **100**, 4197.
85. Brunton, G., Griller, D., Barclay, L.R.C. and Ingold, K.V. (1976) *J. Am. Chem. Soc.*, **98**, 6803.
86. Sprague, E.D. and Williams, F. (1971) *J. Am. Chem. Soc.*, **93**, 787.
87. Le Roy, R.J., Sprague, E.D. and Williams, F. (1972) *J. Phys. Chem.*, **76**, 546.
88. Sprague, E.D. (1977) *J. Phys. Chem.*, **81**, 516.
89. Dubinskaya, A.M. and Butyagin, P.Yu. (1973) *Dokl. AN SSSR*, **211**, 141.
90. Dubinskaya, A.M. (1978) *Usp. Khim.*, **47**, 1169.
91. Dubinskaya, A.M. and Yusupov, N.N. (1975) *Izv. AN SSSR, Khim.*, 2095.

92. Iwasaki, M., Toriyama, K., Muto, H. and Nunome, K. (1978) *Chem. Phys. Lett.*, **56**, 494.
93. Toriyama, K., Nunome, K. and Iwasaki, M. (1980) *J. Phys. Chem.*, **56**, 2374.
94. Wang, H.Y. and Willard, J.E. (1979) *J. Phys. Chem.*, **83**, 2585.
95. Peters, K., Applebury, M.L. and Rentzepis, P.M. (1977) *Proc. Natl. Acad. Sci.* USA, **74**, 3119.
96. Lewis, A. (1978) *Proc. Natl. Acad. Sci.* USA, **75**, 540.
97. Warshel, A. and Deakyne, C. (1978) *Chem. Phys. Lett.*, **55**, 459.
98. Ippen, E.P., Shank, C.V., Lewis, A. and Marcus, R.A. (1978) *Science*, **200**, 1279.
99. Green, B.H., Monger, T.G., Alfano, R.R., Aton, B. and Challenger, R.H. (1977) *Nature*, **269**, 179.
100. Hurley, J.B., Ebrey, T.G., Honig, B. and Ottolenghi, M. (1977) *Nature*, **270**, 540.
101. Doukas, A., Aton, B., Challenger, R.H. and Ebrey, T.G. (1978) *Biochemistry*, **17**, 2430.
102. George, T.F. and Miller, W.H. (1972) *J. Chem. Phys.*, **57**, 2458.
103. Potential Energy Surfaces and Dynamic Calculations (1982) (ed. D.G. Truhlar). Plenum Press, New York, London.
104. Kondrat'ev, V.N., Nikitin, E.E., Reznikov, A.I. and Umanskii, A.I. (1976) *Thermal Biomolecular Reactions in Gases*. (In Russian.) Nauka, Moscow.
105. Jortner, J. and Ulstrup J. (1979) *Chem. Phys. Lett.*, **63**, 236.
106. Klochikhin, V.L. Pshezhetskii, S.Ya. and Trakhtenberg, L.I. (1979) *Vysokomol. Soedin.* **27A**, 2792.
107. Klochikhin, V.L., Livshin, L.I., Pshezhetskii, S.Ya. and Trakhtenberg, L.I. (1980) *Khim. Vysok. Energii*, **14**, 329.
108. Marcus, R.A. (1957) *J. Chem. Phys.*, **26**, 867.
109. Dogonadze, P.P. and Kuznetsov, A.M. (1964) *Izv. AN SSSR, Khim.*, 2140.
110. Perlin, Yu.E. (1963) *Usp. Fiz. Nauk*, **80**, 553.
111. Pekar, S.I. (1951) *A Study on the Electron Theory of Crystals*. (In Russian.) Gostekhizdat, Moscow Leningrad.
112. Ivanov, G.K. and Kozhushner, M.A. (1978) *Fiz. Tverd. Tela*, **20**, 9.
113. Ivanov, G.K. and Kozhushner, M.A. (1982) *Khim. Fiz.*, **1**, 1039.
114. Ovchinnikov, A.A. and Ovchinnikova, M.Ya. (1969) *Zh. Eksp. Teor. Fiz.*, **56**, 1278.
115. Dogonadze, R.R., Kuznetsov, A.M. and Marsagishvili, T.A. (1980) *Electrochim. Acta*, **25**, 1.
116. Ulstrup, J. (1979) *Charge Transfer Processes in Condensed Media*. Springer-Verlag, Berlin.
117. Ovchinnikov, A.A. and Ovchinnikova M.Ya. (1982) *Adv. Quant. Chem.*, **16**, 161.
118. Dogonadze, P.P. and Kuznetsov A.M. (1973) In: *Physical Chemistry*. Kinetics. (In Russian.) VINITI, Moscow. (Itogi Nauki i Tekhniki Vol. 2).
119. Sarai, A. and Kakitani, T. (1981) *Chem. Phys. Lett.*, **77**, 427.
120. Aleksandrov, I.B. and Smedarchina, Z.K. (1982) *Khim. Fiz.*, **1**, 346.
121. Aleksandrov, I.V. and Goldanskii, V.I. (1984) *Chem. Phys.*, **87**, 455.
122. Jäckle, J., Piché L., Arnold, W. and Hunklinger S. (1976) *J. Non-Cryst. Solids*, **20**, 365.
123. Kagan, Yu. and Klinger, M.I. (1976) *Zh. Eksp. Teor. Fiz.*, **70**, 255.
124. Klochikhin, V.L. and Trakhtenberg, L.I. (1981) *Dokl. AN SSSR*, **259**, 399.
125. Benderskii, V.A., Dakhnovskii, Yu.I., Ovchinnikov, A.A. and Philippov, P.G. (1981) *Dokl. AN SSSR*, **261**, 653.
126. Benderskii, V.A., Philippov, P.G., Dakhnovskki, Yu.I. and Ovchinnikov, A.A. (1982) *Chem. Phys.*, **67**, 301.
127. Dakhnovskii, Yu.A., Ovchinnikov, A.A. and Benderskii, V.A. (1982) *Chem. Phys.*, **66**, 93.
128. McKinnon, W.R. and Hurd, C.M. (1983) *J. Phys. Chem.*, **87**, 1283.
129. Siebrand, W., Wildman, T.A. and Zgierski, M.Z. (1983) *Chem. Phys. Lett.*, **98**, 108.

130. Krishtalik, L.I. (1979) *Kinetika i kataliz*, **20**, 847.
131. Ivanov, G.K., Kozhushner, M.A. (1983) *Khim. Fiz.* **1**, 1299.
132. Marcus, R.A. and Coltrin, M.E. (1977) *J. Chem. Phys.*, **67**, 2609.
133. Altkorn, R.I. and Schats, G.C. (1980) *J. Chem. Phys.*, **72**, 3337.
134. Blumenfeld, L.A., Goldanskii, V.I., Podgoretskii, M.I. and Chernavskii, D.S. (1967) *Zh. Strukt.*, **8**, 854.
135. Freed, J.H. (1965) *J. Chem. Phys.*, **43**, 1710.
136. Shteinshneider, N.Ya. (1975) *Teor. Eksp. Khim.*, **11**, 313.

CHAPTER 3
Rate Constant of Solid-Phase Chemical Reactions. Theory

The mathematics used to describe chemical reactions and other radiationless transitions in solids is presented in this chapter. The technique is similar to that used in the conventional theory of radiationless transitions, but in contrast with the latter, it goes beyond the scope of the Franck–Condon approximation, which does not hold for the heavy-particle transfers.

The presentation starts with a discussion of solid-phase chemical reactions; a derivation of the rate constants for the nonadiabatic and adiabatic transitions is given. A double adiabatic approximation is established that allows separation of the nuclear subsystem into two parts: rapid (intramolecular vibrations) and slow (intermolecular vibrations). Criteria for its applicability are given.

The transition rate constant under the double adiabatic approximation is calculated, with oscillations of the potential-barrier parameters due to the intermolecular vibrations, environmental reorganization and energy transfer to intermolecular vibrations being taken into account. Two cases are studied, in which the final intramolecular state is discrete or continuous.

A study of the temperature dependent rate constant is then presented for various models of solids. The transition from tunneling to the activation mechanism is followed.

The role of anharmonic intermolecular vibrations in the elementary processes of solid-phase chemical reactions is discussed.

Charge transfer in the solid-phase is of special interest. Here the importance of the dynamical properties of the potential barrier is emphasized.

Orientational vibrations in a solid lattice are studied; together with translational vibrations, these can force oscillations of the potential barrier crossed by the particle.

3.1. MODEL OF CHEMICAL REACTIONS IN THE SOLID PHASE. NONADIABATIC AND ADIABATIC TRANSITIONS

Similarly to the treatment in [1, 2], a chemical reaction is considered here as a radiationless transition of a particle from an initial bonding term ϵ_i to a final term ϵ_f, which may be either bonding (Fig. 3.1a) or repulsive (Fig. 3.1b). The initial and final terms are solutions of a Schrödinger equation with the electronic–nuclear Hamiltonian H_e^0, obtained by neglecting the nuclear kinetic energy (adiabatic approximation) and the interaction of the terms,

$$H_e^0 \psi(\rho,\mathbf{r}) = \epsilon(\mathbf{r})\,\psi(\rho,\mathbf{r}), \quad H_e^0 = T_e + U_e. \tag{1.1}$$

Here $\rho,\mathbf{r}$ are the electronic and nuclear coordinates; T_e is the electronic kinetic-energy operator; $U_e(\rho,\mathbf{r})$ is the potential energy of the whole system.

As is well-known, the eigenvalues of (1.1) are the initial, $\epsilon_i(\mathbf{r})$, and final, $\epsilon_f(\mathbf{r})$, states to which the electronic wave functions $\psi_{i,f}(\rho,\mathbf{r})$ correspond. Taking account of the interactions between the terms or between the terms and the environment, V, in the vicinity of intersection of the function $\epsilon_i(\mathbf{r})$ and $\epsilon_f(\mathbf{r})$, the electronic wave function may be sought as a

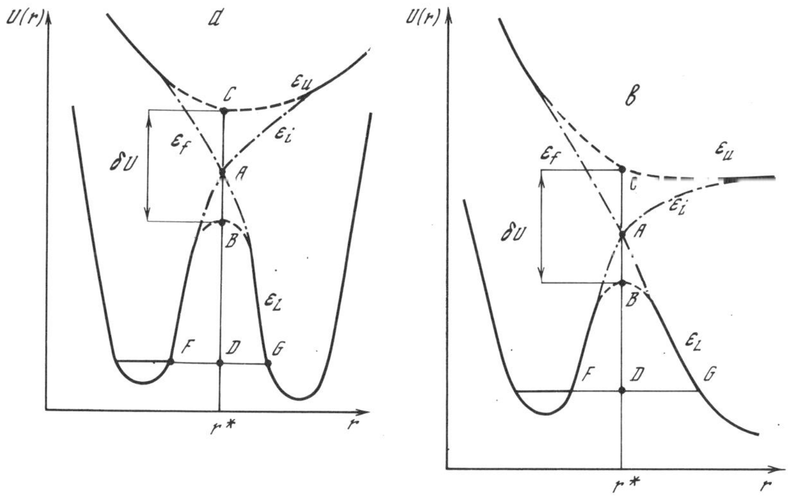

Figure 3.1 A scheme of the electronic terms in the system of reacting molecules. Solid and dashed lines show the upper (ϵ_u) and lower (ϵ_l) adiabatic terms. Solid and dashed-point lines correspond to diabatic terms of the initial (ϵ_i) and final (ϵ_f) states of the system.

linear combination of ψ_i and ψ_f. Substituting such a linear combination into the Schrödinger equation with the Hamiltonian

$$H_e = H_e^0 + V \tag{1.2}$$

results in a secular equation determining the adiabatic terms of the system. Solutions of the secular equation are

$$\epsilon_{u,\ell} = \frac{1}{2}\left\{ H_{ii}^e + H_{ff}^e \pm \left[(H_{ii}^e - H_{ff}^e)^2 + 4(H_{fi}^e)^2 \right]^{1/2} \right\}, \tag{1.3}$$

Only real wave functions will be used below, so the matrix elements may be written omitting the sign of complex conjugation.

Figure 3.1 depicts the upper and lower adiabatic states $\epsilon_u(r)$ and $\epsilon_\ell(\mathbf{r})$. As follows from (1.3), the intersecting terms $\epsilon_i(\mathbf{r})$ and ϵ_f are split and converted into nonintersecting terms $\epsilon_u(r)$ and $\epsilon_\ell(\mathbf{r})$ [3], with the electronic wave functions ψ_u and ψ_ℓ corresponding to them. Here the splitting δU is twice the exchange integral:

$$\delta U = 2\, H_{fi}^e(\mathbf{r}^*). \tag{1.4}$$

Whether the transition is nonadiabatic or adiabatic depends on whether the electronic subsystem has time to rearrange during the particle transition from the initial into the final state. The transition time (which should be distinguished from the transition waiting time, which is equal to the inverse of the transition probability per unit time) is determined by the term-to-term spacing FG and the particle velocity, which in turn depends on the potential-barrier height and the particle mass M:

$$v_0 = \{2\,[E - \epsilon_{i,f}(\mathbf{r}^*)]/M\}^{1/2}. \tag{1.5}$$

Here E and v_0 are the particle energy level and velocity. For overbarrier transitions the velocity is real; but it is imaginary for underbarrier transitions.

The time necessary for the electronic subsystem to rearrange is determined by the relation between the height BD of the potential barrier crossed by the particle (to be more exact, the difference between the energy at which the terms ϵ_i and ϵ_f intersect each other and that of the initial state AD) and the splitting BC.

If the splitting is small compared with the barrier height (BC $\ll$ BD) and the particle goes through the region of quasiintersection of terms fairly rapidly (a more exact criterion will be given below) then the electronic subsystem has no time to rearrange. The nuclear subsystem moves first along the term ϵ_i and there is then a nonadiabatic transition to the term ϵ_f. The functions ψ_i and ψ_f are therefore used as the initial and final electronic wave functions.

If the splitting of the terms is large and the particle under-barrier motion is not rapid then the electronic subsystem has time to rearrange and the transition is of adiabatic type along the term $\epsilon_\ell(\mathbf{r})$, the initial and final states of the electronic subsystem being described by the same function $\psi_\ell(\rho,\mathbf{r})$. As for the initial and final nuclear wave functions, they are not solutions of the Schrödinger equation with the potential energy $\epsilon_\ell(\mathbf{r})$, but either describe particle localization in the left- and right-hand wells (see Fig. 3.1a) or localization in the left-hand potential well together with particle propagation to the right from the barrier (see Fig. 3.1b).

Therefore the rate constants for the adiabatic and nonadiabatic transitions are calculated in different ways. The type of transition is determined by the dimensionless parameter [4]

$$\xi_0 = \frac{2\pi\, \delta U^2}{\hbar |v_0|\, |F_i - F_f|} \,. \tag{1.6}$$

Here $F_{i,f}$ are the slopes of the terms $\epsilon_{i,f}$ at the intersection point; since F_i and F_f are of different signs (i.e. $F_i > 0$, $F_f < 0$), their difference cannot be zero. For $\xi_0 \ll 1$ the transition is nonadiabatic, while for $\xi_0 \gg 1$ it is adiabatic. This criterion holds for both real and imaginary velocities [5].

Those transitions are, as a rule, nonadiabatic that proceed between states with small splittings. That is why a reaction in which the electronic spin of the system changes is nonadiabatic, since the splitting in this case is determined by the weak spin–orbit coupling. If the spin and symmetry of the system do not vary during the reaction then the splitting is usually large and the reaction is adiabatic.

3.2. RATE CONSTANT OF NONADIABATIC TRANSITIONS

An electronic–nuclear system evolves according to the time-dependent Schrödinger equation

$$i\hbar \frac{\partial \Psi(\rho,\mathbf{r},t)}{\partial t} = (T_N + H_e)\, \psi(\rho,\mathbf{r},t), \tag{2.1}$$

where T_N is the kinetic-energy operator of the nuclear subsystem.

In the case of a nonadiabatic transition, the wave function of the system is represented as a linear combination of the initial and final wave functions with coefficients depending on time and nuclear coordinates:

$$\Psi(\rho,\mathbf{r},t) = u_i(\mathbf{r},\mathbf{t})\,\Psi_i(\rho,\mathbf{r}) + u_f(\mathbf{r},t)\,\Psi_f(\rho,\mathbf{r}). \tag{2.2}$$

Substituting (2.2) into (2.1) yields

$$i\hbar\,(\dot{u}_i\Psi_i + \dot{u}_f\Psi_f) - (T_N + H_e)\,(u_i\Psi_i + u_f\Psi_f) = 0 \tag{2.3}$$

Then multiplying (2.3) from the left by the electronic wave functions Ψ_i, Ψ_f and integrating with respect to the electronic coordinates, the system of equations determining the coefficients u_i and u_f is obtained:

$$\left.\begin{aligned}\left[i\hbar\frac{\partial}{\partial t} - (T_N + H^e_{ii} + \Lambda_{ii})\right]u_i - (H^e_{if} + \Lambda_{if})u_f = 0,\\ \left[i\hbar\frac{\partial}{\partial t} - (T_N + H^e_{ff} + \Lambda_{ff})\right]u_f - (H^e_{fi} + \Lambda_{fi})u_i = 0,\end{aligned}\right\} \tag{2.4}$$

$$\Lambda_{jj'} = \int d\rho\,\Psi_j[T_N, \Psi_{j'}], \tag{2.4'}$$

where $[a,b]$ is the commutator of the quantities a and b.

Solutions of (2.4) are sought in the form

$$\begin{aligned}u_i(r,t) &= a_i(t)\,u_i^0(r)\exp\left[-\frac{iE_i^0t}{\hbar}\right],\\ u_f(\mathbf{r},t) &= \sum_f a_f(t)u_f^0(\mathbf{r})\exp\left[-\frac{iE_f^0t}{\hbar}\right]\end{aligned} \tag{2.5}$$

where $u^e_{i,f}(\mathbf{r})$, $E^0_{i,f}$ are the eigenfunctions and eigenvalues of the operators

$$H_i^N = T_N + H^e_{ii} + \Lambda_{ii},\; H_f^N = T_N + H^e_{ff} + \Lambda_{ff}. \tag{2.6}$$

Therefore $u^0_{i,f}(\mathbf{r})$ are the initial and final wave functions of the nuclear subsystem, transitions between these being due to the perturbation

$$U = H^{ee}_{if} + \Lambda_{if}. \tag{2.7}$$

Initial and final potential energies of the nuclear subsystem differ from the $\epsilon_{i,f}(\mathbf{r})$ obtained by solving (1.1). There are corrections due to nonadiabaticity of the system and to the perturbation V (see (1.2)). The splitting is also different; it is now not determined by (1.4), but is

$$\delta U' = 2(H^e_{if}(\mathbf{r}^*) + \Lambda_{if}(\mathbf{r}^*)). \tag{2.8}$$

It should be emphasized that these corrections to the potential energy are usually not large; however, they can be important owing to the very strong dependence of the transition rate constant on the potential-barrier parameters.

The system of equations for the quantities $a_i(t)$ and $a_f(t)$ is

$$\left.\begin{aligned} i\hbar\,\dot{a}_i &= \sum_f \langle u_i^0|U|u_f^0\rangle\, a_f \exp\left[-\frac{i\Delta E_{fi}t}{\hbar}\right], \\ i\hbar\dot{a}_f &= \langle u_f^0|U|u_i^0\rangle\, a_i \exp\left[\frac{i\Delta E_{fi}t}{\hbar}\right], \\ \Delta E_{fi}^0 &= E_f^0 - E_i^0, \quad a_i(0) = 1, \quad a_f(0) = 0. \end{aligned}\right\} \tag{2.9}$$

Similar equations are obtained in quantum electrodynamics on introducing a natural linewidth. This system can be reduced to an integro-differential equation [6], which can be solved by the Laplace technique.

Solutions of (2.9) are

$$a_i(t) = \exp\left[-\frac{\gamma t}{2} - \frac{i\Delta E_i t}{\hbar}\right], \tag{2.10}$$

$$a_f(t) = \frac{\langle u_f^0|U|u_i^0\rangle}{\Delta E_{fi} - \Delta E_L + i\hbar\gamma/2}\left\{\exp\left[\frac{i(\Delta E_{fi} - \Delta E_L)t}{\hbar} - \frac{\gamma t}{2}\right] - 1\right\}, \tag{2.10'}$$

$$\gamma = \frac{2\pi}{\hbar}\sum_f |\langle u_f^0|U|u_i^0\rangle|^2\,\delta(E_f^0 - E_i^0),$$

$$\Delta E_L = -\,⨍_{-\infty}^{\infty} \frac{|\langle u_f^0|U|u_i^0\rangle|^2\,\rho(E_f^0)}{\Delta E_{fi}^0}\,dE_f^0.$$

Here $\rho(E_f^0)$ is the density of final states of the nuclear system; the energy shift ΔE_L is related to the so-called Lamb shift in quantum electrodynamics; $⨍$ denotes the principal value of the integral.

Using (2.10) the probability of finding the system in the initial state is given by

$$|a_i(t)|^2 = \exp[-\gamma t]. \tag{2.11}$$

Thus the quantity γ is the probability (per unit time) of particle transition from the initial state in the term ϵ_i to the term ϵ_f.

The expression for γ is now considered in more detail. If the perturbation U inducing the transition depends weakly on the nuclear coordinates then the value of U at the intersection point of the initial and final terms may be substituted into (2.10′) Taking the quantity $U(\mathbf{r}^*)$ out of the integral over the nuclear coordinates and using (2.7) and (2.8) gives

$$\gamma = \frac{\pi}{2\hbar}\sum_f (\delta U')^2\,|\langle u_f^0|u_i^0\rangle|^2\,\delta(E_f^0 - E_i^0). \tag{2.12}$$

This equation gives the probability per unit time of the transition only from one initial state specified by the wave function u_i^0. In order to get the transition rate constant, an averaging (denoted by the symbol Av_i) of (2.12) over all initial states of the system is necessary:

$$K = \frac{\pi}{2\hbar} Av_i \sum_f (\delta U')^2 |\langle u_f^0 | u_i^0 \rangle|^2 \delta(E_f - E_i). \tag{2.13}$$

Therefore the rate constant of the nonadiabatic transition contains the product of the Franck–Condon factor, $|\langle u_f^0 | u_i^0 \rangle|^2$ and the square of the splitting of the adiabatic terms, the splitting in many cases is determined by the electronic exchange integral.

As a rule, the relaxation rates are much larger than the probability (per unit time) of particle transition between the initial and final states (i.e. the nuclear subsystem equilibrates much more rapidly than the particle is transferred). That is why the distribution of the nuclear subsystem over the energy levels may be treated as an equilibrium one, and the Gibbs distribution may be used when doing the averaging.

It should be pointed out that the overlapping of the nuclear wave functions, which determines the rate constant, depends on the state from which the transition takes place, whereas the probability of finding the system in a certain energy state is governed by the temperature. This is what yields the temperature dependence of the transition rate constant.

3.3. RATE CONSTANT OF ADIABATIC TRANSITIONS

If the parameter $\xi_0 \gg 1$ (see (1.6)) then the electronic subsystem has time to rearrange while the particle passes the quasiintersection of terms and the transition from one potential well into another proceeds along the lower double-well term (see Fig. 3.1). The electronic wave function does not change in the course of the transition, and satisfies the Schrödinger equation with the Hamiltonian (1.2):

$$H_e \Psi_\ell (\rho, \mathbf{r}) = \epsilon_\ell(\mathbf{r}) \Psi_\ell(\rho, \mathbf{r}). \tag{3.1}$$

Similarly to the nonadiabatic transition, the wave function of the whole system should be sought as a linear combination of the electronic wave functions Ψ_ℓ and Ψ_u corresponding to the nuclear subsystem motion in the terms ϵ_ℓ and ϵ_u respectively. If, however, the transition is adiabatic then the upper term is much higher than the lower one and the wave function Ψ_u may be discarded. Therefore the wave function of the electronic–nuclear system is sought as

$$\Psi(\rho, \mathbf{r}, t) = u(\mathbf{r}, t)\, \Psi_\ell(\rho, \mathbf{r}). \tag{3.2}$$

Now the wave function (3.2) is substituted into the time-dependent Schrödinger equation (2.1), which is then multiplied from the left by $\Psi_\ell(\rho, \mathbf{r})$ and integrated over the electronic coordinates:

$$\left(i\hbar \frac{\partial}{\partial t} - H_N\right) u(\mathbf{r}, t) = 0, \quad H_N = T_N + \epsilon_\ell(\mathbf{r}) + \Lambda_{\ell\ell}. \tag{3.3}$$

Here $\Lambda_{\ell\ell}$ is the diagonal matrix element of the nonadiabaticity operator (2.4), which is incorporated, in what follows, into the definition of $\epsilon_\ell(\mathbf{r})$.

If the particle is initially localized in the left-hand potential well and the nuclear subsystem is specified by the wave function u_ℓ then the particle may transfer in the course of time into the right-hand potential well and be specified by the nuclear wave function u_r. It is clear that neither u_ℓ nor u_r are eigenfunctions of the Hamiltonian H_N, for if they were then the initial and final states would be stationary and the transitions in the system would never occur. To be more exact, a transition might take place, but only due to some external perturbations mixing these states (see also Section 1.4).

The wave function u_ℓ and u_r are chosen to be orthonormal; $u(\mathbf{r}, t)$ may then be written as an expansion,

$$\begin{aligned} u(\mathbf{r}, t) &= a_\ell(t)\, u_\ell(\mathbf{r}) \exp[-iH^N_{\ell\ell}t/\hbar] \\ &\quad + \sum_f a_2(t)\, u_r(t) \exp[-iH^N_{rr}t/\hbar], \\ H^N_{aa'} &= \langle a | H_N | a' \rangle. \end{aligned} \tag{3.4}$$

Here the summation is over all possible states of nuclear subsystem with the particle in the right-hand potential well.

Now (3.4) is substituted into (3.3), the equation thus obtained is then multiplied from the left by the wave functions $u_\ell(r)$ and $u_r(r)$ in turn, and an integration over the nuclear coordinates performed. This yields a system of equations determining the coefficients $a_\ell(t)$ and $a_r(t)$:

$$\left.\begin{aligned} i\hbar \dot{a}_i &= \sum_f H^N_{\ell r}\, a_2 \exp\left[-\frac{iE_{r\ell}t}{\hbar}\right], \\ i\hbar \dot{a}_r &= H^N_{r\ell}\, a_i \exp\frac{iE_{r\ell}t}{\hbar}, \\ E_{r\ell} &= H^N_{rr} - H^N_{\ell\ell}, \quad a_\ell(0) = 1, \quad a_r(0) = 0. \end{aligned}\right\} \tag{3.5}$$

In deriving (3.5) account is taken of the fact that the Hamiltonian H_N of the nuclear subsystem within the left-hand well is close to the Hamiltonian to which the wave function $u_\ell(r)$ corresponds, whereas within the right-hand well it is close to that of the wave function $u_r(r)$.

Solutions (3.5) have the same form as the solutions of the system (2.9), so the probability of finding the particle in the left hand well is

$$|a_\ell(t)|^2 = \exp[-\gamma t], \tag{3.6}$$

$$\left.\begin{aligned} \gamma &= \frac{2\pi}{\hbar}\sum_f |H^N_{\ell r}|^2\,\delta(E_\ell - E_r), \\ E_\ell &= H^N_{\ell\ell}, \quad E_r = H^N_{rr} \end{aligned}\right\} \tag{3.6'}$$

Similarly to evaluation of the rate constant of the nonadiabatic transition (2.13), in this case the probability γ per unit time is averaged over the initial states of the system:

$$K = \frac{2\pi}{\hbar}\,\mathrm{Av}_i \sum_f |H^N_{\ell r}|^2\,\delta(E_\ell - E_r). \tag{3.7}$$

In order to calculate the transition matrix element $H^N_{\ell r}$, two orthonormal functions u_1, u_2, corresponding to the wave functions u_ℓ, u_r, are introduced, the former being solutions of the Schrödinger equation with the Hamiltonian H_N:

$$H_N\, u_{1,2}(\mathbf{r}) = E_{1,2}\, u_{1,2}(\mathbf{r}). \tag{3.8}$$

In evaluating the matrix element $H^N_{\ell r}$ in this Section only those nuclear degrees of freedom are taken into account that correspond to the motion of the transfering particle. It will be demonstrated below that, in order to calculate the transition rate constant, knowledge of this matrix element is necessary, as well as of its dependence on the other nuclear coordinates.

The pairs of functions u_ℓ, u_r and u_1, u_2, being orthonormal, are related by the equations

$$u_1 = au_\ell - bu_r, \quad u_2 = bu_\ell + au_r, \quad a^2 + b^2 = 1. \tag{3.9}$$

Proceeding as in [3, 7], where the matrix element is calculated for symmetric potential wells, use is made of the fact that the wave function u_ℓ is an approximate solution of the Schrödinger equation (3.8) within the left-hand potential well ($r < r^*$), whereas u_r is a solution within the right-hand potential well ($r > r^*$). The equations for $u_\ell(r)$ and $u_1(r)$ are

$$\left.\begin{aligned}\frac{\hbar^2}{2M}u_\ell'' + [E_\ell - \epsilon_\ell(\mathbf{r})]\,u_\ell \approx 0 \qquad \text{for} -\infty<r<r^*,\\ \frac{\hbar^2}{2M}u_1'' + [E_1 - \epsilon_\ell(\mathbf{r})]\,u_1 \approx 0 \qquad \text{for} -\infty<r<\infty\end{aligned}\right\} \tag{3.10}$$

The first equation is now multiplied from the left by u_1, and the second by u_ℓ, and their difference is calculated. Integration from $-\infty$ to r^* gives

$$\frac{\hbar^2}{2M}(u_1u_\ell' - u_1'u_\ell)\Big|_{-\infty}^{r^*} = (E_1 - E_\ell)\int_{-\infty}^{r^*} u_1u_\ell dr. \tag{3.11}$$

Using (3.9), which expresses u_1 in terms of u_ℓ and u_r, the equation relating a and b is obtained

$$\frac{\hbar^2}{2M}(u_1u_\ell' - u_\ell u_r')\Big|_{r=r^*} = (E_\ell - E_1)\frac{a}{b}. \tag{3.12}$$

The second equation connecting a and b is obtained by substituting (3.9) for u_1 into the Schrödinger equation (3.8), multiplying it from the left by u_ℓ and integrating over the whole space:

$$H_{\ell r}^N = (E_r - E_1)\,\frac{a}{b}. \tag{3.13}$$

Now equating the left-hand sides of (3.12) and (3.13), the matrix element is obtained:

$$H_{\ell r}^N = -\frac{\hbar^2}{2M}\quad(u_\ell u_r' - u_\ell' u_r)\Big|_{r=r^*}. \tag{3.14}$$

A simple form is acquired by the matrix element $H_{\ell r}^N$ if the quasi-classical approximation for the functions u_ℓ and u_r is applied and the energy levels and angular frequencies of the particle vibrations in the left- and right-hand wells coincide:

$$H_{\ell r}^N = \frac{\hbar\omega}{2\pi}\exp\left\{-\frac{1}{\hbar}\int_{r_1^*}^{r_2^*} dr\,[2M[\epsilon_\ell(r) - E_\ell]]^{1/2}\right\}. \tag{3.15}$$

Here $r_{1,2}^*$ are the boundaries of the potential barrier surmounted by the particle (see Fig. 3.1a); ω is the angular frequency of the particle vibrations in the potential wells.

If the density in the right-hand potential well is assured to be $1/\hbar\omega$, similarly to an oscillator, then (3.15) can be substituted into (3.6) and the summation replaced by an integration, which yields the transition probability per unit time:

$$\gamma = \frac{\omega}{2\pi} \exp\left\{ -\frac{2}{\hbar} \int_{r_1^*}^{r_2^*} dr\, [2M[\epsilon_\ell(r) - E_\ell]]^{1/2} \right\}. \tag{3.16}$$

Equation (3.16), as well as equation (3.17) of Chapter 1, may be represented in the form (3.18) of Chapter 1, which leads to a simple intepretation of the results obtained. The transition probability per unit time is equal to the product of the frequency of particle impacts on the barrier and the probability of penetrating it from one impact. Similar expressions can be obtained for the transition rate constant when the final states are in a continuous spectrum (Fig. 3.1b).

Finally, it should be noted that the rate constants of adiabatic and nonadiabatic transitions may be written using the single formula [8]

$$K = \frac{2\pi}{\hbar} Av_i \sum_f |\langle \varphi_f(\rho,\mathbf{r})|H|\varphi_i(\rho,\mathbf{r})\rangle|^2 \delta(E_f - E_i), \tag{3.17}$$

where H is the Hamiltonian of the whole system and $\varphi_{i,f}$ are the electronic–nuclear functions of the initial and final states of the system. It is easy to see this if the Hamiltonian of the system is written as a sum of the electronic Hamiltonian and the operator for the nuclear kinetic energy, and the matrix element over the electronic wave functions is calculated. For the nonadiabatic transitions

$$\langle \Psi_f|H|\Psi_i\rangle = H_{fi}^\ell + \Lambda_{fi} \equiv U, \tag{3.18}$$

which coincides with the transition-inducing perturbation (2.7). Here the orthogonality of the wave functions $\varphi_{i,f}$ has been used. For the adiabatic transitions, $\Psi_i = \Psi_f = \Psi_\ell$; and using the definition (3.3) of H_N gives

$$\langle \Psi_\ell|H|\Psi_\ell\rangle = H_N, \tag{3.19}$$

which yields the coincidence of (3.7) and (3.17).

3.4. DOUBLE ADIABATIC APPROXIMATION

As we have seen, it is necessary to know the wave function of an electronic–nuclear system in order to calculate the transition rate constant. At the same time, solution (even numerical) of the Schrödinger equation for systems containing a large number of particles, which must be considered when studying chemical reactions in solids, is a very complicated problem. Here we shall consider an approximation that greatly simplifies the problem: the double adiabatic approximation.

The usual adiabatic approximation (see e.g. [9]) is applied to the study of a wide range of problems in the areas of chemical bonding, solid-state physics etc. The validity of this approximation relies upon the large difference in electron and nuclear masses, which results in much greater electron velocity compared with the nuclear velocity. This is why the electronic motion may be dealt with while assuring stationary nuclei.

In order to obtain a zeroth-order approximation, the nuclear kinetic energy is omitted from the Schrödinger equation for the whole system; the electronic wave function and energy depending on the nuclear coordinates. The wave function of the whole system is then represented as a product of the nuclear and electronic wave functions. This leads to a Schrödinger equation for the nuclear subsystem in which the energy of the electronic subsystem plays the role of the potential energy.

The adiabatic approximation therefore allows the division of system containing electrons and nuclei into two subsystems: a rapid one and a slow one. The decoupling of the equations for these subsystems is ensured by the small frequency of the nuclear vibrations compared with the electronic frequencies.

There are also situations when rapid and slow subsystems can be distinguished *within* the nuclear motions. In fact, in many substances intramolecular frequencies are very much greater than intermolecular ones, which allows separate consideration of the electronic, intra- and intermolecular subsystems. Then the wave function of the system, $\Psi_{n,m,\nu}(\rho, \mathbf{r}, \mathbf{R})$, is sought as a product of three wave functions, $\Psi_n(\rho, \mathbf{r}, \mathbf{R})$, $\varphi_{n,m}(\mathbf{r}, \mathbf{R})$, $\chi_{n,m,\nu}(\mathbf{R})$, summed over all quantum numbers. Here n, m, ν and ρ, $\mathbf{r}$, $\mathbf{R}$ are the quantum numbers and coordinates of the electronic, intra- and intermolecular subsystems.

The wave functions Ψ_n and $\varphi_{n,m}$ satisfy the following equations:

$$[T_e(\rho) + U(\rho,\mathbf{r},\mathbf{R})]\Psi_n(\rho,\mathbf{r},\mathbf{R}) = \epsilon_n(\mathbf{r},\mathbf{R})\, \Psi_n(\rho,\mathbf{r},\mathbf{R}), \tag{4.1'}$$

$$[T_N(\mathbf{r}) + \epsilon_n(\mathbf{r},\mathbf{R})]\varphi_{n,m}(\mathbf{r},\mathbf{R}) = \epsilon'_{n,m}(\mathbf{R})\, \varphi_{n,m}(\mathbf{r},\mathbf{R}), \tag{4.1''}$$

where $U(\rho,\mathbf{r},\mathbf{R})$ is the potential energy of the electronic–nuclear interaction and $T_N(\mathbf{r})$ is the kinetic-energy operator of the intramolecular subsystem.

Substituting the wave function $\Psi_{n,m,\nu}(\rho, \mathbf{r}, \mathbf{R})$ into the Schrödinger equation with the Hamiltonian

$$H = T_e(\rho) + T_N(\mathbf{r}) + T_N(\mathbf{R}) + U(\rho,\mathbf{r},\mathbf{R}), \tag{4.2}$$

making use of (4.1) and carrying out standard mathematical manipulations leads to the following system of equations for the intermolecular wave functions:

$$\left.\begin{aligned}
&(T_N(\mathbf{R}) + \epsilon'_{n,m}(\mathbf{R}) - E_{n,m,\nu})\chi_{n,m,\nu}(\mathbf{R}) = \sum_{n',m}(\Lambda^e_{n,m;n',m'} + \Lambda^N_{n,m;n',m'})\chi_{n,m;\nu}(\mathbf{R}),\\
&\Lambda^N_{n,m;n',m'} = -\int d\mathbf{r}\, \varphi_{n,m}[T_N(\mathbf{R}), \varphi_{n'm'}]\,\delta_{nn'},\\
&\Lambda^e_{n,m;n'm'} = -\int d\rho d\mathbf{r}\, \varphi_{n,m}\, \varphi_{n',m'}\Psi_n[T_N(\mathbf{r}) + T_N(\mathbf{R}), \Psi_{n'}],
\end{aligned}\right\} \tag{4.3}$$

where $T_N(R)$ is the kinetic-energy operator of the intermolecular subsystem.

Although the system (4.3) is exact, its solutions can be obtained only under some simplifying assumptions. One of these is the double adiabatic approximation, which allows the nonadiabaticity operators $\Lambda^e_{n,m;n',m'}$ and $\Lambda^N_{n,m;n',m'}$ to be discarded. It is the smallness of the first of these that makes the conventional adiabatic approximation applicable (we shall not discuss here the proof of the conventional adiabatic approximation, and shall assume $\Lambda^e_{n,m;n',m'} = 0$); if the second operator is also small then the intermolecular system may be treated as slow compared with the intramolecular one. In this case the system of coupled equations (4.3) decouples, and independent equations determining the wave functions of the intermolecular subsystem arise:

$$[T_N(\mathbf{R}) + \tilde{\epsilon}_{n,m}(\mathbf{R}) - E^0_{n,m,\nu}]\chi_{n,m,\nu}(\mathbf{R}) = 0 \tag{4.4}$$

The diagonal part of the nonadiabaticity operator in (4.4),

$$\Lambda^N_{n,m;n',m'} = -\left\{\int d\mathbf{r}\, \varphi_{n,m}\, T_N(\mathbf{R})\, \varphi_{n',m'} - \sum_j \frac{\hbar^2}{2M_j}\int d\mathbf{r}\, \varphi_{n,m}\frac{\partial}{\partial \mathbf{R}_j}\varphi_{n',m'}\frac{\partial}{\partial \mathbf{R}_j}\right\}\delta_{nn'}, \tag{4.5}$$

is incorporated into the definition of the potential energy $\tilde{\epsilon}_{n,m}$. Since $n = n'$ and the functions $\varphi_{n,m}$ are normalized ($\int d\mathbf{r}\, \varphi^2_{n,m} = \text{const}$), the second term in (4.5) is zero for $m = m'$. Here M_j and R_j are the mass and coordinate of the jth molecule. Therefore

$$\tilde{\epsilon}_{n,m}(\mathbf{R}) = \epsilon'_{n,m}(\mathbf{R}) + \int d\mathbf{r}\, \varphi_{m,n}\, T_N(\mathbf{R})\, \varphi_{m,n}.$$

Perturbation theory now gives corrections to the solutions of (4.4); these corrections are small compared with the adiabatic solutions. It can easily be seen that the double adiabatic approximation holds if

$$|\langle \chi_{n,m,\nu}|\Lambda^{N}_{n,m;n',m'}|\chi_{n',m',\nu'}\rangle| \ll |E^{0}_{n,m,\nu} - E^{0}_{n',m',\nu'}|.$$

The matrix element of the nonadiabaticity operator in (4.6) is needed in order to find conditions on the system parameters under which the double adiabatic approximation is valid. According to (4.5),

$$\langle \chi_{n,m,\nu}|\Lambda^{N}_{n,m;n',m'}|\chi_{n',m',\nu'}\rangle = \int d\mathbf{r}d\mathbf{R}\,\chi_{n,m,\nu}\,\chi_{n',m',\nu'}\,\varphi_{n,m}\,T_N(\mathbf{R})\varphi_{n',m'}$$

$$+ \sum_j \frac{\hbar^2}{M_j}\int d\mathbf{r}d\mathbf{R}\,\varphi_{n,m}\frac{\partial}{\partial \mathbf{R}_j}\varphi_{n',m'}\chi_{n,m,\nu}\frac{\partial}{\partial \mathbf{R}_j}\chi_{n',m',\nu'}. \qquad (4.7)$$

Thus, in order to evaluate (4.6), it is necessary to know the derivatives of the functions $\varphi_{n,m}(\mathbf{r},\mathbf{R})$ and $\chi_{n,m,\nu}(\mathbf{R})$ with respect to the intermolecular coordinates. The wave function of the intermolecular vibrations varies significantly over a distance of order

$$d_R = (\hbar/M\omega_R)^{1/2},$$

i.e. the estimate

$$\int d\mathbf{R}\,\chi_{n,m,\nu}(R)\frac{\partial}{\partial \mathbf{R}}\chi_{n',m',\nu'}\mathbf{R}) \approx (M\omega_R/\hbar)^{1/2} \qquad (4.8)$$

may be used; here ω_R is the characteristic intermolecular frequency.

As far as the intermolecular function is concerned, its derivative, may be estimated, as in (4.8), as

$$\int d\mathbf{r}\,\varphi_{n,m}(\mathbf{r},\mathbf{R})\frac{\partial}{\partial \mathbf{r}}\chi_{n',m'}(\mathbf{r},\mathbf{R}) \approx (M_r\omega_r/\hbar)^{1/2}, \qquad (4.9)$$

where M_r and ω_r are the mass and frequency of the oscillator corresponding to the intramolecular coordinate over which the differentiation is carried out.

Equation (4.9) allows estimation of the derivative of the function $\varphi_{n,m}(\mathbf{r},\mathbf{R})$ over the intermolecular coordinate, provided that the combination of intra- and intermolecular coordinates on which the intramolecular wave function depends is known. As an example, we consider the case of two coupled oscillators whose potential energy is

$$\epsilon(r,R) = \tfrac{1}{2}\gamma_r(r-r_0)^2 + \tfrac{1}{2}\gamma_R(R-R_0)^2 + \gamma_0(r-r_0)(R-R_0). \qquad (4.10)$$

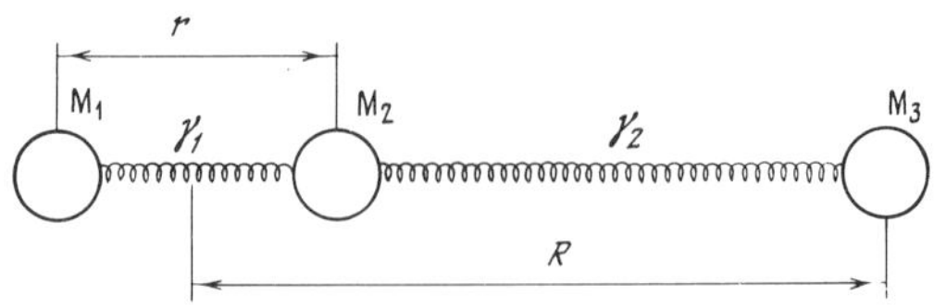

Figure 3.2 A scheme illustrating weak interaction of two particles M_1 and M_2 strongly bound with the third practicle M_3.

Here r_0, R_0 and γ_r, γ_R are the equilibrium coordinates and elastic constants for the intra- and intermolecular subsystems; γ_0 is the coupling constant, which depends in general on the interaction between the molecules in the system. However, estimates may be obtained assuming $\gamma_0 \approx \gamma_R$. Then the potential energy (4.10) corresponds to the model depicted in Fig. 3.2, where $\gamma_r = \gamma_1 + \gamma_2$, $\gamma_R = \gamma_2$ and $\gamma_0 = \gamma_2/2 = \gamma_R/2$

The potential energy (4.10) is diagonalized,

$$\epsilon(r,R) = \frac{\gamma_r}{2}\left[r - r_0 + \frac{\gamma_0}{\gamma_r}(R - R_0)\right]^2 + \frac{1}{2}\left(\gamma_R - \frac{\gamma_0^2}{\gamma_r}\right)(R - R_0)^2$$

and substituted into the (4.1)″); the argument of the wave function $\varphi_{n,m}$ is then

$$r - r_0 + \frac{\gamma_0}{\gamma_r}(R - R_0). \tag{4.11}$$

It follows from (4.11) that the derivative of the intramolecular wave function over the intermolecular coordinate is

$$\frac{\partial}{\partial R}\varphi_{n,m}(r,R) = \frac{\gamma_0}{\gamma_r}\frac{\partial}{\partial r}\varphi_{n,m}(r,R), \tag{4.12}$$

which leads to the following estimate for the matrix element of the non-adiabaticity operator (4.7):

$$\langle \chi_{n,m,\nu} | \Lambda^N_{n,m;n;m'} | \chi_{n;m;\nu'} \rangle \approx \hbar\omega_r \frac{\gamma_0}{2\gamma_r}\left[\frac{\gamma_0 M_r}{\gamma_r M} + 2\left(\frac{M_r\omega_R}{M\,\omega_r}\right)^{1/2}\right]. \tag{4.13}$$

Recalling that for $m \neq m'$

$$|E_{n,m} - E_{n,m'}| \geqslant \hbar\omega_r,$$

the inequality (4.6) can be rewritten [10] as

$$\frac{1}{2}\frac{\gamma_0}{\gamma_r}\left[\frac{\gamma_0 M_r}{\gamma_r M} + 2\left(\frac{M_r\omega_R}{M\omega_r}\right)^{1/2}\right] \ll 1. \tag{4.14}$$

Therefore, if the mass of the transferred particle is much smaller than the average molecular mass, or the elasticity of the intramolecular subsystem is substantially greater than that of the intermolecular one, then the double adiabatic approximation applies and the wave function of the system may be represented as the product

$$\Psi_{n,m,\nu}(\rho,\mathbf{r},\mathbf{R}) = \Psi_n(\rho,\mathbf{r},\mathbf{R})\Psi_{n,m}(\mathbf{r},\mathbf{R})\,\chi_{n,m,\nu}(\mathbf{R}). \tag{4.15}$$

It should be kept in mind that the condition (4.14) as it is ensures only that the wave function of the whole system can be written as a product of the wave functions of three subsystems (4.15), but it is *not* sufficient for calculating the transition matrix elements by means of these functions. As shown in [11], another requirement arises in the study of particle transfer — the characteristic frequency of potential-barrier fluctuations must be small compared with the inverse transversal time. This time can be estimated as the distance transversed by the particle divided by its velocity (1.5).

3.5. RATE CONSTANT OF SOLID-PHASE CHEMICAL REACTIONS IN THE DOUBLE ADIABATIC APPROXIMATION WHEN THE INTRAMOLECULAR STATES OF THE REACTION PRODUCTS ARE DISCRETE

We now calculate the rate constant of a chemical reaction in the solid state. As mentioned above, chemical reactions are simulated by radiationless transitions, so the results to be presented below are general enough and may be used for various radiationless transitions in solids.

The rate constant of both adiabatic and nonadiabatic transitions is given by (3.17). If the double adiabatic approximation holds and the wave function of the electronic-nuclear system is (4.15), then (3.17) transforms as

$$K = \frac{2\pi}{\hbar}\,\mathrm{Av}_i \sum_f |\langle \chi_{n_f,m_f,\nu_f} | V_{n_f m_f;n_i,m_i}(\mathbf{R}) | \chi_{n_i,m_i,\nu_i}\rangle|^2 \delta(E_f - E_i), \tag{5.1}$$

$$V_{n_f,m_f;n_i,m_i}(\mathbf{R}) = \int d\rho d\mathbf{r}\, \Psi_{n_f} \varphi_{n_f,m_f} H\, \varphi_{n_i m_i} \Psi_{n_i}. \tag{5.1$'$}$$

If the matrix element (5.1′) depends only weakly on the intermolecular coordinates then it may be taken out of the integral, so that the transition rate constant is determined by the Franck–Condon factor of the intermolecular subsystem. This approximation is valid if a light particle (e.g. an electron) is transferred, its localization radius greatly exceeding the amplitude of the intermolecular vibrations. The same aproach is often used when studying transitions of heavy particles, i.e.

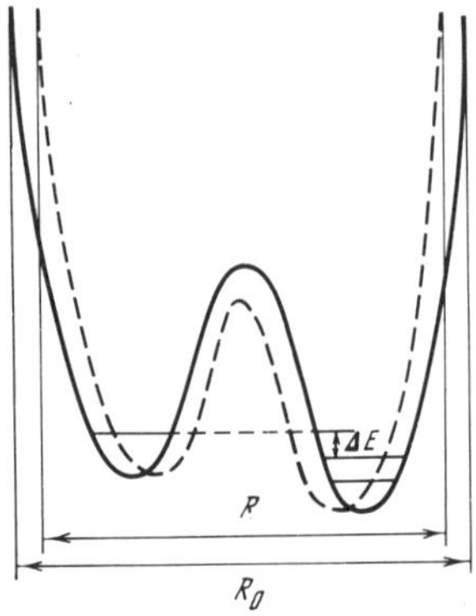

Figure 3.3 A scheme of the potential energy of the tunneling particle. The dashed line shows the energy after a rapprochment of the molecules.

the Franck–Condon principle is carried over to the case of separate intra- and intermolecular subsystems [2, 12–14]. However, calculations [10] show (see also Chapter 5) that the matrix element (5.1′) in this case is a very steep function of the intermolecular coordinates and may not be taken out of the integral.

The dependence of the matrix element (5.1′) on the coordinate R is studied using the example of hydrogen abstraction from molecules by radicals [15–17]. Hydrogen transitions in these reactions proceed adiabatically and the potential energy has a double-well form (Fig. 3.3). The radical–molecule distance, which determines the parameters of the potential barrier transversed by the particle in the course of the transition, depends on the intermolecular coordinates; therefore the matrix element (5.1′) (see also (3.15)) is an exponential function of the coordinate R. It can be estimated that, for a parabolic potential barrier of 1 eV height, a zero-point amplitude decrease of the barrier width leads to an increase in the matrix element (5.1′) of more than an order of magnitude.

For the nonadiabatic transitions (5.1′) is rewritten, using (2.13) as

$$K = \frac{2\pi}{\hbar} Av_i \sum_f (\delta U')^2 \,|\langle \chi_{n_f,m_f,\nu_f}(\mathbf{R})|\, S_{fi}(\mathbf{R})\,|\chi_{n_i,m_i,\nu_i}\rangle|^2 \,\delta(E_f - E_i), \qquad (5.2)$$

where $S_{fi}(R)$ is the overlap integral of the wave functions corresponding to the intramolecular initial and final terms. The overlap integral depends exponentially on the minimum positions of the terms between which the transition occurs. Thus the minimum positions of the initial and final terms, and hence the overlap integral, depend on the intermolecular coordinates. This dependence arises for different reasons in mono- and bimolecular reactions. In the case of the monomolecular reaction the minimum positions and the shapes of the terms depend on the intermolecular coordinates as the result of interaction of the molecule in which the transition occurs with the environment [18, 19]. For bimolecular reactions the initial and final terms relate to different molecules and the overlap integral obviously depends on the intermolecular coordinates.

The transition rate constant is now calculated for the case when the intramolecular states of the reaction products are discrete. The problem is to carry out the summation in (5.1), taking account of the dependence of the matrix element (5.1′) on the intermolecular coordinates. Here the mathematical problem lies in how to incorporate the potential-barrier fluctuations discussed in the previous chapter.

In the theory of radiationless transitions the sums in the (5.1) are calculated under the assumption that $V_{n_f m_f; n_i m_i}$ is constant (the Franck–Condon principle) using the techniques of generating polynomials [20], operator calculus [21] and the density matrix [22]. It is the latter technique that seems to be best when the Condon approximation is not used. It should also be kept in mind that the problem is not simply to consider several terms of the expansion of the rapid subsystem matrix element over the slow subsystem coordinates [23]; the matrix element is now extremely sensitive to the variations of the intermolecular coordinates, and the usual expansion is no longer sufficient.

It is convenient to denote the matrix element (5.1′) by V(R); then, introducing the Boltzmann factor, (5.1) can be written as

$$\left.\begin{aligned} K &= \frac{2\pi}{\hbar}\sum_{i,f} \exp[-\beta E_i]\,|\langle \nu_f|V(\mathbf{R})|\nu_i\rangle|^2\,\delta(E_f - E_i) \Big/ \sum_i \exp[-\beta E_i], \\ \beta &= 1/k_B T. \end{aligned}\right\} \tag{5.3}$$

All the states of both the rapid and slow subsystems are summed in (5.3). It should be recalled that the rapid subsystem comprises both electrons and intramolecular oscillations, while the slow subsystem comprises the intermolecular oscillations. In the double adiabatic approximation the energy of the system is equal to the sum of the energies of the rapid and slow subsystems: $E_{i,f} = E_{n_{i,f}} + E_{\nu_{i,f}}$. Here the index n stands for the rapid-subsystem states.

If temperatures are considered that are low enough such that the thermal energy is much less than the energy quantum of the rapid subsystem then the summation over n_i in (5.3) may be omitted [24], i.e. the transition proceeds mainly from the lowest level. In addition to this, it is assured that the transition is to a *single* final state of the rapid subsystem n_f. It will be shown in Chapter 5 that in the radical reactions it is the transition to the nearest (from below) energy state that gives the main contribution to the reaction. Therefore summations only over the slow subsystem states ν_i, ν_f are retained in (5.3). Here the argument of the δ-function can conveniently be rewritten in terms of the energies of the slow and rapid subsystems:

$$E_f - E_i = -\Delta E + E_{\nu_f} - E_{\nu_i}, \quad \Delta E = E_{n_i} - E_{n_f},$$

where ΔE is the energy defect in the rapid subsystem (see Fig. 3.3). If $\Delta E > 0$ then the energy flows from the rapid subsystem to the slow one, and the process is exothermic; the opposite case $\Delta E < 0$ corresponds to the endothermic process. The following property of the energy defect in the rapid subsystem will be used below: although the quantities E_{n_i} and E_{n_f} depend on the intermolecular coordinate, nevertheless as computations show, their difference ΔE depends only weakly on the intermolecular coordinates.

Similarly to [22], the integral representation of the δ-function,

$$\delta(E_f - E_i) = \frac{1}{2\pi} \int_{-\infty}^{\infty} ds \exp[-i(\Delta E + E_{\nu_i} - E_{\nu_f})s], \tag{5.4}$$

is used. Then (5.4) is substituted into (5.3) and the energy defect in the rapid subsystem is treated as a constant

$$K = \frac{1}{\hbar} \int_{-\infty}^{\infty} ds \exp[-i\Delta Es] \sum_{\nu_i,\nu_f} \langle \nu_f | V | \nu_i \rangle \exp[-(\beta + is)E_{\nu_i}]$$
$$\times \langle \nu_i | V | \nu_f \rangle \exp[-(-is)E_{\nu_f}] \Big/ \sum_i \exp[-\beta E_{\nu_i}]. \tag{5.5}$$

The density matrix of the intermolecular subsystem is now introduced; it is defined as

$$\rho_\lambda(\mathbf{R},\mathbf{R}') = \sum_\nu |\nu\rangle \exp[-\lambda E_\nu]\langle \nu|. \tag{5.6}$$

Distinguishing the combinations of the type (5.6) in (5.5), the latter is transformed as

$$K = \frac{\dfrac{1}{\hbar}\displaystyle\int_{-\infty}^{\infty} ds \exp[-i\Delta Es]\mathrm{Tr}\,\{V(R)\,\rho^{i}_{\beta+is}(\mathbf{R},\mathbf{R}')\,V(\mathbf{R}')\rho^{f}_{-is}(\mathbf{R},\mathbf{R}')\}}{\mathrm{Tr}\,\{\rho^{i}_{\beta}(\mathbf{R},\mathbf{R}')\}}. \quad (5.7)$$

Therefore the calculations in (5.7) present no difficulties if the density matrices of the intermolecular subsystem and the R dependence of the matrix elements V(R) are known.

It is usual in the theory of solids to apply a harmonic approximation, which allows the introduction of normal coordinates for the intermolecular subsystem. Then its energy is presented as a sum of the energies of independent oscillators (phonons), and summation is carried out over all possible normal lattice modes. Therefore the density matrix of the intermolecular (phonon) subsystem is the product of the density matrices of harmonic oscillators corresponding to these modes. The density matrix of a one-dimensional harmonic oscillator is (see e.g. [25])

$$\rho_\lambda(q_\mu, q'_\mu) = \{2\pi \sinh\lambda\hbar\omega_\mu\}^{-1/2}\exp\left\{-\frac{1}{4}(q_\mu + q_{\mu'} - 2q^0_\mu)^2 \tanh \tfrac{1}{2}\lambda\hbar\omega_\mu \right.$$
$$\left. -\frac{1}{4}(q_\mu - q'_\mu)^2 \coth \tfrac{1}{2}\lambda\hbar\omega_\mu\right\}. \quad (5.8)$$

Here q_μ is the dimensionless normal coordinate of the phonon mode μ; ψ_μ is the frequency of this mode; q^0_μ is the equilibrium value of the normal coordinate.

It should be pointed out that the rate constant is given by (5.7) even for transitions in isolated molecules, for example rupture of one bond and creation of another. If one or several vibration modes of the molecule are low frequency compared with the remaining ones then the former may be treated as a slow subsystem and the matrix element V(R) may be calculated using the wave functions of the rapid (i.e. electronic subsystem and high-frequency modes).

However, if it is not possible to separate the vibrational modes into low- and high-frequency modes then the transition rate constant in such a molecule may be calculated by means of (5.7), but now the double adiabatic approximation does not apply and it has to be assumed that V(R) is the electronic matrix element and ρ^i_λ, ρ^f_λ are the density matrices of the whole vibrational system of the molecule. This is why the rate constants to be calculated below apply to radiationless transitions both in solids and complicated isolated molecules. In both cases the normal coordinates allow representation of the density matrices ρ^i_λ, ρ^f_λ as products of the density matrices of linear harmonic oscillators (5.8) corresponding to the slow-subsystem vibrations:

$$\rho_\lambda(\{q_\mu\}, \{q'_\mu\}) = \prod_{\mu,\mu'} \{2\pi \sinh\lambda\hbar\omega_\mu\}^{-1/2}$$
$$\exp\left[-\frac{1}{4}(q_\mu + q'_\mu - 2q^0_\mu)t_{\mu\mu'}(q_{\mu'} + q'_{\mu'} - 2q^0_{\mu'}) - \frac{1}{4}(q_\mu + q'_\mu)c_{\mu\mu'}(q_{\mu'} - q'_{\mu'})\right], \tag{5.9}$$
$$t_{\mu\mu'} = \tanh(\tfrac{1}{2}\lambda\hbar\omega_\mu)\,\delta_{\mu\mu'}, \quad c_{\mu\mu'} = \coth(\tfrac{1}{2}\lambda\hbar\omega_\mu)\,\delta_{\mu\mu'}.$$

Equation (5.7) contains the density matrices of both initial and final states of the intermolecular subsystems. It should be kept in mind that the equilibrium coordinates, eigenfrequencies and directions of the principal axes do not necessarily coincide in the initial and final states; hence the normal coordinates might be different in these states. Since the density matrices $\rho^{i}_{\beta+is}$ and ρ^{f}_{-is} are to be written in the same coordinates, only one of them, say $\rho^{i}_{\beta+is}$, can be represented, as in (5.9), in terms of the diagonal matrices $t_{\mu\mu'}$ and $c_{\mu\mu'}$, whereas for ρ^{f}_{-is} these matrices are nondiagonal. This is why the transition rate constant is calculated using an arbitrary coordinate frame for the initial and final states. Coordinates Q are introduced that are normalized such that the Hamiltonian of the intermolecular subsystem in the harmonic approximation takes the form

$$\mathscr{H}_{i,f} = \frac{1}{2}\left[(\mathbf{Q} - \mathbf{Q}^{i,f}_0)(\Omega_{i,f})^2(\mathbf{Q} - \mathbf{Q}^{i,f}_0) - \frac{\partial^2}{\partial \mathbf{Q}^2}\right], \tag{5.10}$$

where $\mathbf{Q}^{i,f}_0$ and $\Omega_{i,f}$ are the equilibrium coordinate and frequencies tensors in the initial and final states.

Now the density matrix of the intermolecular subsystem to which the Hamiltonian (5.10) corresponds is represented as

$$\rho^{i,f}_\lambda(\mathbf{Q},\mathbf{Q}') = \{\det(2\pi\Omega^{-1}_{i,f})\sinh(\lambda\Omega_{i,f})\}^{-1/2}$$
$$\exp\left\{-\frac{1}{4}[\mathbf{Q} + \mathbf{Q}' - 2\mathbf{Q}^{i,f}_0]\,\mathbf{T}_{i,f}[\mathbf{Q} + \mathbf{Q}' - 2\mathbf{Q}^{i,f}_0] - \frac{1}{4}[\mathbf{Q} - \mathbf{Q}']\,\mathbf{C}_{i,f}[\mathbf{Q} - \mathbf{Q}']\right\}, \tag{5.11}$$
$$\mathbf{T}_{i,f} = \Omega_{i,f}\tanh\tfrac{1}{2}\lambda\Omega_{i,f}, \quad \mathbf{C}_{i,f} = \Omega_{i,f}\coth\tfrac{1}{2}\Omega_{i,f}.$$

It should be emphasized that both the Hamiltonian (5.10) and the exponent in (5.11) are invariant with respect to linear transformations (rotations and displacements). Analytical functions of $\Omega_{i,f}$ are expressed as usual as power series. The trace of the density matrix ρ^{i}_{β} in the denominator of (5.7) is

$$\mathrm{Tr}\,\{\rho^{i}_{\beta}(\mathbf{R},\mathbf{R}')\} = \det[2\sinh(\beta\Omega_i/2)]^{-1}. \tag{5.12}$$

New coordinates transform (5.7) into

$$K = \frac{1}{\hbar}\det[2\sinh(\tfrac{1}{2}\beta\Omega_i)]\int ds\,d\mathbf{Q}\,d\mathbf{Q}'\,V(\mathbf{Q})\,\rho^{i}_{\beta+is}(\mathbf{Q},\mathbf{Q}')V(\mathbf{Q}')\,\rho^{f}_{-is}(\mathbf{Q},\mathbf{Q}'). \tag{5.13}$$

The integrals over $\mathbf{Q}$ and $\mathbf{Q}'$ can be performed if the matrix elements $V(\mathbf{Q})$ are known. As shown above, these matrix elements depend exponentially on the intermolecular coordinates; therefore $V(\mathbf{Q})$ is conveniently represented as

$$V(\mathbf{Q}) = V(\mathbf{Q}_0)\exp\left[-\frac{1}{2}J(\mathbf{Q})\right],\quad \mathbf{Q}_0 = \tfrac{1}{2}(\mathbf{Q}^{i}_{0}+\mathbf{Q}^{f}_{0}). \tag{5.14}$$

Then the function $J(\mathbf{Q})$ is expanded in a power series in the displacements from the equilibrium positions, and only two terms are retained:

$$V(\mathbf{Q}) = V(\mathbf{Q}_0)\exp[-\tfrac{1}{2}\mathbf{J}'(\mathbf{Q}_0)(\mathbf{Q}-\mathbf{Q}_0) - \tfrac{1}{4}(\mathbf{Q}-\mathbf{Q}_0)\,\mathbf{J}'(\mathbf{Q}_0)\,(\mathbf{Q}-\mathbf{Q}_0)]. \tag{5.15}$$

Here $\mathbf{J}'$ is a vector and $\mathbf{J}''$ is a tensor. For the nonadiabatic transitions between two parabolic terms, the approximate expression (5.15) is exact; $\mathbf{J}'$ and $\mathbf{J}''$ are expressed by means of the term parameters.

It is easy to calculate the integrals in (5.13) over the intermolecular coordinates, since the integrand is an exponential with an exponent of quadratic form:

$$\begin{aligned} B = &-\frac{1}{4}(\mathbf{Q}+\mathbf{Q}'-2\mathbf{Q}^{i}_{0})\mathbf{T}_i(\mathbf{Q}+\mathbf{Q}'-2\mathbf{Q}^{i}_{0}) - \frac{1}{4}(\mathbf{Q}-\mathbf{Q}')\mathbf{C}_i(\mathbf{Q}-\mathbf{Q}') \\ &- \frac{1}{4}(\mathbf{Q}+\mathbf{Q}'-2\mathbf{Q}^{f}_{0})\mathbf{T}_f(\mathbf{Q}+\mathbf{Q}'-2\mathbf{Q}^{f}_{0}) - \frac{1}{4}(\mathbf{Q}-\mathbf{Q}')\mathbf{C}_f(\mathbf{Q}-\mathbf{Q}') \\ &- \frac{1}{2}\mathbf{J}'(\mathbf{Q}-\mathbf{Q}_0) - \frac{1}{4}(\mathbf{Q}-\mathbf{Q}_0)\,\mathbf{J}''(\mathbf{Q}-\mathbf{Q}_0) \\ &- \frac{1}{2}\mathbf{J}'(\mathbf{Q}'-\mathbf{Q}_0) - \frac{1}{4}(\mathbf{Q}'-\mathbf{Q}_0)\,\mathbf{J}''(\mathbf{Q}'-\mathbf{Q}_0). \end{aligned} \tag{5.16}$$

To carry out the integration in (5.13), it is necessary to group the terms in (5.16) to obtain two bilinear forms of independent linear combinations of $\mathbf{Q}$ and $\mathbf{Q}'$. The required transformations give

$$
\begin{aligned}
B &= \frac{1}{4}(\mathbf{Q}+\mathbf{Q}'-\mathbf{Q}_1)[\mathbf{T}_i+\mathbf{T}_f+\tfrac{1}{2}\mathbf{J}''](\mathbf{Q}+\mathbf{Q}'-\mathbf{Q}_1) \\
&- \frac{1}{4}(\mathbf{Q}-\mathbf{Q}')[\mathbf{C}_i+\mathbf{C}_f+\tfrac{1}{2}\mathbf{J}''](\mathbf{Q}=\mathbf{Q}') \\
&+ \frac{1}{4}[(\mathbf{T}_i-\mathbf{T}_f)\Delta+\mathbf{J}'][\mathbf{T}_i+\mathbf{T}_f+\tfrac{1}{2}\mathbf{J}'']^{-1}[\mathbf{T}_i-\mathbf{T}_f)\Delta+\mathbf{J}'] \\
&- \frac{1}{4}\Delta(\mathbf{T}_i+\mathbf{T}_f)\Delta, \\
\mathbf{Q}_1 &= -[\mathbf{T}_i+\mathbf{T}_f+\tfrac{1}{2}\mathbf{J}'']^{-1}[(\mathbf{T}_i-\mathbf{T}_f)\Delta+\mathbf{J}']+2\mathbf{Q}_0,\ \Delta=\mathbf{Q}_0^f-Q_0^i.
\end{aligned}
\tag{5.17}
$$

Now the formula

$$
\int dX \exp\left[-\sum_{i,j} a_{ij}X_iX_j\right] = [\det(a/\pi)]^{-1/2}, \tag{5.18}
$$

is used and the transition rate constant is obtained as

$$
\begin{aligned}
K &= \frac{1}{\hbar}|V_{n_f n_i}(R_0)|^2 \int_{-\infty}^{\infty} ds\{\det\phi\}^{-1/2}\exp\Big\{-i\Delta Es-\frac{1}{4}\Delta(\mathbf{T}_i+\mathbf{T}_f)\Delta \\
&+\frac{1}{4}[(\mathbf{T}_i-\mathbf{T}_f)\Delta+\mathbf{J}'][\mathbf{T}_i+\mathbf{T}_f+\tfrac{1}{2}\mathbf{J}'']^{-1}[(\mathbf{T}_i-\mathbf{T}_f)\Delta+\mathbf{J}'], \\
\phi &= [2\sinh(\beta\Omega_i/2)]^{-2}\,\Omega_i^{-1}\sinh[(\beta+is)\Omega_i]\,(\mathbf{T}_i+\mathbf{T}_f+\tfrac{1}{2}\mathbf{J}'') \\
&\times(\mathbf{C}_i+\mathbf{C}_f+\tfrac{1}{2}\mathbf{J}'')\,\Omega_f^{-1}\sinh(-is\Omega_f), \\
&\{\det\phi\}^{-1/2} \equiv \exp\left[-\frac{1}{2}\mathrm{Tr}\,(\ell n\phi)\right].
\end{aligned}
\tag{5.19}
$$

It follows from (5.19) that the magnitude and temperature dependence of the transition rate constant are largely influenced by three factors: fluctuations of the potential-barrier parameters (whose contribution is specified by $\mathbf{J}'$ and $\mathbf{J}''$), rearrangement of the environment due to the transition, which can be described by equilibrium coordinate shifts Δ, and the energy defect of the rapid subsystem, ΔE.

The transition rate constant is simplified if the intermolecular frequencies and normal coordinates do not change in the course of the transition. Even if the normal frequencies in the initial and final states do not coincide, this is usually neglected, since such a frequency effect only appears in higher orders of the adiabatic approximation than the effects of the equilibrium coordinate shifts [26]. In fact, as follows from (5.19), the frequency variation mainly affects the preexponential factor (for $\Omega_i = \Omega_f$ and small $\mathbf{J}''$ the matrix $\boldsymbol{\Phi}$ is equal to unity), whereas the exponent does not vary significantly. Therefore it may be assumed that $\Omega_i = \Omega_f = \Omega$ in (5.19); then the values $\mathbf{T}_i \pm \mathbf{T}_f$, on which the transition rate constant depends, are

$$\mathbf{T}_i \pm \mathbf{T}_f = \frac{2\Omega \sinh[\frac{1}{2}(\beta + is \mp is)\,\Omega]}{\cosh(\frac{1}{2}\beta\Omega) + \cosh[\frac{1}{2}\beta + is)\Omega]}. \tag{5.20}$$

If the normal coordinates of the intermolecular subsystem in the initial and final states coincide then the tensors $\mathbf{T}_{i,f}$, $\mathbf{C}_{i,f}$ and Ω are diagonal in this frame, and the vectors Δ and $\mathbf{J}'$ are $\Delta = \{\ldots \Delta_\mu \ldots\}$, $\mathbf{J}' = \{\ldots \partial J/\partial Q_\mu \ldots\}$. The tensor $\mathbf{J}''$ as a rule remains nondiagonal even in normal coordinates. The transition rate constant is the simplest if dimensionless coordinates q_μ are used that are normalized such that the initial and final phonon (intermolecular) Hamiltonians are

$$\mathscr{H}^{i,f} = \frac{\hbar}{2}\sum_\mu \omega_\mu \left[(q_\mu - q_\mu^{0i,f})^2 - \frac{\partial^2}{\partial q_\mu^2}\right]. \tag{5.21}$$

Here $q_\mu^{0i,f}$ are the equilibrium initial and final normal coordinates.

The equations relating the coordinates

$$q_\mu = \Omega_\mu^{1/2} Q_\mu, \quad Q_\mu = q_\mu/(\hbar\omega_\mu)^{1/2} \tag{5.22$'$}$$

and frequencies

$$\omega_\mu = \Omega_\mu/\hbar, \quad \Omega_\mu = \hbar\omega_\mu \tag{5.22$''$}$$

can be found by comparing the Hamiltonians (5.10) and (5.21). The quantities

$$\Delta_\mu = \Delta q_\mu/(\hbar\omega_\mu),^{1/2} \frac{\partial J}{\partial Q_\mu} = (\hbar\omega_\mu)^{1/2} \frac{\partial J}{\partial q_\mu}, \tag{5.22$'''$}$$

$$\frac{\partial^2 J}{\partial Q_\mu \partial Q_{\mu'}} = \hbar\,(\omega_\mu\omega_{\mu'})^{1/2} \frac{\partial^2 J}{\partial q_\mu \partial q_{\mu'}}$$

are also transformed. Now substituting (5.20) and (5.22) into (5.19), the transition rate constant is

$$K = \frac{1}{\hbar} \,|V_{n_f,n_i}(\mathbf{R}_0)|^2 \int_{-\infty}^{\infty} ds\{\det\phi\}^{-1/2}$$
$$\exp\left\{ -i\Delta E s + \frac{1}{4}\sum_{\mu\mu'} a_\mu (b^{-1})_{\mu\mu'} a_{\mu'} \right.$$
$$\left. - \frac{1}{2}\sum_\mu \frac{(\Delta q_\mu)^2 \sinh(\frac{1}{2}\beta\hbar\omega_\mu)}{\cosh(\frac{1}{2}\beta\hbar\omega_\mu) + \cosh[(\frac{1}{2}\beta + is)\hbar\omega_\mu]}\right\}, \tag{5.23}$$

$$a_\mu = \frac{\partial J}{\partial q_\mu} + 2\,\frac{2\Delta q_\mu \sinh[(\frac{1}{2}\beta + is)\hbar\omega_\mu]}{\cosh(\frac{1}{2}\beta\hbar\omega_\mu) + \cosh[(\frac{1}{2}\beta + is)\hbar\omega_\mu]},$$

$$b_{\mu\mu'} = \frac{1}{2}\frac{\partial^2 J}{\partial q_\mu \partial q_{\mu'}} + \frac{2\delta_{\mu\mu'}\sinh(\frac{1}{2}\beta\hbar\omega_\mu)}{\cosh(\frac{1}{2}\beta\hbar\omega_\mu) + \cosh[(\frac{1}{2}\beta + is)\hbar\omega_\mu]}.$$

The substitution $s \to s + \frac{1}{2}i\beta$ is then made in the integral (5.23) and the integration contour is shifted upwards through a distance $\frac{1}{2}\beta$. This results in an equation that will be used below in studying the temperature dependence of the transition rate constant:

$$K = \frac{1}{\hbar}\,|V_{n_f n_i}(\mathbf{R}_0)|^2 \int_{-\infty}^{\infty} ds\{\det\phi\}^{-1/2}\exp\left\{\frac{\beta\Delta E - i\Delta E s}{2} + \frac{1}{4}\sum_{\mu\mu'} A_\mu (B^{-1})_{\mu\mu'} A_{\mu'} - \frac{1}{2}\sum_\mu \frac{(\Delta q_\mu)^2 \sinh(\frac{1}{2}\beta\hbar\omega_\mu)}{\cosh(\frac{1}{2}\beta\hbar\omega_\mu) + \cos(s\hbar\omega_\mu)}\right\}, \quad (5.24)$$

$$A_\mu = \frac{\partial J}{\partial q_\mu} + \frac{2i\Delta q_\mu \sin(s\hbar\omega_\mu)}{\cosh(\frac{1}{2}\beta\hbar\omega_\mu) + \cos(s\hbar\omega_\mu)},$$

$$B_{\mu\mu'} = \frac{1}{2}\frac{\partial^2 J}{\partial q_\mu \partial q_{\mu'}} + \frac{2\delta_{\mu\mu'}\sinh(\frac{1}{2}\beta\hbar\omega_\mu)}{\cosh(\frac{1}{2}\beta\hbar\omega_\mu) + \cos(s\hbar\omega_\mu)}.$$

Now an example of a single-mode intermolecular subsystem is considered whose frequency is Ω_0. This demonstrates qualitative features that are specific to the temperature dependence of the transition rate constant. Only the low-temperature region is studied, for the sake of simplicity, when the transitions between two terms proceed via tunneling; at low temperatures the intermolecular amplitude is not large and the quadratic term in the expansion (5.15) may be neglected,† i.e. $\partial^2 J/\partial q_\mu \partial q_{\mu'} = 0$ is assumed in (5.24).

Obviously, $J(\{q_\mu\})$ depends only on the distance between terms, hence it is the deviation from the equilibrium distance that serves as the normal coordinate in the single-mode approximation. Now it should be kept in mind that the equilibrium distance $R_0^{i,f}$ is not necessarily the same initially and finally, and the expansion

$$\exp[a\cos(s\hbar\Omega_0)] = \sum_n I_n(a)\cos[ins\hbar\Omega_0]$$

† However, this cannot be done for the zero-point vibrations, whose amplitude does not depend on temperature. So this approximation affects only the absolute value of the rate constant, without changing its temperature dependence.

should be used, where $I_n(a)$ is the Bessel function of imaginary argument. Then (5.24) for the transition rate constant becomes

$$
\begin{aligned}
K &= \frac{2\pi}{\hbar}|V(\mathbf{R}_0)|^2 \left|\frac{\tilde{\alpha}_0+\Delta\xi}{\tilde{\alpha}_0-\Delta\xi}\right|^{\Delta\nu} \exp\left\{\frac{\hbar\Omega_0\Delta\nu}{2k_BT}\right. \\
&\quad \left. + \frac{\tilde{\alpha}_0^2-\Delta\xi^2}{2}\coth\left(\frac{\hbar\Omega_0}{2k_BT}\right)\right\} I_{\Delta\nu}\left\{\frac{|\tilde{\alpha}_0^2-\Delta\xi^2|}{2\sinh(\hbar\Omega_0/2k_BT)}\right\}\rho_f, \\
\tilde{\alpha}_0 &= \left(\frac{\hbar}{M\Omega_0}\right)^{1/2}\frac{J_0'}{2}, \quad \Delta\xi = \left(\frac{M_0\Omega_0}{\hbar}\right)^{1/2}\Delta R_0, \\
\Delta R_0 &= R_0^f - R_0^i, \quad \Delta\nu = \left[\frac{\Delta E}{\hbar\Omega_0}\right], \quad J_0' = \frac{\partial J}{\partial R_{12}}.
\end{aligned}
\tag{5.25}
$$

Here ρ_f is the density of final states, M_0 is the reduced mass of the reactants, R_{12} is the distance between the reactants, [. . .] means integer part; energy conservation allows the retention of only one term in the Fourier series.

If the condition $\tilde{\alpha}_0 \gg \Delta\xi$ is satisfied, i.e. the transition does not cause any appreciable rearrangement of the environment, then the rate constant becomes especially simple:

$$
K = \frac{2\pi}{\hbar}|V(\mathbf{R}_0)|^2 \exp\left[\frac{\hbar\Omega_0\Delta\nu}{2k_BT} + \frac{\tilde{\alpha}_0^2}{2}\coth\frac{\hbar\Omega_0}{2k_BT}\right] I_{\Delta\nu}\left\{\frac{2_0^2}{2\sinh(\hbar\Omega_0/2k_BT)}\right\}\rho_f.
\tag{5.26}
$$

This equation allows analysis of the temperature dependence of the tunneling reaction rate constant. For temperatures $T \ll \hbar\Omega_0/2k_B$, when the intermolecular vibrations are quantum (i.e. observe quantum statistics) the low-temperature limit of the rate constant is obtained:

$$
\begin{aligned}
K &= \frac{2\pi}{\hbar}|V(R_0)|^2 \frac{1}{|\Delta\nu|!}\left\{\frac{\hbar(J_0')^2}{8M_0\Omega_0}\right\}^{|\Delta\nu|} \exp\left\{\frac{\hbar\Omega_0(\Delta\nu-|\Delta\nu|)}{2k_BT}\right. \\
&\quad \left. + \frac{\hbar(J_0')^2}{8M_0\Omega_0}\right\}\rho_f.
\end{aligned}
\tag{5.26'}
$$

Now it can be seen that the fluctuations of the potential-barrier parameters caused by the intermolecular vibrations are important even at $T = 0$. These are the zero-point vibrations, which effectively decrease the potential barrier; in the present case, the zero-point vibrations in the intermolecular subsystem. Calculations show that the zero-point vibrations lead to an 4–6 order-of-magnitude increase in the transition rate constant for hydrogen in radical reactions [27].

It also follows from (5.26′) that the rate constant depends on the energy defect ΔE of the intramolecular subsystem; the effective potential barrier becomes smaller with increasing quantum number describing the intermolecular subsystem in the final state. This yields the factor $(\hbar(J_0')^2/8M_0\Omega_0)^{|\Delta\nu|}$. On the other hand, the larger the difference in the quantum numbers of the intermolecular subsystem in the initial and final states the more oscillations are made by the integrand in the matrix element $\langle \nu_f | V(R) | \nu_i \rangle$ in (5.5); this results in a factorial dependence on $\Delta\nu$. Hence there exists an optimal value of $\Delta\nu$ at which the rate constant is a maximum.

Moreover, if $\Delta\nu < 0$ (i.e. $\Delta E < 0$) and the energy flows from the intermolecular into intramolecular subsystem then, as follows from (5.26), the temperature dependence of the transition rate constant is activation-like, the activation energy being equal to the energy absorbed by the intramolecular subsystem from the intermolecular one. Therefore a final low-temperature limit exists only in nonendothermic reactions. If $\Delta E \geqslant 0$ then it can easily be seen that the transition rate constant at $T \to 0$ does not depend on temperature.

At higher temperatures, when the argument of the Bessel function becomes large compared with $\Delta\nu$, the rate constant takes the form

$$K = \frac{2\pi}{\hbar} |V(\mathbf{R}_0)|^2 \left[\frac{\pi\tilde{\alpha}_0^2}{\sinh\left(\dfrac{\hbar\Omega_0}{2k_B T}\right)} \right]^{-1/2} \times \exp\left[\frac{\hbar\Omega_0}{2k_B T}\Delta\nu + \frac{\tilde{\alpha}_0^2}{2}\coth\left(\frac{\hbar\Omega_0}{4k_B T}\right)\right]. \tag{5.27}$$

"Defreezing" of the rate constant occurs at temperatures $T \approx \hbar\Omega_0/4k_B$; in the high-temperature domain, $T \gg \hbar\Omega_0/4k_B$, when the intermolecular vibrations may be treated as classical, the rate constant becomes

$$K = \frac{2\pi}{\hbar} |V(\mathbf{R}_0)|^2 \left[\frac{2M_0\Omega_0^2}{\pi(J_0')^2 k_B T} \right]^{1/2} \exp\left[\frac{(J_0')^2 k_B T}{2M_0\Omega_0} \right] \rho_f, \tag{5.28}$$

so that its logarithm becomes a linear function of temperature.

At even higher temperatures, (5.26) does not apply, since the intermolecular amplitude is no longer small and the quadratic term in the expansion (5.15) must be taken into account.

3.6. RATE CONSTANT OF SOLID-PHASE CHEMICAL REACTIONS IN THE DOUBLE ADIABATIC APPROXIMATION WHEN THE INTRAMOLECULAR STATES OF THE REACTION PRODUCT ARE IN THE CONTINUOUS SPECTRUM

We now consider the role of the environment in the elementary process of solid-phase chemical reactions when the intramolecular states of one (or several) reaction product lie in the continuous spectrum. These processes include reactions producing vibrationally excited molecules (in strongly exothermal reactions the product states might lie in the quasicontinuum region [28]), dissociation of electronically excited molecules (the final state of the system is unstable) and other processes.

There are a number of differences from the case when the intramolecular subsystem has discrete final states, since the passage from the starting formula (5.3) for the rate constant in the double adiabatic approximation to (5.5) assumed that only one initial state and the final state of the rapid subsystem closest to it need to be taken into account.

There is still no need to sum over n_i in (5.3), even in the case to be considered below, but, as far as n_f is concerned, the situation is completely different. When the final energies of the rapid subsystem lie in the continuous spectrum, they are fixed and determined only by the equation

$$E_{n_f} = E_{n_i} + E_{\nu_i} - E_{\nu_f}. \tag{6.1}$$

Thus the transitions to the states whose energies differ from the initial energy may contribute significantly to the rate constant.

It is convenient to chose an energy-normalized wave function for the final state; then transforming the summation over n_f in (5.3) into an integration over E_{n_f} yields

$$\begin{gathered} K = \frac{\pi}{\hbar Z_i} \sum_{\nu_i \nu_f} \mathrm{ex}[-\beta E_{\nu_i}] \langle \nu_i | V(E_{n_f}, \mathbf{R}) | \nu_f \rangle \langle \nu_f | V(E_{n_f}, \mathbf{R}') | \nu_i \rangle, \\ Z_i = \sum_{\nu_i} \exp[-\beta E_{\nu_i}], \end{gathered} \tag{6.2}$$

the final state energy of the intramolecular subsystem E_{n_f} being determined by (6.1).

If the matrix element $V(E_{n_f}, R)$ depends only weakly on the energy transferred into the intermolecular subsystem, $E_{\nu_f} - E_{\nu_i}$, then the transition rate constant may be written in terms of diagonal elements of the density matrix of the initial slow (intermolecular) subsystem. In fact, (5.6) for the density matrix and the equality $\sum_\nu |\nu\rangle\langle\nu| \equiv 1$, following from the completeness of the intermolecular wave-function set, allows conversion of (6.2) into

$$K = \frac{2\pi}{\hbar \, \mathrm{Tr}\{\rho^{i}_{\beta}(\mathbf{R};\mathbf{R}')\}} \mathrm{Tr}\{|V(E_{n_i},\mathbf{R})|^2 \rho^{i}_{\beta}(\mathbf{R},\mathbf{R}')\}, \tag{6.3}$$

which is used to calculate the tunneling rate constant in [29, 30]. This equation has a simple physical interpretation: the transition rate constant is equal to the transition probability per unit time at a given phonon configuration R multiplied by the probability of this configuration, all being averaged over the possible realizations.

An arbitrary frame may be used in (6.2), as well as in the (5.5), to carry out all summations. In most cases, however, the frame of normal coordinates describing the environment does not change in the course of the transition. That is why only normal coordinates are used here in studying the rate constant; this consideration can be easily generalized to an arbitrary frame.

As mentioned above, the matrix element $V(E_{n_f}, \{q_\mu\})$ is an exponentially steep function of the intermolecular coordinates, its dependence on E_{n_f} also being, as a rule, exponential [31]. Because of this, the matrix element can be conveniently represented as

$$\left.\begin{aligned} V(E_{n_f}, \{q_\mu\}) &= V(E_{n_i}, \{q^0_\mu\}) \exp\left[-\frac{1}{2} J(E_{n_f}, \{q_\mu\})\right], \\ V(E_{n_i}, \{q^0_\mu\}) &\equiv V(E_{n_i}, \mathbf{R}_0),\ q^0_\mu = \tfrac{1}{2}(q^{oi}_\mu + q^{of}_\mu). \end{aligned}\right\} \tag{6.4}$$

Then the function $J(E_{n_f}, \{q_\mu\})$ can be expanded in a power series over the energy transferred into the slow subsystem and the phonon coordinates; here only three terms are retained:

$$\begin{aligned} J(E_{n_f}, \{q_\mu\}) = &-J'_E(E_{\nu_i} - E_{\nu_f}) + J'_\mu(q_\mu - q^0_\mu) \\ &+ \tfrac{1}{2}(q_\mu - q^0_\mu)\, J''_{\mu\mu'}(q_{\mu'} - q^0_{\mu'}). \end{aligned} \tag{6.5}$$

Equations (6.4) and (6.5) are substituted into (6.2), and (5.6) for the density matrix of phonons is used, giving [8]

$$\begin{aligned} K = \frac{2\pi}{\hbar} &|V(E_{n_i}, \mathbf{R}_0)|^2 \, [\mathrm{Tr}\{\rho^{i}_{\beta}(q_\mu, q'_\mu)\}]^{-1} \\ &\times \mathrm{Tr}\{\rho^{i}_{\beta - J'_E}(q_\mu, q'_\mu) \exp[-\tfrac{1}{2} J'_\mu(q_\mu + q'_\mu - 2q^0_\mu) \\ &- \tfrac{1}{4}(q_\mu - q^0_\mu)\, J''_{\mu\mu'}(q_{\mu'} - q^0_{\mu'}) \\ &- \tfrac{1}{4}(q'_\mu - q^0_\mu) J''_{\mu\mu'}(q'_{\mu'} - q^0_{\mu'})]\, \rho^{f}_{J'_E}(q_{\mu'} q'_\mu)\}. \end{aligned} \tag{6.6}$$

The trace in (6.6) is calculated in a fashion similar to that used in deriving (5.17) and (5.19). Straightforward but tedious calculations result in the following transition rate constant [8]:

$$
\begin{aligned}
\mathrm{K} &= \frac{2\pi}{\hbar}\,|\mathrm{V}(\mathrm{E}_{\mathrm{n_i}},\mathbf{R}_0)|^2\{\det\tilde{\phi}\}^{-1/2}\exp\left\{\frac{1}{4}\sum_{\mu\mu'}\tilde{\mathrm{A}}_\mu(\tilde{\mathrm{B}}^{-1})_{\mu\mu'}\tilde{\mathrm{A}}_{\mu'}\right.\\
&\quad\left.-\frac{1}{4}\sum_{\mu}\frac{(\Delta \mathrm{q}_\mu)^2\sinh(\beta\hbar\omega_\mu/2)}{\cosh[\frac{1}{2}(\beta-\mathrm{J}'_{\mathrm{E}})\hbar\omega_\mu]\cosh(\frac{1}{2}\mathrm{J}'_{\mathrm{E}}\hbar\omega_\mu)}\right\},\\
\tilde{\mathrm{A}}_\mu &= \mathrm{J}'_\mu + \frac{\Delta \mathrm{q}_\mu\sinh[\frac{1}{2}\beta-\mathrm{J}'_{\mathrm{E}})\hbar\omega_\mu]}{\cosh[\frac{1}{2}(\beta-\mathrm{J}'_{\mathrm{E}})\hbar\omega_\mu]\cosh(\frac{1}{2}\mathrm{J}'_{\mathrm{E}}\hbar\omega_\mu)},\\
\tilde{\mathrm{B}}_{\mu\mu'} &= \frac{1}{2}\mathrm{J}''_{\mu\mu'} + \frac{\delta_{\mu\mu'}\sinh(\frac{1}{2}\beta\hbar\omega_\mu)}{\cosh[\frac{1}{2}(\beta-\mathrm{J}'_{\mathrm{E}})\hbar\omega_\mu]\cosh(\frac{1}{2}\mathrm{J}'_{\mathrm{E}}\hbar\omega_\mu)},\\
\tilde{\Phi} &= \left\{\delta_{\mu\mu'} + \frac{\mathrm{J}''_{\mu\mu'}\cosh[\frac{1}{2}(\beta-\mathrm{J}'_{\mathrm{E}})\hbar\omega_\mu]\cosh(\frac{1}{2}\mathrm{J}'_{\mathrm{E}}\hbar\omega_\mu)}{2\sinh(\frac{1}{2}\beta\hbar\omega_\mu)}\right\}\\
&\quad\times\left\{\delta_{\mu\mu'} + \frac{\mathrm{J}''_{\mu\mu'}\sinh[\frac{1}{2}(\beta-\mathrm{J}'_{\mathrm{E}})\hbar\omega_\mu]\sinh(\frac{1}{2}\mathrm{J}'_{\mathrm{E}}\hbar\omega_\mu)}{2\sinh(\frac{1}{2}\beta\hbar\omega_\mu)}\right\}
\end{aligned}
\tag{6.7}
$$

To conclude this section, we give the formula that describes the transition rate constant in the single-mode approximation at $\mathrm{J}''_{\mu\mu'} = \Delta \mathrm{q}_\mu = 0$, which corresponds to low temperatures and small environment rearrangement (cf. (5.26)):

$$
\mathrm{K} = \frac{2\pi}{\hbar}\,|\mathrm{V}(\mathrm{E}_{\mathrm{n_i}},\mathbf{R}_0)|^2\exp\left\{\frac{\hbar(\mathrm{J}'_0)^2\cosh(\frac{1}{2}\mathrm{J}'_{\mathrm{E}}\hbar\Omega_0)\cosh[\frac{1}{2}(\beta-\mathrm{J}'_{\mathrm{E}})\hbar\Omega_0]}{4\mathrm{M}_0\Omega_0\sinh(\frac{1}{2}\beta\hbar\Omega_0)}\right\} \tag{6.8}
$$

The low-temperature limit of the rate constant is

$$
\mathrm{K} = \frac{2\pi}{\hbar}\,|\mathrm{V}(\mathrm{E}_{\mathrm{n_i}},\mathbf{R}_0)|^2\exp\left\{\frac{\hbar\mathrm{J}'_0}{4\mathrm{M}_0\Omega_0}\exp[-\tfrac{1}{2}\mathrm{J}'_{\mathrm{E}}\hbar\mathrm{M}_0]\cosh(\tfrac{1}{2}\mathrm{J}'_{\mathrm{E}}\hbar\Omega_0)\right\}. \tag{6.9}
$$

This equation predicts that for $\mathrm{J}'_{\mathrm{E}} \ll (\hbar\Omega_0)^{-1}$ the phonon zero-point vibrations for the continuous final intermolecular states of the reaction products contribute twice as much as those for the case of the discrete final states.

In the high-temperature region, when the intermolecular vibrations may be treated as classical, the rate constant coincides to (5.28) to within the preexponential factor. Therefore if the intermolecular vibrations are classical then the features of the final term do not affect the exponent, although the preexponential factor is affected very slightly.

3.7. TEMPERATURE DEPENDENCE OF THE TUNNELING RATE CONSTANT. THE DEBYE MODEL OF SOLIDS

We have seen that transition rate constants derived in the previous sections depend on the parameters Δq_μ, J'_μ and $J''_{\mu\mu'}$. The form of these parameters is influenced by the choice of the coordinate system and the model describing the solid. If the reaction does not yield charge transfer or appreciable conformational rearrangement of the reactants then the environmental rearrangement may be neglected and the assumption $\Delta q_\mu = 0$ in (5.24) and (6.7) may be made. We shall study this particular case, which will reveal those specific features of solid-phase chemical reactions that are caused by the potential barrier fluctuations due to intermolecular vibrations; environmental change will be considered later when studying charge transfer (Section 3.10).

Since $J(\{q_\mu\})$ depends on the term-to-term spacing, the derivation of this function is largely determined by how the reactant displacements from their equilibrium position, $\xi_{1,2}$, depend on the normal coordinates:

$$\left.\begin{aligned} J'_\mu &= J'\left(\frac{\partial\xi_1}{\partial q_\mu}-\frac{\partial\xi_2}{\partial q_\mu}\right), \quad J''_{\mu\mu'} = J''\left(\frac{\partial\xi_1}{\partial q_\mu}-\frac{\partial\xi_2}{\partial q_\mu}\right)\left(\frac{\partial\xi_1}{\partial q_{\mu'}}-\frac{\partial\xi_2}{\partial q_{\mu'}}\right), \\ J' &= \frac{\partial J}{\partial R_{12}}\left|\frac{\partial R_{12}}{\partial\xi_1}\right|, \quad J'' = \frac{\partial^2 J}{\partial R_{12}^2}\left|\frac{\partial R_{12}}{\partial\xi_1}\right|^2 \end{aligned}\right\} \tag{7.1}$$

It has been mentioned in Section 3.5 that $\partial J/\partial R_{12}$ depends on the process and is different for mono- and bimolecular reactions.

The expressions (5.24) and (6.7) for the transition rate constants are valid over a wide temperature range and describe both tunneling and overbarrier chemical reactions. Here only the tunneling transition mechanism is of interest; it is realized at sufficiently low temperatures [32, 33] (see (1.3) in Chapter 2). If the condition $T < \gamma/J''k_B$ (see (4.5) in Chapter 2) also holds then another simplification may be made — the value $J''_{\mu\mu'}$ may be neglected. Then (5.24) and (6.7) are transformed, and for discrete intramolecular states of the reaction products (distinctions due to the product states lying in the continuous spectrum will be pointed out below):

$$\begin{aligned} K = \frac{1}{\hbar}\,|V(\mathbf{R}_0)|^2\exp\left[\frac{\Delta E}{2k_BT}+\frac{1}{8}\sum_\mu (J'_\mu)^2\coth\left(\frac{\hbar\omega_\mu}{2k_BT}\right)\right] \\ \times\int_{-\infty}^{\infty} ds\,\exp\left[-i\Delta Es+\frac{1}{8}\sum_\mu\frac{(J'_\mu)^2\cos(s\hbar\omega_\mu)}{\sinh(\hbar\omega_\mu/2k_BT)}\right]. \end{aligned} \tag{7.2}$$

If the second term in the exponent of the integrand is small, which is the case only at temperatures much lower than the Debye temperature, then the exponential may be expanded, and the first nonvanishing term with $\Delta E \neq 0$ at $|\Delta E| \gg k_B T$ yields [10, 34]:

$$K = \frac{\pi}{4\hbar} |V(\mathbf{R}_0)|^2 (J'_{\mu_0})^2 \exp\left[\frac{\Delta E - |\Delta E|}{2} + \right.$$
$$\left. + \frac{1}{8} \sum_{\mu} (J'_{\mu})^2 \coth\left(\frac{\hbar\omega_\mu}{2k_B T}\right)\right] \rho(\Delta E). \tag{7.3}$$

Here $\rho(\Delta E)$ is the phonon density of states, the index μ_0 is determined by $\hbar\omega_0 = |\Delta E|$. It can easily be seen that the expansion of the exponential holds if the inequality

$$\tfrac{1}{2} (J'_{\mu_0})^2 |\Delta E| \rho(\Delta E) \exp[-|\Delta E|/2k_B T] \ll 1$$

is valid.

For $|\Delta E| > \hbar\omega_D$ (ω_D is the Debye frequency) the expansion of the exponential should also include higher-order terms describing multiphonon processes. Such a correction to (7.3) might, however, influence only the weakly dependent preexponential factor. As far as the rate-constant temperature dependence is concerned, it is completely governed by the exponential in the (7.3) for any relation between ΔE and $\hbar\omega_D$. Therefore we assume hereinafter, without loss of generality that $|\Delta E| < \hbar\omega_D$.

For $\Delta E > 0$ (exothermic process) the first term in the exponent turns out to be zero, and for $\Delta E < 0$ (endothermic process) it results in an activation-like temperature dependence of the rate constant (cf.(5.26′)). The second term is also temperature-dependent; however, it can only be analysed by choosing a more specific model for the solid. When the reactant mass and interaction constant differ only slightly from those of the surrounding molecules the medium may be treated as an ideal crystalline lattice. Normal phonon coordinates are now introduced to represent the longitudinal displacement of the nth molecule from the equilibrium position in terms of these coordinates:

$$\left.\begin{aligned} \xi_n &= \left(\frac{\hbar}{MN}\right)^{1/2} \sum_{\mathbf{k},\sigma} (\omega_{\mathbf{k},\sigma})^{-1/2} \mathbf{e}_{\mathbf{k},\sigma} \ell_{\mathbf{k},\sigma} \exp[i\mathbf{k}\mathbf{a}_n], \\ \mathbf{a}_n &= \{a_x n_x, a_y n_y, a_z n_z\}. \end{aligned}\right\} \tag{7.4}$$

Here N is the number of molecules in the crystalline lattice; $\mathbf{a}_{x,y,z}$ are the intermolecular spacings in the directions x, y, z; $\omega_{\mathbf{k},\sigma}$ and $\mathbf{e}_{\mathbf{k},\sigma}$ are the frequency and the polarization vector of the phonon with the wave vector **k** and branch number σ.

Substituting (7.1) and (7.4) into (7.3) gives for the transition rate constant

$$K = \frac{\pi}{4\hbar} |V(\mathbf{R}_0)|^2 (J'_\mu)^2 \exp\left[\frac{\Delta E - |\Delta E|}{2k_B T} + \right.$$

$$\left. + \frac{\hbar (J')^2}{2NM} \sum_{\mathbf{k}} \sin^2\left(\frac{\mathbf{ka}}{2}\right) \sum_\sigma \frac{1}{\omega_{\mathbf{k}\sigma}} \coth\left(\frac{\hbar\omega_{\mathbf{k}\sigma}}{2k_B T}\right) \cos^2\theta_{\mathbf{k}\sigma}\right] \rho(\Delta E), \tag{7.5}$$

$$\cos\theta_{\mathbf{k}\sigma} = \ell_{\mathbf{k}\sigma} J'/|J'|.$$

Here a is the vector connecting the centers of mass of the molecules whose vibrations give rise to the barrier fluctuations.

Equation (7.5) can be simplified by means of the Debye model for the solid. The principal assumptions of this model are as follows: (i) the phonon frequencies for all acoustic modes coincide and depend only on the modulus of the wave vector; a stronger assumption is usually used: (ii) $\omega_{\mathbf{k}\sigma} = s|\mathbf{k}|$, where s is the sound velocity; (iii) the true Brillouin zone is replaced by a sphere with maximal wave vector k_D chosen in such a way that the volumes of the true and effective Brillouin zones coincide.

Using assumptions (i)–(iii) a temperature-dependent rate constant arises. Since coth x = 1 + exp(−x)/sinh x, the second term in the exponent (7.5) is separated into two parts; one of which does not depend on temperature and is due to the zero-point contribution, whereas the other one is temperature-dependent. The transition rate constant in the Debye approximation is thus

$$K \sim \exp\left[(\Delta E - |\Delta E|)/2k_B T + \alpha' T^4\right],$$

$$\alpha' = \frac{\pi^6}{60} \frac{(J')^2 k_B^4}{\hbar^3 \omega_D^5}. \tag{7.6}$$

At higher temperatures the exponent in (7.2) is simplified in another way [35]. The function $\cos(s\hbar\omega_\mu)$ is expanded and two terms are retained (the applicability of this approximation will be discussed below). Then the integral in (7.2) is easily performed, and the transition rate constant becomes

$$K = \frac{4}{\hbar} |V(\mathbf{R}_0)|^2 \left[2\pi \Big/ \sum_\mu \frac{(\hbar\omega_\mu)^2 (J'_\mu)^2}{\sinh(\hbar\omega_\mu/2k_B T)}\right]^{1/2}$$

$$\times \exp\left[\frac{\Delta E}{2k_B T} + \frac{1}{8}\sum_\mu (J'_\mu) \coth\left(\frac{\hbar\omega_\mu}{4k_B T}\right)\right]. \tag{7.7}$$

A neighborhood of the point

$$s_0 = i8\,\Delta E \Big/ \sum_\mu \frac{(\hbar\omega_\mu)^2 (J'_\mu)^2}{\sinh(\hbar\omega_\mu/2k_BT)}. \tag{7.8}$$

makes the major contribution to the integral (7.7). It is then obvious that the expansion of $\cos(s\hbar\omega_\mu)$ is valid if the inequality

$$|\hbar\omega_D s_0| \ll 1 \tag{7.9}$$

holds. The relation

$$\sum_\mu \frac{(\hbar\omega_\mu)^2 (J'_\mu)^2}{\sinh(\hbar\omega_\mu/2k_BT)} > \frac{\hbar^2\omega_D}{\sinh(\hbar\omega_D/2k_BT)} \sum_\mu \omega_\mu (J'_\mu)^2 = \frac{2\hbar^3\omega_D (J')^2}{M\sinh(\hbar\omega_D/2k_BT)} \tag{7.10}$$

and (7.8) convert the inequality (7.9) into

$$\sinh\left(\frac{\hbar\omega_D}{2k_BT}\right) \ll \frac{\hbar^2 (J')^2}{4M|\Delta E|}. \tag{7.11}$$

The summation in the exponent in (7.7) for the Debye model of the solid lattice is now carried out in the same fashion as in the derivation of (7.6). It is necessary only that the inequality (7.11) should be consistent with the conditions of applicability of the Debye model.

At high enough temperatures ($T > \hbar\omega_D/4k_B$) the function $\coth(\hbar\omega_\mu/4k_BT)$ in (7.7) may be expanded and the first term retained. The rate constant is then given by

$$K = \frac{2}{\hbar^2}|V(\mathbf{R}_0)|^2 \left(\frac{2\pi M}{(J')^2 k_BT}\right)^{1/2} \exp\left[\frac{\Delta E}{2k_BT} + \alpha'' T\right],$$

$$\alpha'' = \frac{k_B}{2\hbar}\sum_\mu (J'_\mu)^2/\omega_\mu. \tag{7.12}$$

If it is also borne in mind that $|\Delta E| < \hbar\omega_D$ then the temperature dependence of the rate constant is due mainly to the linear temperature term in the exponent. A similar result arises by first considering the particle transition at fixed intermolecular coordinates and then averaging the transition probability over various reactant separations, assuming the molecular vibrations to be classical (cf. (4.6) in Chapter 2).

If the states of the reaction products lie in the continuous spectrum then (6.8) predicts that for $J'_E \ll \beta,\ \hbar\Omega_0$ the Debye approximation yields (7.6) (in which $\Delta E = 0$ should be assumed) but with α' doubled; the contribution of the zero-point vibrations is also doubled.

At high temperatures the rate constant is described by (7.12); however, "defreezing" of the rate constant as well as of the internal energy of the solid occurs at $T \approx \hbar\omega_D/2k_B$ rather than at $T \approx \hbar\omega_D/4k_B$ as in the case of discrete states of the reaction products.

3.8. TUNNELING AND ACTIVATION MECHANISMS OF REACTIONS. THE EINSTEIN MODEL OF SOLIDS

We now consider the transition rate constant (5.24) at temperatures of order $\hbar\omega_D/4k_B$ and higher. The quadratic term in the $J(\mathbf{Q})$ expansion may then become necessary (see (5.15)), and transitions from higher levels of the intramolecular (rapid) system may contribute appreciably to the transition probability. As will be shown below, the higher levels contribute only at such temperatures when the intermolecular vibrations are classical. At the same time, the quadratic term could be of importance in the temperature range where the intermolecular vibrations are still of quantum character.

This is why the quadratic term will be examined first, assuming that transitions proceed from the ground level of the intramolecular subsystem; only after this will transitions from higher levels be investigated, treating the intermolecular vibrations as classical.

We cannot see how to calculate the temperature-dependent rate constant in the general case, taking account of $J''_{\mu\mu'}$; since it is not clear how to invert the Nth-rank matrix $B_{\mu\mu'}$ (see (5.24)). In this temperature range, however, the solid lattice vibrations are described satisfactorily by the Einstein approximation, in which the molecular displacements from their equilibrium positions are the normal coordinates (vibrations of different molecules are assumed to be independent). Indicating the reactant coordinates by suffices 1 and 2,

$$\left.\begin{aligned} &J'_1 = -J'_2 = J'(\hbar/M\Omega)^{1/2} \equiv \bar{J}', \quad J''_{11} = J''_{22} = -J''_{12} = -J''_{21} = J''\frac{\hbar}{M\Omega} \equiv \bar{J}'', \\ &J'_\ell = J''_{\ell\ell'} = 0 \text{ for } \ell, \ell' \neq 1, 2 \end{aligned}\right\} \tag{8.1}$$

where Ω is the Einstein frequency.

In this approximation the matrices A_μ and $B_{\mu\mu'}$ are quite simple (as before, we assume that $\Delta q_\mu = 0$):

$$A_\mu = \frac{\partial J}{\partial q_\mu} = \bar{J}' \begin{pmatrix} 0 \\ \vdots \\ 0 \\ 1 \\ -1 \\ 0 \\ \vdots \\ 0 \end{pmatrix}, \quad B_{\mu\mu'} = \frac{1}{2} J'' \left(\begin{array}{c|cc|c} 0 & 0 & & 0 \\ \hline & 1 & -1 & \\ 0 & & & 0 \\ & -1 & 1 & \\ \hline 0 & 0 & & 0 \end{array} \right) + cI, \tag{8.2}$$

$$c = \frac{2\sinh(\hbar\Omega/2k_BT)}{\cosh(\hbar\Omega/2k_BT) + \cos(s\hbar\Omega)},$$

and the inverse matrix $(B^{-1})_{\mu\mu'}$ can be analytically calculated as

$$(B^{-1})_{\mu\mu'} = \frac{1}{c}\left(I + \frac{\frac{1}{2}\bar{J}''}{\bar{J}'' + c} \left(\begin{array}{c|cc|c} 0 & 0 & & 0 \\ \hline & 1 & -1 & \\ 0 & & & 0 \\ & -1 & 1 & \\ \hline 0 & 0 & & 0 \end{array} \right) \right), \tag{8.3}$$

where I is the unit matrix. Multiplying the matrix (8.2) and (8.3) and substituting the result into (5.24) at $\Delta q_\mu = 0$ gives

$$K = \frac{1}{\hbar} |V(\mathbf{R}_0)|^2 \left[\left(1 + \frac{\bar{J}''}{2} \coth \frac{\hbar\Omega}{4k_BT} \right) \left(1 + \frac{\bar{J}''}{2} \tanh \frac{\hbar\Omega}{4k_BT} \right) \right]^{-1/2}$$

$$\times \int_{-\infty}^{\infty} ds \exp \Bigg[-i\Delta Es + \frac{\Delta E}{2k_BT} +$$

$$+ \frac{(J')^2}{4} \left(\frac{\sinh(\hbar\Omega/2k_BT)}{\cosh(\hbar\Omega/2k_BT) + \cos(s\hbar\Omega)} + \frac{\bar{J}''}{2} \right)^{-1} \Bigg]. \tag{8.4}$$

At low temperatures $\cosh(\beta\hbar\Omega/2) \gg 1$, whereas at high temperatures the value of $\cos(s\hbar\Omega)$ at the saddle point s_0 is close to unity (see (7.8)). This permits the expansion of the exponent in (8.4) in terms of the parameter

$$(1 - \cos(s\hbar\Omega))/(1 + \cosh(\tfrac{1}{2}\beta\hbar\Omega)),$$

which yields the exponent in the form

$$-i\Delta Es + \frac{\Delta E}{2k_BT} + \frac{(\bar{J}')^2}{4} \left[\tanh \frac{\hbar\Omega}{4k_BT} + \frac{\bar{J}''}{2} \right]^{-1}$$

$$- \frac{(J')^2}{8} \frac{\sinh(\hbar\Omega/4k_BT)/\cosh^3(\hbar\Omega/4k_BT)}{[\tanh(\hbar\Omega/4k_BT) + \frac{1}{2}\bar{J}'']^2} (1 - \cosh(s\hbar\Omega)).$$

Now Fourier analysis of the function exp $[a \cos(s\hbar\Omega)]$ (see the derivation of (5.25)) can be used to obtain the following transition rate constant [19, 36]:

$$K = \frac{2\pi}{\hbar} |V(\mathbf{R}_0)|^2 \varphi_3^{-1/2} \exp\left[\frac{\Delta E}{2k_B T} + \varphi_1(1-\varphi_2)\right] I_{\frac{|\Delta E|}{\hbar\Omega}}(\varphi_1\varphi_2)\rho_f,$$

$$\varphi_1 = \frac{(\bar{J}')^2}{4}\left[\tanh\left(\frac{\hbar\Omega}{4k_B T}\right) + \frac{\bar{J}''}{2}\right]^{-1},$$

$$\varphi_2 = \frac{\sinh(\hbar\Omega/4k_B T)/\cosh^3(\hbar\Omega/4k_B T)}{2[\tanh(\hbar\Omega/4k_B T) + \frac{1}{2}\bar{J}'']}, \tag{8.5}$$

$$\varphi_3 = \left[1 + \frac{J''}{2}\coth\left(\frac{\hbar\Omega}{4k_B T}\right)\right]\left[1 + \frac{J''}{2}\tanh\left(\frac{\hbar\Omega}{4k_B T}\right)\right].$$

If the argument of the Bessel function $I_{\frac{|\Delta E|}{\hbar\Omega}}(a)$ is large then

$$K = \frac{2\pi}{\hbar} |V(\mathbf{R}_0)|^2 \varphi_3^{-1/2} \exp\left[\frac{\Delta E}{2k_B T} + \varphi_1\right]. \tag{8.6}$$

At high temperatures, when the intermolecular vibrations may be treated as classical, the exponent becomes

$$K \sim \exp\left[-\frac{(J')^2}{2J''}\frac{M\Omega^2}{2J''k_B T + M\Omega^2}\right], \tag{8.7}$$

and coincides with the result of averaging of the transition probability over reactant displacements from the equilibrium positions [30, 34] (see (4.7) in Chapter 2). If the inequality $T > M\Omega^2/2J''k_B$ also holds then the transition rate constant becomes activation-like:

$$K \sim \exp[-E_a/k_B T], \quad E_a = \frac{M\Omega^2}{4}\left(\frac{J'}{J''}\right)^2. \tag{8.8}$$

In order to understand the nature of the activation energy and the reasons for the activation-like dependence, a transition between two parabolic terms should be examined:

$$U_i = \tfrac{1}{2}\gamma_0(x - x_i^0)^2, \quad U_f = \tfrac{1}{2}\gamma_0(x - x_f^0)^2. \tag{8.9}$$

The quantities J' and J'' are expressed in terms of the parameters

$$J' = \frac{m_0\omega_0\Delta_0}{\hbar}, \; J'' = \frac{m_0\omega_0}{\hbar}, \gamma_0 = m_0\omega_0^2, \Delta_0 = x_f^0 - x_i^0. \tag{8.10}$$

Here m_0 and ω_0 are the mass and frequency of the oscillator corresponding to the intramolecular motion of the terms (8.9). Substituting (8.10) into (8.8), the activation energy is

$$E_a = \tfrac{1}{4} M\Omega^2 \Delta_0^2 = \tfrac{1}{2}\gamma_2 \Delta_0^2, \tag{8.11}$$

which means that the activation energy equals the work performed against the elastic forces in order to merge the initial and final terms [36, 37]. Therefore if the transitions from the higher levels of the rapid subsystem are not taken into account then it is not the particle that should be activated over the barrier but rather the slow subsystem is activated in such a way that the barrier vanishes.

In the other limiting case (the distance between the terms is assumed to be fixed), taking account of the higher level occupation, the transition rate constant between two parabolic terms is (see e.g. [13]):

$$K \sim \exp\left[-\frac{m_0\omega_0}{2\hbar}\Delta_0^2 \coth\left(\frac{\hbar\omega_0}{2k_BT}\right)\right] I_0\left(\frac{m_0\omega_0\Delta_0^2}{2\hbar\sinh(\hbar\omega_0/2k_BT)}\right). \tag{8.12}$$

For the sake of simplicity, the energy defect of the intramolecular subsystem is assumed to be zero. It can easily be seen that the rate constant (8.12) "defreezes" at much higher temperatures than when account is taken of potential-barrier fluctuations (see (8.5)).

If the condition

$$\frac{m_0\omega_0\Delta_0^2}{2\hbar\sinh(\hbar\omega_0/2k_BT)} \gg 1$$

holds for (8.12) then the asymptotics of the Bessel function yields

$$K \sim \exp\left[-\frac{m_0\omega_0}{2\hbar}\Delta_0^2 \tanh\left(\frac{\hbar\omega_0}{4k_BT}\right)\right], \tag{8.13}$$

and for $T > \hbar\omega_0/4k_B$ the rate constant is of Arrhenius type, with an activation energy that is equal to the potential barrier height, $\frac{1}{2}\gamma_0\Delta_0^2$, i.e. in this case the transferring particle really overcomes the barrier.

Since at temperatures in the "defreezing" region, $T \approx \hbar\omega_0/4k_B$ the intermolecular vibrations are classical, (8.13) is averaged over displacements from the equilibrium positions:

$$K \sim \exp\left[-\frac{m_0\omega_0}{2\hbar}\Delta_0^2 \tanh\left(\frac{\hbar\omega_0}{4k_BT}\right) \times \right.$$
$$\left. \times \frac{M\Omega^2}{2(m_0\omega_0/\hbar)\tanh(\hbar\omega_0/4k_BT)k_BT + M\Omega^2}\right]. \tag{8.14}$$

In the temperature range $M\Omega^2/2k_BJ'' < T < \frac{1}{4}\hbar\omega_0$ the rate constant is given by (8.8) and (8.11), whereas for $T > \hbar\omega_0/4k_B$ the activation energy is determined by the parameters of both environment and terms:

$$E_a = \frac{\gamma_0\Delta_0^2}{2} \quad \frac{\gamma_2}{\gamma_0+\gamma_2}.$$

Therefore (5.24) takes account both of the dependence of the rate constant of the solid-phase reaction on the potential barrier parameters and of the medium properties over a wide temperature range — from helium temperatures, where the reaction proceeds via the tunneling mechanism, to high temperatures (but certainly not higher than the melting point), where the transition is activation-like.

3.9. INFLUENCE OF ANHARMONICITY OF THE INTERMOLECULAR VIBRATIONS ON THE RATE OF SOLID-PHASE CHEMICAL REACTIONS

The above study of chemical reactions has been carried out assuming that the intermolecular vibrations are harmonic, which gives an analytical solution. If the intermolecular potential is arbitrary then the rate constant can be found numerically [38].

At low temperatures, when the intermolecular potential amplitude is small, the harmonic approximation is quite sufficient, but at higher temperatures corrections due to anharmonicity may be essential and cannot be neglected. It is interesting to learn what are the consequent deviations from anharmonicity result, and at what temperatures, calculating the rate constant of solid-phase chemical reaction, there is a need to take account of anharmonicity of intermolecular vibrations.

If the anharmonicity is not large then it can be treated using similar mathematics as in the harmonic approximation, all the above formulae for the transition rate constant remaining the same; however, the equilibrium phonon coordinates, X_0 and frequencies Ω_0 now become temperature-dependent.

The starting formula for the transition rate constant (5.7) is general and holds for arbitrary systems that are not necessarily harmonic. Hence if the density matrix of the intermolecular subsystem under study is known then to find the rate constant it is only necessary to calculate the integrals.

In order to find the density matrix of an anharmonic intermolecular subsystem, a technique is applied using the principle of minimum free energy [25]:

$$F \leqslant F_0 + Av_0(H - H_0), \tag{9.1}$$

where F and H are the true Gibbs free energy and Hamiltonian; F_0 and H_0 are trial Gibbs free energy and Hamiltonian; Av_0 means averaging over the system characterized by the Hamiltonian H_0. For an arbitrary attracting potential, U(X) the true Hamiltonian is

$$H = \frac{p^2}{2M} + U(X). \tag{9.2}$$

Now the trial Hamiltonian is chosen in the form

$$H_0 = \frac{p^2}{2M_0} + \frac{M_0\Omega_0^2}{2}(X - X_0)^2; \tag{9.3}$$

then the density matrix used in averaging retains its previous form (5.11), with the only difference being that the values M_0, X_0 and Ω_0 vary so as to fit the Gibbs-free-energy minimum (9.1). Then substituting (9.2) and (9.3) into (9.1) and fitting M_0, X_0 and Ω_0 yields the following equations:

$$M = M_0, \tag{9.4$'$}$$

$$Av_0 \frac{\partial U}{\partial X} = 0, \tag{9.4$''$}$$

$$Av_0 X \frac{\partial U}{\partial X} - Av_0 \frac{p^2}{2M}. \tag{9.4$'''$}$$

Equations (9.4) allow calculation of the temperature dependences of the quantities X_0 and Ω_0. Then using the virial theorem, i.e. (9.4′), and the notation $Y = X - X_0$, (9.4″) can be rewritten more conveniently as

$$Av_0 \frac{\partial^2 U(Y + X_0)}{\partial Y^2} = M\Omega_0^2. \tag{9.5}$$

The potential U(X) is now separated into a harmonic part and an anharmonicity correction:

$$U(X) \equiv U(Y + X_0) = \tfrac{1}{2}M\Omega^2(Y + X_0)^2 + U_a(Y + X_0). \tag{9.6}$$

The potential $U(Y + X_0)$ is expanded in X_0 up to quadratic terms; then using the relations

$$\frac{\partial U}{\partial X} = \frac{\partial U}{\partial Y} = \frac{\partial U}{\partial X_0},$$

(9.4″) and (9.5) give

$$Av_0[U_a'(Y)] + X_0\{M\Omega^2 + Av_0[U''(Y)]\} = 0, \tag{9.7}$$

$$M\Omega^2 + Av_0[U_a''(Y)] = M\Omega_0^2. \tag{9.8}$$

Substituting (9.8) into (9.7), the following equations for X_0 and Ω_0 are obtained:

$$X_0 M\Omega_0^2 + Av_0[U_a''(Y)] = 0, \tag{9.7'}$$

$$M(\Omega^2 - \Omega_0^2) + Av_0[U_a''(Y)] = 0. \tag{9.8'}$$

These equations have a simple physical interpretation the first determines the oscillator equilibrium, i.e. the additional force $Av_0[U'(Y)]$ is balanced by an elastic force equal to the elasticity constant multiplied by the displacement; the second equation indicates that the effective elasticity is the elasticity due to the harmonic part of the true potential plus an additional elasticity $Av_0[U''(Y)]$.

As an example, we consider the potential

$$U(X) = \tfrac{1}{2} M\Omega^2 Y^2 + \gamma_3 Y^3 + \gamma_4 Y^4 \tag{9.9}$$

of an anharmonic oscillator. It includes the anharmonic correction

$$U_a(Y) = \gamma_3 Y^3 + \gamma_4 Y^4. \tag{9.10}$$

Substituting (9.10) into (9.7') and (9.8'), the functions $X_0(T)$ and $\Omega_0^2(T)$ are found as

$$X_0(T) = \frac{3\hbar\gamma_3}{2M\Omega_0^3} \coth \frac{\hbar\Omega_0}{2k_B T}, \tag{9.11}$$

$$\Omega_0^2(T) = \Omega^2 + \frac{6\hbar\gamma_4}{M^2\Omega_0} \coth \left(\frac{\hbar\Omega_0}{2k_B T}\right). \tag{9.12}$$

These equations easily yield the conditions on the temperature for which the anharmonicity may be neglected. The third-order correction is small if $X_0(T)$ is small compared with $(J')^{-1}$; this means that the smallness of $X_0(T)$ compared with the thermal vibration amplitude is not a sufficient condition for neglecting the anharmonicity, since, as a rule

$$(J')^{-2} \ll Av_0[Y^2] = \frac{\hbar^2}{M\Omega_0} \coth \left(\frac{\hbar\Omega_0}{2k_B T}\right). \tag{9.13}$$

The fourth-order correction results in a variation of the intermolecular frequency; this variation depends on the sign of the correction. If $\gamma_4 > 0$ then the phonon frequency increases with temperature, but if $\gamma_4 < 0$ it decreases. This can significantly affect the temperature-dependent rate constant of solid-phase reactions. According to (5.28), the slope of the logarithm of the tunneling rate constant versus temperature in the high-temperature range is inversely proportional to the squared frequency. Hence if $\gamma_4 > 0$ then the slope decreases with increasing temperature, whereas it increases if $\gamma_4 < 0$. When

$$3(J')^2 \gamma_4 k_B^2 T^2 / M^3 \Omega^6 \ll 1$$

the fourth-order anharmonicity may be neglected.

This study of anharmonic oscillations allows conclusions to be drawn concerning the behaviour of the rate constant of solid-phase chemical reactions near a phase-transition point. It follows from (9.12) that the intermolecular frequency may become as small as necessary, i.e. a soft mode appears near the phase transition [39]. When the frequency decreases, the intermolecular amplitude increases, which results in a rapid increase of the rate constant of solid-phase chemical reactions near phase transitions [30, 40–42].

3.10. SOME FEATURES OF HEAVY CHARGED-PARTICLE TRANSFER IN THE SOLID PHASE

Charge transfer is known to be an important stage of the radiation–chemical and electrochemical reactions. Because of this, a large number of experimental and theoretical papers (e.g. [2, 12, 36, 43, 44]) have dealt with the principal features of charge transfer in the solid phase. These studies have shown that the charge-transfer rate depends essentially on the environmental properties. The effect that has mainly been taken into account is the charge in the equilibrium phonon coordinates after the transition. Environmental polarization varies, yielding an additional potential barrier (compared with that caused by the electrostatic and exchange interactions between the reactants) to be overcome by the system in the course of the transition.

If electron transfer is studied then in fact the environmental effect is reduced to repolarization. As far as the transfer of heavy particles is concerned, then it is intermolecular vibration, causing fluctuations of the potential barrier overcome by the particle, that is most important additional factor.

In studying the principal features of charge transfer in the solid phase, we shall proceed from (5.24) for the processes with discrete final states and from (6.7) for those with continuous final intermolecular states. These formulae allow the simultaneous consideration of both factors: the change in equilibrium phonon coordinates during transition (repolarization of the medium) and the potential-barrier fluctuations caused by intermolecular vibrations.

First (5.24) is considered. At low temperatures, when the quadratic term in (5.15) may be neglected (i.e. $\partial^2 J/\partial q_\mu \partial q_{\mu'} = 0$), we have [34]

$$K = \frac{1}{\hbar} \, |V(\mathbf{R}_0)|^2 \int_{-\infty}^{\infty} ds \exp\left\{ - i\Delta E \left(s + \frac{i}{2k_BT}\right) \right.$$
$$+ \frac{1}{2}\sum_{\mu} [\tfrac{1}{4}(J'_\mu)^2 - (\Delta q_\mu)^2] \coth\left(\frac{\hbar\omega_\mu}{2k_BT}\right)$$
$$+ \frac{1}{2}\sum_{\mu} [\tfrac{1}{2}(J'_\mu)^2 + (\Delta q_\mu)^2] \frac{\cos(s\hbar\omega_\mu)}{\sinh(\hbar\omega_\mu/2k_BT)}$$
$$\left. + \frac{i}{2}\sum_{\mu} J'_\mu \Delta q_\mu \frac{\sin(s\hbar\omega_\mu)}{\sinh(\hbar\omega_\mu/2k_BT)} \right\}. \tag{10.1}$$

The exponential in the integral is expanded in a series similar to that used in deriving (7.3); the rate constant for single-phonon processes is thus obtained as

$$K - \frac{\pi}{\hbar} \quad |V(\mathbf{R}_0)|^2(\tfrac{1}{2}J'_{\mu_0} \pm \Delta q_{\mu_0})^2 \exp\left\{ - \frac{(|\Delta E| \mp |\Delta E|}{2k_BT} \right.$$
$$\left. + \frac{1}{2}\sum_{\mu} [\tfrac{1}{4}(J'_\mu)^2 - (\Delta q_\mu)^2] \coth\left(\frac{\hbar\omega_\mu}{2k_BT}\right) \right\}. \tag{10.2}$$

The upper sign is for phonon emission ($\Delta E > 0$); the lower sign is for absorption ($\Delta E < 0$).

The low-temperature limit of the charge-transfer rate constant in the solid phase follows from (10.2) on putting $\coth(\hbar\omega_\mu/2k_BT)$ equal to one. It is clear that only nonendothermic reactions possess a low-temperature limit; this assertion following from the first term in the exponent. The next term leads to a decrease of the potential barrier as a result of zero-point lattice vibrations, while the final term causes a tunneling rearrangement of the medium owing to particle transfer. This term, particularly in polar matrices, can be rather large ~40–50 [45], which makes observation of low-temperature heavy charged-particle transfer in polar matrices doubtful.

At higher temperatures the exponent in the integrand cannot be expanded as a series, and (10.1) has to be simplified in another fashion, similar to the transformation of (7.2) into (7.7). Both functions $\cos(s\hbar\omega_\mu)$ and $\sin(s\hbar\omega_\mu)$ in the exponent in (10.1) are expanded up to quadratic terms; the integral is then easily performed to give

$$K \sim \exp\left[\frac{\Delta E}{2k_BT} + \frac{1}{8}\sum_{\mu} (J'_\mu)^2 \coth\left(\frac{\hbar\omega_\mu}{4k_BT}\right) \right.$$
$$\left. - \frac{1}{2}\sum_{\mu} (\Delta q_\mu)^2 \tanh\left(\frac{\hbar\omega_\mu}{4k_BT}\right)\right]. \tag{10.3}$$

The condition for applicability of (10.3) is obtained by generalizing the condition (7.11):

$$\sinh(\hbar\omega_D/2k_BT) \ll \hbar\omega_D \left\{ \frac{\hbar(J')^2}{4M\omega_D} + \sum_\mu (\Delta q_\mu)^2 \right\} \Big/ \Delta E. \qquad (10.4)$$

In addition to the condition (10.4), there is an upper bound on the temperature, since the quadratic term in the expansion of the exponent of the matrix element (6.4) has been neglected. At the same time, this quadratic term must be included in the expansion when considering the rate constant of heavy charged-particle transfer at temperatures comparable to the Debye temperature, where activation-like behaviour might occur (as we have seen for reactions without any charge in the medium). Therefore, as in the $\Delta q_\mu = 0$ case, the Einstein model of the solid is employed. Using (8.1), the column vector A_μ is

$$A_\mu = \bar{J}' \begin{pmatrix} 0 \\ \vdots \\ 0 \\ 1 \\ -1 \\ 0 \\ \vdots \\ 0 \end{pmatrix} + 2i \frac{\sin(s\hbar\Omega)}{\cosh \dfrac{\hbar\Omega}{2k_BT} + \cos(s\hbar\Omega)} \begin{pmatrix} 0 \\ \vdots \\ 0 \\ \Delta q_1 \\ \Delta q_2 \\ 0 \\ \vdots \\ 0 \end{pmatrix}, \qquad (10.5)$$

the matrix $B^{-1}_{\mu\mu'}$ retaining its form (8.3). The product of the matrices in the exponent of (5.24) is then found and transformations are performed similar to those involved in deriving (8.5). The result [36] is

$$K = \frac{2\pi}{\hbar} |V(\mathbf{R}_0)|^2 \varphi_3^{-1/2} \exp\left[\frac{\Delta E}{2k_BT} + \varphi_1(1-\varphi_2) - \frac{E_r}{\hbar\Omega}\coth\left(\frac{\hbar\Omega}{2k_BT}\right)\right] I_{\frac{|\Delta E|}{\hbar\Omega}}\left(\varphi_1\varphi_2 + \frac{E_r}{\hbar\Omega}\sinh^{-1}\left(\frac{\hbar\Omega}{2k_BT}\right)\right)\rho_f, \qquad (10.6)$$

$$E_r = \tfrac{1}{2}\hbar\Omega \sum_\mu (\Delta q_\mu)^2.$$

Here E_r is the energy of environmental rearrangement. To obtain (10.6), the smallness of the quantity $J'(\Delta q_1 - \Delta q_2)$ compared with $E_r/\hbar\Omega$ has been used. The ratio of these quantities can easily be found if it is taken into account that the transition changes the equilibrium positions of a large number of molecules, and the rearrangement energy is due to all these displacements. As for the quantity $J'(\Delta q_1 - \Delta q_2)$, it is determined only by the reactant displacements. In fact, calculations [45] have shown that the value $E_r/\hbar\Omega$ may be as high as ~ 50, whereas $J'(\Delta q_1 - \Delta q_2)$ is only ~ 5 (in the evaluations it was assumed that $|\Delta q| \approx 0.1$ Å and $J' \approx 50$ Å^{-1}).

We now consider some limiting cases of (10.6). At low temperatures, when the argument of the Bessel function is small ($T \ll \hbar\Omega/2k_B$),

$$K = \frac{2\pi}{\hbar}|V(\mathbf{R}_0)|^2(1+\tfrac{1}{2}\bar{J}'')^{-1/2}\left(\frac{|\Delta E|}{\hbar\Omega}!\right)^{-1}\left[\frac{E_r}{\hbar\Omega} + \frac{E_J}{\hbar\Omega}(1+\tfrac{1}{2}\bar{J}'')^{-2}\right]^{|\Delta E|/\hbar\Omega} \exp\left\{\frac{\Delta E - |\Delta E|}{2k_BT} + \frac{E_J}{\hbar\Omega}(1+\tfrac{1}{2}\bar{J}'')^{-1} - \frac{E_r}{\hbar\Omega}\right\}\rho_f, \tag{10.7}$$

$$E_J = \hbar^2(J')^2/4M.$$

the terms in the exponent having the same physical meaning as in (10.2).

The Einstein model allows investigation of the case $|\Delta E| > \hbar\Omega$. As mentioned above, the ratio of ΔE and $\hbar\Omega$ does not affect the temperature dependence of the rate constant, but can influence its absolute value. If $|\Delta E| \gg \hbar\Omega$ then Stirling's formula can be used. Equation (10.7) then becomes

$$K = \frac{2\pi}{\hbar}|V(\mathbf{R}_0)|^2\left(\frac{2\pi|\Delta E|}{\hbar\Omega}\right)^{1/2}\left(\frac{E_r}{|\Delta E|}\right)^{\Delta E/\hbar\Omega} \times \exp\left[\frac{\Delta E - |\Delta E|}{2k_BT} + \frac{E_J}{\hbar\Omega} - \frac{E_r}{\hbar\Omega} + \left(1 + \frac{E_J}{E_r}\right)\frac{|\Delta E|}{\hbar\Omega}\right]\rho_f. \tag{10.8}$$

It can be seen that the last term in the exponent (10.7), which is due to the environmental rearrangement is compensated at the expense of the energy defect of the rapid subsystem.

In the high-temperature region, when the argument of the Bessel function becomes large, its asymptotic expansion may be used and the transition rate constant now becomes

$$K = |V(\mathbf{R}_0)|^2\left[2\pi\Omega\sinh\left(\frac{\hbar\Omega}{2k_BT}\right)\Big/\hbar E_r\right]^{1/2}\varphi_3^{-1/2} \times \exp\left[\frac{\Delta E}{2k_BT} + \varphi_1 - \frac{E_r}{\hbar\Omega}\tanh\left(\frac{\hbar\Omega}{4k_BT}\right) - \frac{1}{2}\left(\frac{\Delta E}{\hbar\Omega}\right)^2\left(\frac{E_r}{\hbar\Omega}\sinh^{-1}\left(\frac{\hbar\Omega}{2k_BT}\right) + \varphi_1\varphi_2\right)^{-1}\right]\rho_f. \tag{10.9}$$

This formula is convenient for analyses treating the intermolecular vibrations as classical:

$$K = \frac{2\pi}{\hbar} |V(\mathbf{R}_0)|^2 \left[4\pi E_r k_B T \left(1 + \frac{J'' k_B T}{\gamma_R}\right)\right]^{-1/2}$$

$$\times \exp\left[-\frac{(\Delta E - E_r)^2}{4E_r k_B T} + \frac{(J')^2 k_B T}{2(\gamma_R + J'' k_B T)}\right]\rho_f, \qquad (10.10)$$

$$\varphi_1\varphi_2 \ll E_r/\hbar\Omega \sinh\left(\frac{\hbar\Omega}{2k_B T}\right), \quad \rho_f = (\hbar\Omega)^{-1}.$$

Thus the contributions to the temperature-dependent transition rate constant caused by changes in the medium and reactant thermal vibrations separate. The first term in the exponent (10.10) is for the environmental change, while the second accounts for the influence of the medium on the potential barrier overcome by the particle. If $\Delta E \approx E_r$ (i.e. the activation energy caused by the environmental rearrangement and the energy defect of the rapid subsystem is small) then only the second term leads to a temperature dependence of the transition rate constant. For $T < \gamma_R/J'' k_B$ it yields a linearly temperature-dependent logarithm of the rate constant characteristic of tunneling particle transfer [10, 34, 46].

At higher temperatures $T > \gamma_R/J'' k_B$ the temperature dependence of the rate constant is activation-like [30, 34], the activation energy being the sum of two terms [36, 37]. The first of these is due to the environment reorganization and energy defect of the rapid subsystem; the second one is the energy necessary to draw the potential well together until barrier disappears. Hence, transitions in this temperature region occur when the potential barrier to be overcome by the particle disappears completely owing to thermal fluctuations, and the rate constant is the product of the transition rate constant in the absence of the barrier and the probability of such a fluctuation.

Figure 3.4 shows the temperature dependence of the rate constant for proton transfer (curve 1) for typical parameters. It has been calculated [36] using (10.6) at $J'' = 0$, $E_r = 1\text{eV}$, $\Delta E = 0$, $E_J = \frac{1}{6}$ eV, $\Omega = 3 \times 10^{13}\,\text{s}^{-1}$ and compared with the curve (2) calculated at $E_J = 0$. It can easily be seen that the fluctuations of the potential barrier due to reactant thermal vibrations are extremely important when calculating the rate constant of heavy charged-particle transfer. A calculation of the rate constant at $E_r = 0$ (curve 3) is also shown; here there is no environmental rearrangement due to the particle transition (this happens, for example, in the case of hydrogen-atom transfer). The curve 3 lies rather higher than the curves 1 and 2, indicating that if there

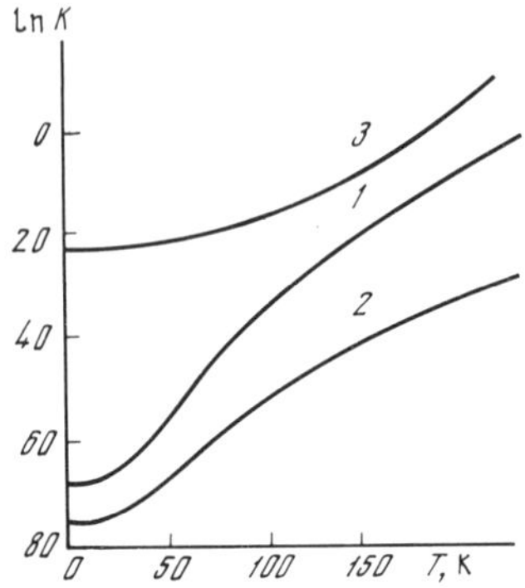

Figure 3.4 Temperature dependent rate constant of the particle transfer in solid phase at low temperatures.

are several competing reaction channels then the one with the lowest environmental rearrangement energy wins. Hence, the reaction mechanism depends on the medium in which it proceeds. The mechanism can even change when going from the gaseous phase to the solid one. Also, the environmental-rearrangement contribution to the transition rate constant depends on temperature; hence the reaction mechanism is determined by the temperature. Thus the ion–molecule reactions

$$RH^+ + RH \rightarrow R + HRH^+$$

are known to proceed in the gaseous phase and in the condensed phase at high temperatures via proton transfer, whereas at low temperature the formation of the same products is expected via hydrogen-atom rather than proton transfer:

$$RH^+ + RH \rightarrow HRH^+ + R.$$

The formulae obtained in this section also lead to a description of the kinetic isotope effect for solid-phase chemical reactions. Since this effect can easily be seen to be independent of the environmental-rearrangement energy, the influence of the medium is confined to potential-barrier fluctuations caused by intermolecular vibrations. The increasing amplitude of thermal vibrations with higher temperatures results in a decrease of the isotope effect. Keeping in mind that J', $J'' \sim M^{1/2}$ (M is the mass of the transferring particle), (10.10) yields, on substituting deuterium for hydrogen,

$$\ell n \left(\frac{K_H}{K_D}\right) \approx \text{const} - \frac{(J')^2 k_B T \left[\gamma_R + 2\left(1 - \frac{1}{\sqrt{2}}\right) J'' k_B T\right]}{2(\gamma_R + J'' k_B T)(\gamma_R + \sqrt{2} J'' k_B T)}, \tag{10.11}$$

i.e. there is a term that leads to a decrease in the isotope effect at higher temperatures. Here J′ and J″ are determined by the transition matrix element for the hydrogen atom; K_H and K_D are the transition rate constants for normal and deuteriated molecules.

It should be noted that intermolecular vibrations are not the only cause of the diminishing isotope effect with temperature — the semi-classical kinetic isotope effect mentioned in Chapter 2 has a similar effect.

We now consider charged-particle transfer when the final states of the rapid subsystem lie in the continuous spectrum. At low temperatures, when the quadratic term in the expansion (6.5) may be neglected, the exponent in (6.7) simplifies and $\tilde{\phi} = 1$, the transition rate constant becomes

$$\begin{aligned} K = {} & \frac{2\pi}{\hbar} |V(E_{n_i}, \mathbf{R}_0)|^2 \\ & \times \exp\left\{ \frac{1}{4} \sum_\mu (J'_\mu)^2 \frac{\cosh(\frac{1}{2} J'_E \hbar\omega_\mu)\cosh[\frac{1}{2}(\beta - J'_E)\hbar\omega_\mu]}{\sinh(\frac{1}{2}\beta\hbar\omega_\mu)} \right. \\ & - \sum_\mu \frac{(\Delta q_\mu)^2 \sinh(\frac{1}{2} J'_E \hbar\omega_\mu)\sinh[\frac{1}{2}(\beta - J'_E)\hbar\omega_\mu]}{\sinh(\frac{1}{2}\beta\hbar\omega_\mu)} \\ & \left. + \frac{1}{2} \sum_\mu \frac{J'_\mu \Delta q_\mu \sinh[(\frac{1}{2}\beta - J'_E)\hbar\omega_\mu]}{\sinh(\frac{1}{2}\beta\hbar\omega_\mu)} . \right\} \end{aligned} \tag{10.12}$$

If $J'_E = 0$, i.e. if the matrix element (6.4) depends only weakly on the energy transferred to the phonon subsystem (see also (6.3)), then the rate constant is given by

$$K = \frac{2\pi}{\hbar} |V(E_{n_i}, \mathbf{R}_0^i)|^2 \exp\left[\frac{1}{4} \sum_\mu (J'_\mu)^2 \coth(\tfrac{1}{2}\beta\hbar\omega_\mu) \right]. \tag{10.13}$$

The last term in the exponent in (10.12) at $J'_E = 0$ shifts the point near which the matrix element (6.4) is expanded from R_0 to R_0^i, since the equality

$$|V(E_{n_i}, \mathbf{R}_0)|^2 \exp\left[\frac{1}{2} \sum_\mu J'_\mu \Delta q_\mu \right] \approx |V(E_{n_i}, \mathbf{R}_0^i)|^2$$

holds. This is quite natural, since, in applying (6.3) the expansion should be made in the vicinity of the point R_0^i. At the same time, it can be shown (in a way similar to that used in deriving (10.6)) that this term in the exponent of (10.12) is smaller than the second one. This is why, when studying the temperature dependence of solid-phase chemical reactions, we shall restrict ourselves to the first two terms in the exponent of (10.12).

We now consider the low-temperature limit of the chemical reaction rate, using (10.12). Assuming T = 0 in (10.12), we get

$$K = \frac{2\pi}{\hbar}\,|V(E_{n_i}, \mathbf{R}_0)|^2 \exp\left\{\frac{1}{4}\sum_\mu \exp[-\tfrac{1}{2}J'_E\hbar\omega_\mu]\cosh(\tfrac{1}{2}J'_E\hbar\omega_\mu)(J')^2 - \sum_\mu \exp[-\tfrac{1}{2}J'_E\hbar\omega_\mu]\sinh(\tfrac{1}{2}J'_E\hbar\omega_\mu)\,\Delta q_\mu^2\right\}. \tag{10.14}$$

The first term in the exponent yields an increase in the rate constant as a result of zero-point oscillations of the potential barrier; the second one describes the environmental rearrangement. It can easily be seen that the contributions of these terms depend both on the environmental properties and the potential-barrier parameters.

Comparing (10.14) and (10.13) as $T \to 0$, conditions are obtained that must be satisfied by J'_E (and hence the matrix element (6.4)) so that the transition rate constant is determined by (6.3):

$$J'_E \ll (\hbar\omega_D)^{-1}, \quad J'_E \ll E_r^{-1}. \tag{10.15}$$

We now consider the transition rate constant at temperatures where the Einstein model is valid. The column vector $\tilde{A}_\mu$ in (6.7) then becomes

$$\tilde{A}_\mu = \bar{J}'\begin{pmatrix} 0 \\ \vdots \\ 0 \\ 1 \\ -1 \\ 0 \\ \vdots \\ 0 \end{pmatrix} + \frac{\sinh[\frac{1}{2}(\beta - J'_E)\hbar\Omega]}{\cosh[\frac{1}{2}(\beta - J'_E)\hbar\Omega]\cosh(\frac{1}{2}J'_E\hbar\Omega)}\begin{pmatrix} 0 \\ \vdots \\ 0 \\ \Delta q_1 \\ \Delta q_2 \\ 0 \\ \vdots \\ 0 \end{pmatrix}. \tag{10.16}$$

The matrix $(\tilde{B}^{-1})_{\mu\mu'}$ can be represented in the form (8.3), with the only difference being that the quantity

$$\tilde{C} = \frac{\sinh(\frac{1}{2}\beta\hbar\Omega)}{\cosh[\frac{1}{2}(\beta - J'_E)\hbar\Omega]\cosh(\frac{1}{2}J'_E\hbar\Omega)}$$

is substituted for c. Then insertion of $\tilde{A}_\mu$ and $(\tilde{B}^{-1})_{\mu\mu'}$ into (6.7) and neglect of small terms $\sim J'\Delta q_{1,2}$, gives [8]

$$K = \frac{2\pi}{\hbar}|V(E_{n_i}, \mathbf{R}_0)|^2\{\det\tilde{\phi}\}^{-1/2}$$

$$\times \exp\left\{\frac{1}{2}(J')^2 \Big/ \left[J'' + \frac{\sinh(\frac{1}{2}\beta\hbar\Omega)}{\cosh(\frac{1}{2}J'_E\hbar\Omega)\cosh[\frac{1}{2}(\beta - J'_E)\hbar\Omega]}\right]\right.$$
$$\left. - \frac{2E_r \sinh(\frac{1}{2}J'_E\hbar\Omega)\sinh[(\beta - J'_E)\hbar\Omega]}{\hbar\Omega \sinh(\frac{1}{2}\beta\hbar\Omega)}\right\}. \tag{10.17}$$

For the classical intermolecular vibrations at temperatures $T \gg \hbar\Omega/2k_B$ (but for $2J''k_BT/M\Omega^2 \ll 1 + \hbar^2J''J'_E/2M$) the rate constant is simplified and the exponent in (10.17) is a linear function of temperature; but if the latter inequality does not hold then the rate constant is

$$K = \frac{2\pi}{\hbar}|V(E_{n_i}, \mathbf{R}_0)|^2\{\det\tilde{\phi}\}^{-1/2}$$
$$\times \exp\left\{\frac{(J')^2}{2J''} - (J')^2/J''\left(1 + \frac{\hbar^2J'_EJ''}{2M} + \frac{2J''k_BT}{M\Omega^2}\right)\right.$$
$$\left. + J'_EE_r(J'_Ek_BT - 1)\right\}. \tag{10.18}$$

It follows that the environment also influences the isotope effect on the chemical reaction rate; substituting deuterium for hydrogen, the isotope kinetic effect is

$$\ell n\left(\frac{K_H}{K_D}\right) \approx (\sqrt{2} - 1)\left\{2|\ell n|V(E_{n_i}, \mathbf{R}_0)|| - \frac{(J')^2}{2J''}\right.$$
$$\left. - J'_EE_r[(\sqrt{2} + 1)J'_Ek_BT - 1]\right\}. \tag{10.19}$$

Here J', J'' and J'_E are those relevant to the nondeuteriated molecules; (10.19) is obtained under the assumption that

$$\frac{2J''k_BT}{M\Omega^2} \gg 1 + \frac{J''J'_E\hbar^2}{2M}.$$

The kinetic isotope effect, as given by (10.19), differs markedly from the isotope effect for discrete final intramolecular states (see 10.11), since (10.19), in contrary with (10.11), depends on environmental-rearrangement energy.

The temperature dependence of chemical reactions with continuous final intramolecular states also differ markedly from those with discrete final states. Thus the rate constant (10.3) "defreezes" at $T \approx \hbar\omega_D/4k_B$, while (10.12) states that for the continuous spectrum of the final intramolecular states, the "defreezing" of the rate constant as well as of the internal energy occurs at $T \approx \hbar\omega_D/2k_B$. This conclusion can be verified experimentally. Also at variance are the contributions of the zero-point vibrations.

Finally, the consideration of this section applies not only to charge transfer but is also valid for any chemical reaction with an environmental change. If, however, no such change occurs then it must be assumed that $\Delta q_\mu = 0$, and the results presented above (see Sections 3.7 and 3.8) must be used.

The results of this section may also be used in the study of electron transfer in solids. However, the electron localization radius in the potential well is much larger than the amplitudes of intermolecular vibrations i.e. the relative reactant displacements, which are very important for heavy particle transfers, do not affect either the value or the temperature dependence of the rate constant of the electron tunneling transfer. Formally, the parameters J' and J'' are proportional to $m^{1/2}$, and hence these values for the electron transfer are nearly forty times smaller than for the hydrogen transfer. Therefore it may be assumed that $J' = J'' = 0$ in (10.1)–(10.19) in order to obtain formulae describing the electron tunneling transfer in solids.

3.11. ORIENTATIONAL INTERMOLECULAR VIBRATIONS IN SOLID PHASE TUNNELING REACTIONS

As demonstrated above, the temperature-dependent rate constant of atomic tunneling in solids is generally affected by intermolecular vibrations. At liquid-helium temperatures account must be taken of the zero-point vibrations of the solid lattice, whose amplitudes are inversely proportional to the square root of the Debye frequency, their contribution to the exponent being $\sim \omega_D^{-1}$. At higher temperatures the vibrations become "unfrozen", and at $T \approx \hbar\omega_D/4k_B$ the rate constant becomes temperature-dependent.

When studying the influence of intermolecular vibrations on solid-phase chemical reactions, only translational vibrations were taken into account. However, since the orientational vibrations are characterized by lower frequencies then the translational ones, they are capable of affecting both the zero-point contribution (increasing it) and the "defreezing" temperature (decreasing it). Moreover, the dispersions of the orientational and translational modes are at variance [47], which can also lead to some peculiar features in the temperature dependence of the tunneling rate constant.

This section deals with the way in which the orientational vibrations in a solid lattice influence the transition rate of tunneling atoms. Special attention is paid to the case of a one-dimensional lattice — an analytical solution is found for this case and all the main features of the rate-constant temperature dependence relevant to the orientational vibrations are retained. Situations when all three dimensions are important are dealt with and the results obtained are compared with those for the translational vibrations of a one-dimensional solid lattice.

As before, chemical reactions are treated as radiationless transitions of particles from one potential well into another. The states of the tunneling particle in the potential wells are described by intramolecular wave functions, with only single energy level being assumed in each well; the fluctuations of the potential barrier are described by phonon states. The double adiabatic approximation is used, i.e. the intramolecular frequencies are assumed to be much greater then the orientational Debye frequency. This allows separation of the intra- and interatomic vibrations (see Section 3.4). The transition rate constant is then given by (5.24), which (neglecting environmental rearrangement ($\Delta q_\mu = 0$) and the quantity $J''_{\mu\mu'}$) becomes at low temperatures

$$K = \frac{1}{\hbar}\,|V(\mathbf{R}_0)|^2 \exp\left[\frac{\Delta E}{2k_BT} + F_1(T)\right]\int_{-\infty}^{\infty} ds\, \exp[-i\Delta Es + F_2(T)],$$

$$F_i = \sum_\lambda (J'_\lambda)^2 f_i(\omega_\lambda, T),\ f_1 = \coth\left(\frac{\hbar\omega_\lambda}{2k_BT}\right),\quad f_2 = \frac{\cos(s\hbar\omega_\lambda)}{\sinh(\hbar\omega_\lambda/2k_BT)}.$$

In a colinear model of the reaction, the translational modes yield, for a one-dimensional lattice,

$$\begin{aligned}(J'_\lambda)^2 &= \left(\frac{\partial J}{\partial X_1}\right)^2 \left|\frac{\partial X_1}{\partial q_\lambda} - \frac{\partial X_2}{\partial q_\lambda}\right|^2 \\ &= \frac{(J')^2\hbar\,|e^{i\lambda X_1} - e^{i\lambda X_2}|^2}{8NM\omega_\lambda} = \frac{(J')^2\hbar(1-\cos\lambda\tilde{a})}{4NM\omega_\lambda},\end{aligned} \tag{11.2}$$

where X_1 and X_2 are the molecule coordinates between which the tunneling occurs; $\tilde{a} = |X_1 - X_2| = R_0$ is the equilibrium distance between these molecules (lattice spacing); N is the number of molecules in the lattice. The translational modes have the following dispersion:

$$\omega_\lambda = \Omega_1(1-\cos(\lambda\tilde{a}))^{1/2},\quad 0 \leqslant \lambda < \pi/\tilde{a},\quad 0 \leqslant \omega_\lambda < \sqrt{2}\Omega_1. \tag{11.3}$$

In order to calculate the rate constant for the orientational phonons, equations are also required for $(J'_\lambda)^2$ and ω_λ. A one-dimensional model is used for the orientational phonons [47]:

$$\left.\begin{aligned} \omega_\lambda &= \Omega_2(1+\tilde{q}\cos(\lambda\tilde{a}))^{1/2}, \quad 0 \leqslant \lambda < \pi/\tilde{a}, \quad -1 \leqslant \tilde{q} \leqslant 1, \\ (1-|\tilde{q}|)^{1/2}\Omega_2 &= \omega_{\min} \leqslant \omega_\lambda \leqslant \omega_{\max} = (1+|q|)^{1/2}\Omega_2, \quad \Delta\omega = \omega_{\max}-\omega_{\min}. \end{aligned}\right\} \tag{11.4}$$

There are two important differences between the orientational and translational phonons: first, the orientational vibrations have lower frequencies (Ω_2 is known [48] to lie within the range 10–1000 cm^{-1}) and hence have greater amplitudes; secondly, since $\tilde{q} \neq -1$, the spectrum (11.4) starts from a finite frequency $\omega_{\min} > 0$ at $\tilde{q} > -1$ rather than from zero frequency.

Since the cases $\tilde{q} \neq -1$ are of importance for us, the interpretation of the value $\tilde{q}$ is discussed in more detail. The potential energy of two neighboring interacting molecules is

$$\begin{aligned} U(\varphi_1,\varphi_2) &= \tfrac{1}{2}\gamma_\varphi\,(\varphi_1^2+\varphi_2^2+2\tilde{q}\varphi_1\varphi_2) \\ &= \tfrac{1}{2}\gamma_\varphi\,[\tfrac{1}{2}(1+\tilde{q})\,(\varphi_1+\varphi_2)^2 + \tfrac{1}{2}(1-\tilde{q})\,(\varphi_1-\varphi_2)^2]. \end{aligned} \tag{11.5}$$

At $\tilde{q} = -1$ the potential energy (similarly to the translational case) would depend on the difference in molecule rotations, $\varphi_1-\varphi_2$, whereas at $\tilde{q} = 1$ it depends on the sum, $\varphi_1+\varphi_2$. Therefore the quantities $\frac{1}{2}(1+\tilde{q})$ and $\frac{1}{2}(1-\tilde{q})$ specify the relative weights of the contributions to the potential energy from the rotational sum and difference of the rotational shifts respectively. The sum and difference vary with the frequencies

$$\Omega_- = [\gamma_\varphi(1-\tilde{q})/I]^{1/2} = \Omega_2[\tfrac{1}{2}(1-\tilde{q})]^{1/2}, \; \Omega_+ = \Omega_2[\tfrac{1}{2}(1+\tilde{q})]^{1/2}. \tag{11.6}$$

These follow from the equations of motion of two molecules with the angular momenta I in the potential (11.5). Bearing in mind that the effective force constants in a one-dimensional chain of molecules are twice those for two interacting molecules, it can easily be seen that the frequencies (11.4) lie between the frequencies ($\sqrt{2}\Omega_-$, $\sqrt{2}\Omega_+$) of the rotational counter- and covibrations†. This also holds for translational vibrations, since for these $\tilde{q} = -1$ and the frequency of the "covibrations" is $\Omega_+ = 0$. Examples of $\tilde{q} \neq -1$ cases are $\tilde{q} = 1/2$ for the dipole potential energy and $\tilde{q} = 0$ for independent molecules (Einstein model, $\Delta\omega \ll \omega_\lambda$). It should also be pointed out that long-wave orientational phonons,

$$\begin{aligned} \omega_\lambda &\approx \Omega_2(1-|\tilde{q}|)^{1/2} + \frac{|\tilde{q}|}{4(1-|\tilde{q}|)^{1/2}}\,\Omega_2\tilde{a}^2\lambda^2 \\ &= \omega_{\min} + (\overline{\omega}_{\max}-\overline{\omega}_{\min})\left(\frac{\tilde{a}\lambda}{\pi}\right)^2 = \omega_{\min} + \Delta\overline{\omega}\left(\frac{\tilde{a}\lambda}{\pi}\right)^2 \end{aligned} \tag{11.7}$$

† Here we use the terms counter- and covibrations to denote the vibrations when two molecules rotate in opposite or in the same directions, respectively.

are similar to long-wave translational optical phonons, but the frequencies are much lower.

The normalized density of states of the orientational phonons is

$$g(\omega) = \frac{2\,\omega}{[(\omega^2 - \omega_{min}^2)(\omega_{max}^2 - \omega^2)]^{1/2}}. \tag{11.8}$$

Using Fig. 3.5, $(J_\lambda')^2$ is given for orientational vibrations by

$$\left.\begin{aligned} (J_\lambda')^2 &= \left(\frac{\partial J}{\partial R_0}\right)^2 \left|\frac{\partial R}{\partial \varphi_1}\frac{\partial \varphi_1}{\partial \varphi_\lambda} + \frac{\partial R}{\partial \varphi_2}\frac{\partial \varphi_2}{\partial \varphi_\lambda}\right|^2 \\ &= \frac{\hbar \bar{s}^2 (J')^2}{8NI\omega_\lambda}\,|\alpha_1 e^{i\lambda X_1} + \alpha_2 e^{i\lambda X_2}|^2 = \frac{\hbar (J')^2}{4N(I/\bar{s}^2)\omega_\lambda}\,(1 - \bar{q}\cos(\lambda\tilde{a})), \\ \alpha_i &= \left(\frac{\partial R}{\partial \varphi_i}\right) \Big/ \bar{s}, \quad \bar{s}^2 = \tfrac{1}{2}[(R'_{\varphi_1})_0^2 + (R'_{\varphi_2})_0^2], \end{aligned}\right\} \tag{11.9}$$

$$\alpha_1^2 + \alpha_2^2 = 2, \quad -1 \leqslant \bar{q} = \alpha_1\alpha_2 \leqslant 1, \tag{11.10}$$

where φ_λ are normal coordinates with wavenumber λ; the index 0 indicates equilibrium values. The third important difference between the orientational and translational phonons is the presence of the parameters $\bar{q} \neq -1$ and $\bar{s}^2 \approx r_{1,2}^2$ (here $I/\bar{s}^2 \neq M$) in the expression for $(J_\lambda')^2$. The values $\bar{s}^2$ and $\bar{q}$ depend on the geometry of the molecules and the equilibrium positions of the tunneling atoms inside them, in contrast with the value $\tilde{q}$ in (11.4), which is determined by the potential energy of the interacting molecules. The tunnel length (see Fig. 3.5) is

$$R^2 = \tilde{a}^2 + r_1^2 + r_2^2 - 2r_1r_2\cos(\varphi_1 - \varphi_2) + 2\tilde{a}(r_1\sin\varphi_1 - r_2\sin\varphi_2). \tag{11.11}$$

Some examples of different values of $\bar{s}$ and $\bar{q}$ for $r_1 = r_2 = r_0$ are as follows.

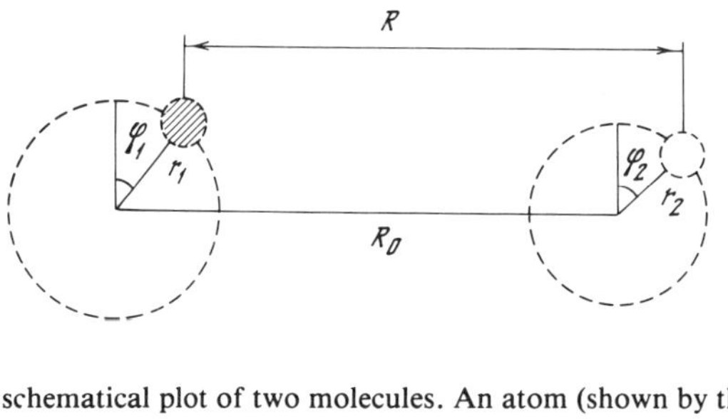

Figure 3.5 A schematical plot of two molecules. An atom (shown by the shaded circle) with the orientation ϕ_1 in one of the molecules transfers to the place (shown by the open circle) in the other molecule with orientation ϕ_2, the distance R being passed.

(i) $\varphi_1^0 = \varphi_2^0 = 0$: vibrations of both molecules are effectively changed; $R - \bar{s} = r_0$, $\bar{q} = -1$, $I/\bar{s}^2 = I/r_0^2 \approx M$;

(ii) $\varphi_1^0 = \varphi_2^0 + \pi$: the molecules at equilibrium are "antiphase"; $\bar{s} = r_0|\cos\varphi_1^0|$, $\bar{q} = +1$, $I/\bar{s}^2 \geqslant M$.

(iii) $\varphi_1^0 = 0$, $\varphi_2^0 = \frac{1}{2}\pi$: only the φ_1 rotation is effectively changed;

$R - \bar{s} \approx r_0/\sqrt{2}$, $\bar{q} \approx 0$, $I/\bar{s}^2 \approx 2M$, and the dependence of $(J')^2$ on the wavenumber λ (see (11.9)) is now absent.

We now examine the consequences of the above peculiarities of orientational vibrations. Using (11.4), (11.8) and (11.9), we find that $F_i(T)$ is given by

$$F_i(T) = \frac{\hbar(J')^2}{4N(I/\bar{s}^2)\Omega_2} \sum_\lambda \frac{1+\bar{q}\cos(\lambda\tilde{a})}{\omega_\lambda/\Omega_2} f_i(\omega_\lambda, T)$$

$$= \frac{\hbar(J')^2}{4(I/\bar{s}^2)\Omega_2} \int_{\omega_{min}}^{\omega_{max}} d\omega_\lambda\, g(\omega_\lambda) \frac{1+\bar{q}\cos(\lambda\tilde{a})}{\omega_\lambda/\Omega_2} f_i(\omega_\lambda, T). \tag{11.12}$$

It follows from (11.12) that, owing to the inequality $\Omega_2 < \Omega_1$, the value of the factor Ω_2^{-1} before the sum (integral) in (11.12), together with the relation

$$\frac{F_i^{or}}{F_i^{tr}} \sim \frac{M\Omega_1}{(I/\bar{s}^2)\Omega_2} \approx \frac{\Omega_1}{\Omega_2} > 1, \tag{11.13}$$

ensures that $F_i(T)$ in the exponents (11.1) for the orientational phonons exceed the corresponding quantities for the translational phonons.

At low temperatures $T \ll \Delta E/2k_B$, (11.1) becomes simplified, since $\exp[F_2(T)]$ can then be expanded as a series. The first nonvanishing term with $\Delta E \neq 0$ is (see (7.3))

$$K = \frac{1}{\hbar^2} |V(\mathbf{R}_0)|^2 (J'_\lambda)^2 g\left(\frac{\Delta E}{\hbar}\right) \exp\left[\frac{\Delta E - |\Delta E|}{2k_B T} + F_i(T)\right], \tag{11.14}$$

where the wavenumber λ_0 is given by $\hbar\omega_{\lambda_0} = |\Delta E|$.

If $|\Delta E| < \hbar\omega_{min}$ then the energy defect can be compensated by a translational phonon absorption (or — which is less probable — by the absorption of several orientational phonons), i.e. in this case translational phonons should be also taken into account. For $\Delta E > \hbar\omega_{min}$ the expansion of the exponent should also include higher-order terms corresponding to the absorption of several phonons (if $|\Delta E| < \hbar\Omega_1$ then it is sufficient to account for the translational phonons). However, these corrections to (11.14) can influence only the weakly temperature-dependent preexponential factor. As far as the temperature dependence of the rate constant is concerned, it is mainly determined by the exponential (11.14) for any ratio of the quantities ΔE and $\hbar\omega_{min}$, $\hbar\omega_{max}$. That is why, without losing the general character of our results, we may assume that $\hbar\omega_{min} < \Delta E < \hbar\omega_{max}$.

The rate constant of tunneling transitions depends on ΔE in the same fashion as in the case of translational vibrations. So we need not pay much attention to this problem.

It follows from (11.14) that all necessary information on the principal features of the temperature-dependent rate constant for rotational vibrations is incorporated in the function $F_1(T)$. This is analyzed in the long-wave approximation, which holds at low temperatures:

$$F_1(T) = \frac{\hbar(J')^2}{4(I/\bar{s}^2)\omega_{min}} \int_0^1 \frac{1+\bar{q}+\bar{q}\pi^2\bar{\lambda}^2/2}{1+\bar{\lambda}^2\overline{\Delta\omega}/\omega_{min}} \coth\left(\frac{\hbar\omega_{min}}{2k_BT} + \frac{\hbar\Delta\omega}{2k_BT}\bar{\lambda}^2\right) d\bar{\lambda} \tag{11.15}$$

$$= \frac{\hbar(J')^2}{4(I/\bar{s}^2)\omega_{min}} \left\{ \left[\frac{\bar{q}\pi^2\omega_{min}}{2\overline{\Delta\omega}} + \frac{1+\bar{q}-\bar{q}\pi^2\omega_{min}/2\Delta\omega}{(\Delta\omega/\omega_{min})^{1/2}} \arctan\left(\frac{\Delta\omega}{\omega_{min}}\right) \right] + 2\int_0^1 \frac{1+\bar{q}+\frac{1}{2}\bar{q}\pi^2\bar{\lambda}^2}{1+\bar{\lambda}^2\Delta\omega/\omega_{min}} \frac{d\bar{\lambda}}{\exp[(\hbar\omega_{min}/2k_BT)+(\hbar\overline{\Delta\omega}/2k_BT)\bar{\lambda}^2]-1} \right\} \tag{11.15'}$$

The first term in the braces (11.15′) does not depend on temperature and reflects the contribution of the zero-point vibrations to the exponent. It is proportional to $1/\overline{\Delta\omega}$, i.e. it is markedly larger than $1/\Omega_1$, ensuring a large contribution of the rotational zero-point vibrations compared with the translational ones.

We now consider the second term in (11.15′). It is easy to see that the features of the rate constant relevant to the orientational modes are most pronounced at temperatures $T < \hbar\omega_{min}/k_B$. The integral in (11.15′) is easily calculated if it is kept in mind that the principal contribution comes from the region of small $\bar{\lambda}$ and the first factor in the integrand varies more slowly than the second. Therefore, putting $\bar{\lambda} = 0$ in the first factor, taking into account that $\exp[(\hbar\omega_{min}/2k_BT) + (\hbar\overline{\Delta\omega}/2k_BT)\bar{\lambda}^2] \gg 1$ and extending the range of integration to infinity, we get

$$F_1(T) = \text{const} + \frac{\hbar(J')^2(1+\bar{q})}{4(I/\bar{s}^2)\omega_{min}}\left(\frac{\pi k_BT}{\hbar\overline{\Delta\omega}}\right)^{1/2} \exp\left[-\frac{\hbar\omega_{min}}{k_BT}\right], \tag{11.16}$$

It is obvious that such a form of the temperature dependence is caused by a deviation of the value $\tilde{q}$ from -1, and hence by the influence of the covibrations.

Therefore, in contrast with the translational phonons, which give a power-like dependence of the function $F_1(T)$ on temperature ($F_1(T) = \text{const} + AT^2$ for a linear lattice and $F_1(T) = \text{const} + BT^4$ for a three-dimensional lattice — see (7.6)), the orientational vibrations do not yield any noticeable temperature dependence of the rate constant at low temperatures.

Experimental studies of tunneling reactions [15–17] show that at $T \leqslant 40$ K the rate constant of hydrogen-atom abstraction from molecules by methyl radicals practically does not depend on temperature. To explain this, variations of the reactant-to-reactant distances have been assumed [10, 49] to be due to a local mode; then $F_1 = \text{const} + \alpha_0^2 \exp[-\hbar\Omega_0/k_BT]$ (see (5.26)).

It follows from (11.16) that at $T < \hbar\omega_{min}/k_B$ the reason why the tunneling rate constant does not depend on temperature might be that the principal influence on the potential barrier to be transversed by the particle is exerted by orientational vibrations.

At temperatures $\hbar\omega_{min} < k_BT < \hbar\omega_{max}$ (the case $\overline{\Delta\omega}/\omega_{min} \gg \pi^2\bar{q}/2(1+\bar{q})$ is considered) the second factor in the integrand (11.15) varies more slowly than the first, and may be taken out of the integral at $\bar{\lambda} = 0$. Then

$$F_1(T) = \frac{\hbar\,(J')^2}{4(I/\bar{s}^2)\omega_{min}}\left[\frac{\bar{q}\,\pi^2\omega_{min}}{2\overline{\Delta\omega}} + \frac{1+\bar{q}-\bar{q}\,\pi^2\omega_{min}/2\overline{\Delta\omega}}{(\overline{\Delta\omega}/\omega_{min})^{1/2}}\right]\coth\left(\frac{\hbar\omega_{min}}{2k_BT}\right). \tag{11.17}$$

If $\overline{\Delta\omega}/\omega_{min} \leqslant \bar{q}\,\pi^2/2(1+\bar{q})$ at temperatures satisfying both $k_BT \gg \hbar\omega_{min}$ and $k_BT < \hbar\overline{\Delta\omega}$ then the integral (11.15) is proportional to $k_BT/\hbar\omega_{min}$. This large term, however, disappears in three dimensions. Taking into account that the main contribution to the integral (11.15′) comes from $\bar{\lambda} \approx (k_BT/\hbar\overline{\Delta\omega})^{1/2}$, we find that

$$F_1(T) \sim \text{const} + (k_BT/\hbar\overline{\Delta\omega})^{3/2}.$$

Finally, for $k_BT > \hbar\omega_{max}$

$$F_1(T) \sim (J')^2\,k_BT/4(I\bar{s}^2)\Omega_2^2,$$

i.e. $F_1(T)$ is a linear function of temperature, similarly to the case of translational phonons, but with much larger coefficient. The ratio of the rotational and translational coefficients is $(\Omega_1/\Omega_2)^2$.

At higher temperatures the single-phonon approximation (7.3) for the rate constant is no longer valid, and (7.7) should be used. It is conveniently employed in the form

$$K = \frac{1}{\hbar} |V(\mathbf{R}_0)|^2 \left(\frac{2\pi}{F_3(T)}\right)^{1/2} \exp\left[\frac{\Delta E}{2k_BT} + F_4(T)\right],$$

$$F_3(T) = \frac{1}{2}\sum_{\mu} \frac{(J'_\lambda)^2(\hbar\omega_\lambda)^2}{\sinh(\hbar\omega_\lambda/2k_BT)}, \tag{11.18}$$

$$F_4(T) = \sum_{\lambda} (J'_\lambda)^2 \coth\frac{\hbar\omega_\lambda}{2k_BT}.$$

The functioin $F_3(T)$ is now calculated by means of the same approximations as those used in obtaining (11.16) and (11.17). It is also clear that $F_4(T) = F_1(2T)$.

In the high-temperature range, where the orientational amplitudes are large, the second term in the expansion of the function $J(\{q_\mu\})$ (see (5.15)) must be taken into account. Equation (5.24) is therefore used. At these temperatures the Einstein model for solids is valid ($\omega_\lambda = \Omega_2$); also the inequality $R_0 = \tilde{a} \gg r_{1,2}$ (see Fig. 3.5) enables us to take $J''_{12} = 0$ ($J''_{12} = \partial J/\partial\varphi_1\,\partial\varphi_2$). So, proceeding similarly to derivation of (8.2) for the matrix $(B^{-1})_{\mu\mu'}$ in (5.24), we arrive at

$$(B^{-1})_{\mu\mu'} = \begin{pmatrix} \ddots & & & & \\ & 1/C & & & 0 \\ & & 1/(\frac{1}{2}J''_{11}+C) & & \\ & & & 1/(\frac{1}{2}J''+C) & \\ & 0 & & & 1/C \\ & & & & & \ddots \end{pmatrix} \tag{11.19}$$

Equation (11.19) now gives the transition rate constant in the form

$$\begin{aligned} K = {} & \frac{2\pi}{\hbar} |V(R_0)|^2 \left[\left(1 + \frac{\hbar(J_{11}+J_{22})}{4I\Omega_2}\coth\frac{\hbar\Omega_2}{4k_BT}\right)\right. \\ & \times \left.\left(1 + \frac{\hbar(J_{11}+J_{22})}{4I\Omega_2}\tanh\frac{\hbar\Omega_2}{4k_BT}\right)\right]^{-1/2} \\ & \times \exp\left[\frac{\Delta E}{2k_BT} + \sum_i \phi_i(1-\phi'_i)\right] I_{\frac{|\Delta E|}{\hbar\Omega_2}}\left(\sum_i \phi_i\phi'_i\right)\rho_f, \\ \phi_i = {} & \frac{\hbar\,(J'_i)^2}{8I\Omega_2}\left[\tanh\frac{\hbar\Omega_2}{4k_BT} + \frac{\hbar J''_{ii}}{4I\Omega_2}\right]^{-1}, \\ \phi'_i = {} & \frac{\sinh(\hbar\Omega_2/4k_BT)/\cosh(\hbar\Omega_2/4k_BT)}{2[\tanh(\hbar\Omega_2/4k_BT) + J''_{ii}/4I\Omega_2]}, \\ J'_i = {} & \frac{\partial J}{\partial\varphi_i},\ J''_{ii} = \frac{\partial^2 J}{\partial\varphi_i^2},\ i = 1, 2. \end{aligned} \tag{11.20}$$

When the orientation vibrations in the medium may be treated classically ($T > \hbar\Omega_2/4k_B$), (11.20) is simplified to become

$$K \sim \exp\left[\sum_i \frac{\frac{1}{2}(J_i')^2 k_B T}{I\Omega_2^2 + J_{ii}'' k_B T}\right]. \tag{11.21}$$

It follows from (11.21) that the rate constant tends to a constant with increasing temperature, which can easily be understood if it is recalled that the orientational vibrations promote the approach of the reactants only in a restricted range of angles (see Fig. (3.5)). So a further increase of the rotational amplitude does not increase the rate constant. If the inequality $J_{ii}'' k_B T \gg I\Omega_2^2$ holds then the exponent may be expanded in a series and the effective activation energy E_a introduced:

$$K \sim \exp\left(-\frac{E_a}{k_B T}\right) \; ; E_a = \frac{1}{2} I\Omega_2^2 \sum_i \left(\frac{J_i'}{J_{ii}''}\right)^2. \tag{11.22}$$

This energy depends on the system geometry, in contrast with the translational vibrations, for which the activation energy is the work performed against the elastic forces in order to merge the potential wells corresponding to the initial and final states of the transferring particle (see (8.8)).

Orientational vibrations therefore result in a temperature-dependent tunneling rate constant that differs significantly from that of translational vibrations, especially at low temperatures. Orientational vibrations may be of importance owing to their large amplitude. The reasons for the activation-like dependences of the tunneling rate constants for orientational and translational vibrations are different.

It should also be pointed out that it is possible to have situations where steric conditions cannot be met without rotation of the molecule (i.e. if the transferring atom is blocked by other reactant and environment atoms). Then an appreciable part of the activation energy is work against the elastic forces, which must be performed in order to obtain the desired rotation, i.e. not only does the preexponential factor depend on the steric conditions, but so also does the activation energy.

3.12. ELECTRON TUNNELING TRANSFER IN A SOLID MATRIX

The results of this Chapter may also be used in the study of electron transfer in solid matrices. However, the localization radius of the electron in a potential well is much greater than the intermolecular amplitude, so the relative reactant displacements, which are of fundamental importance for heavy-particle transitions, do not influence the magnitude and temperature dependence of the rate constant of electron tunneling transfer. Formally, this follows from the proportionality of the parameters J' and J'' to $m^{1/2}$, which means that for electron transfer these quantities are approximately forty times smaller than for hydrogen transfer. Hence the formulae for electron tunneling transfer in a solid matrix are obtained by assuming $J' = J'' = 0$ in (10.1)–(10.19).

If the electron is transferred in a nonpolar matrix and neither donor nor acceptor are markedly altered then the rate constant is completely determined by the potential barrier through which the electron tunnels. In studying long-range tunneling (e.g. several tens of Ångströms) when the features of electron tunneling transfer are exhibited (e.g. logarithmic kinetics [50, 51]), the barrier may be treated as being rectangular, and a simple rate constant can be calculated:

$$K = \nu_0 \exp\left[-\frac{2}{\hbar}(2m\, V_0)^{1/2}\Delta X\right], \tag{12.1}$$

where V_0 is the potential-barrier height and ΔX is the tunneling distance (see also (1.2) in Chapter 2).

It is possible to find the temperature below which electron tunneling transitions prevail over activation-like ones. The Arrhenius exponential is compared with (12.1) in a fashion similar to [32], in which this temperature is found for the parabolic barrier (see Section 2.1). Thus we obtain

$$T_t = \hbar V_0^{1/2}/2k_B\, \Delta X(2m)^{1/2}. \tag{12.2}$$

The values $V_0 = 2$eV and $\Delta X = 30$Å yield a tunneling temperature $T_t = 500$ K, i.e. electron transitions below this temperature proceed mainly via tunneling.

Only electron transitions from the zero-point energy level need be taken into account, since transitions from higher levels contribute only weakly to the rate constant. In fact, higher level occupation need be considered only if the condition (see Section 2.4)

$$\rho_n W_n \geqslant \rho_0 W_0 \tag{12.3}$$

holds. Assuming the Gibbs distribution ($p_n \sim \exp[-E_n/k_B T]$, E_n is the energy of the nth level measured from the ground one) in the inequality (12.3) gives a condition for the higher-level transitions to be important:

$$T \geqslant \hbar \frac{\hbar[V_0^{1/2} + (V_0 - E_n)^{1/2}]}{2k_B\Delta X(2m)^{1/2}} = T_t + \frac{\hbar(V_0 - E_n)^{1/2}}{2k_B\Delta X(2m)^{1/2}} . \qquad (12.4)$$

Thus the temperature at which the higher-level occupation needs to be taken into account is higher than the tunneling temperature; hence the higher levels become important only at temperatures where electron transfer proceeds mainly via the overbarrier mechanism.

If the electron is transferred in a polar medium then the matrix is altered in the course of the transition and a more detailed treatment is necessary. The environmental change was first studied in [52, 53] and there is an extensive review literature on this topic (see e.g. [2, 12, 44, 54]), so only a brief summary of the results is presented here.

According to the theory of radiationless transitions, the electron transition rate constant, taking account of the environmental rearrangement is

$$K = \frac{1}{\hbar} |V_{n_f n_i}(\mathbf{R}_0)|^2 \int_{-\infty}^{\infty} ds \exp\left[\frac{\Delta E}{2k_B T} - i\Delta E s - \frac{1}{2}\sum_{\mu} \Delta q_\mu^2 \coth\left(\frac{\hbar\omega_\mu}{2k_B T}\right) + \frac{1}{2}\sum_{\mu} \frac{\Delta q_\mu^2 \cos(s\hbar\omega_\mu)}{\sinh(\hbar\omega_\mu/2k_B T)}\right] . \qquad (12.5)$$

At low temperatures ($T \ll \hbar\omega_D/2k_B$)

$$K = \frac{\pi}{\hbar} \Delta q_{\mu_0}^2 |V_{n_f n_i}(\mathbf{R}_0)|^2 \exp\left[\frac{\Delta E - |\Delta E|}{2k_B T} - \frac{1}{2}\sum_{\mu} \Delta q_\mu^2 \coth\left(\frac{\hbar\omega_\mu}{2k_B T}\right)\right], \qquad (12.6)$$

$$V_{n_i n_f}(\mathbf{R}_0) \sim \exp\left[-\frac{1}{\hbar}(2m\, V_0)^{1/2}\Delta X\right],$$

i.e. apart from the term in the exponent, which is associated with electronic wave-function decay inside the potential barrier (see (12.1)), a term appears that corresponds to environmental rearrangement. The electron transition rate constant is thus the product of the probability per unit time of electron underbarrier tunneling and the probability of environmental change.

In an Einstein solid the rate constant of electron transfer is rather simple:

$$K = \frac{2\pi}{\hbar} |V_{n_f n_i}(\mathbf{R}_0)|^2 \exp\left[\frac{\Delta E}{2k_B T} - \frac{E_r}{\hbar\Omega} \coth\left(\frac{\hbar\Omega}{2k_B T}\right)\right] \times I_{\frac{\Delta E}{\hbar\Omega}}\left(E_r/\hbar\Omega \sinh\left(\frac{\hbar\Omega}{2k_B T}\right)\right)\rho_f, \tag{12.7}$$

and it can easily be seen that its temperature dependence is determined by the energy of the environmental rearrangement and the energy defect of the rapid subsystem.

At high temperatures ($T > \hbar\omega_D/4k_B$) the temperature dependence of the rate constant becomes activation-like, with activation energy

$$E_a = (\Delta E - E_r)^2/4E_r. \tag{12.8}$$

Equations (12.5)–(12.7) are obtained from (5.24), (10.2), (10.6), (10.10) at $J'_\mu = J''_{\mu\mu'} = J' = J'' = 0$. They hold if the electron transfers to a discrete acceptor level. When the final intramolecular state of the electron lies in the continuum then (10.12) should be used. Then, assuming as before $J'_\mu = J''_{\mu\mu'} = 0$, we have

$$K = \frac{2\pi}{\hbar} |V(E_{n_i}, \{q^0_\mu\})|^2 \exp\left\{ - \sum_\mu \frac{\Delta q^2_\mu \sinh(\frac{1}{2} J'_E\, \hbar\omega_\mu)\sinh[\frac{1}{2}(\beta - J'_E)\hbar\omega_\mu]}{\sinh(\frac{1}{2}\beta\hbar\omega_\mu)} \right\},$$

which readily yields all limits (cf. (10.14)–(10.18)).

References

1. Goldanskii, V.I., Frank-Kamenetskii, M.D. and Barkalov I.M. (1973) *Science*, **182**, 1344.
2. Dogonadze, R.R. and Kuznetsov, A.M. (1973) In: *Physical Chemistry*. Kinetics. (In Russian.) VINITI, Moscow, (Itogi Nauki i Tekhniki; vol. 2).
3. Landau, L.D. and Lifshitz, E.M. Quantum Mechanics. Nonrelativistic Theory. Pergamon, Oxford.
4. Kondrat'ev, V.N. and Nikitin, E.E. (1974) *Kinetics and Mechanism of Gaseous Reactions*. (In Russian.) Nauka, Moscow.
5. Bykhovskii, V.K., Nikitin, E.E. and Ovchinnikova, M.Ya. (1964) *Zh. Eksp. Teor. Fiz.*, **47**, 750.
6. Louisell, W.H. (1964) *Radiation and Noise in Quantum Electronics*. McGrow-Hill, New York.
7. Fain, B. (1979) *Chem. Phys. Lett.*, **67**, 267.
8. Trakhtenberg, L.I. (1983) *Khim. Fiz.*, **1**, 1689.
9. Davydov, A.C. Quantum Mechanics, 2nd edn. Pergamon, Oxford.

10. Trakhtenberg, L.I., Klochokhin, V.L. and Pshezhetskii S.Ya. (1982) *Chem. Phys.*, **69**, 121.
11. Ivanov, G.K. and Kozhushner, M.A. (1983) *Khim. Fiz.*, **1**, 1299.
12. Ulstrup, J. (1979) *Charge Transfer Process in Condensed Media*. Springer-Verlag, New York, 1979.
13. Jortner, J. and Ulstrup, J. (1979) *Chem. Phys. Lett.*, **63**, 236.
14. Buhks, E. and Jortner, J. (1980) *J. Phys. Chem.*, **84**, 3370.
15. Hudson, R.L., Shiotany, M. and Williams, F. (1977) *Chem. Phys. Lett.*, **48**, 193.
16. Sprague, E.D. (1977) *J. Phys. Chem.*, **81**, 516.
17. Le Roy, R.J., Murai, H. and Williams, F. (1980) *J. Am. Chem. Soc.*, **102**, 2325.
18. Makshantsev, B.I. and Fleurov, V.N. (1979) *Chem. Phys.*, **37**, 271.
19. Trakhtenberg, L.I. (1983) *Dokl. AN SSSR*, **269**, 1131.
20. Pekar, S.I. (1953) *Usp. Fiz. Nauk*, **50**, 197.
21. Feynman, R. (1951) *Phys. Rev.*, **84**, 108.
22. Kubo R. and Toyozawa, Y. (1955) *Prog. Theor. Phys.*, **13**, 160.
23. Bikbaev, N.Kh., Ivanov, A.I., Lomakin, G.S. and Ponomarev, O.A. (1978) *Zh. Eksp. Teor. Fiz.*, **74**, 2154.
24. Klochikhin, V.L., Livshin, L.I., Pshezhetskii, S.Ya. and Trakhtenberg, L.I. (1980) *Khim. Vysok. Energii*, **14**, 329.
25. Feynman, R. (1972) *Statistical Mechanics*. Benjamin, Reading, Mass.
26. Perlin, Yu.E. (1963) *Usp. Fiz. Nauk*, **80**, 553.
27. Trakhtenberg, L.I., Klochikhin, V.L. and Pshezhetskii, S.Ya. (1981) *Chem. Phys.*, **59**, 191.
28. Bagratishvili, V.N., Letokhov, V.S., Makarov, A.A. and Ryabov, E.A. (1981) In: *Physics of Atoms and Molecules*. Optics. (In Russian.) VINITI, Moscow, (Itogi Nauki i Tekhniki, Vol. 2).
29. Klochikhin, V.L., Pshezhetskii, S.Ya. and Trakhtenberg, L.I. (1979) *Vysokomol. Soedin.*, **27A**, 2792.
30. Benderskii, V.A., Goldanskii, V.I. and Ovchinnikov, A.A. (1980) *Chem. Phys. Lett.*, **73**, 492.
31. Plotnikov, V.G. and Ovchinnikov, A.A. (1978) *Usp. Khim.*, **47**, 444.
32. Goldanskii, V.I. (1959) *Dokl. AN SSSR*, **124**, 1261.
33. Goldanskii, V.I. (1959) *Dokl. AN SSSR*, **127**, 1037.
34. Klochikhin, V.L. and Trakhtenberg, L.I. (1981) *Dokl. AN SSSR*, **259**, 399.
35. Trakhtenberg, L.I. and Fleurov, V.N. (1982) *Zh. Eksp. Teor. Fiz.*, **83**, 1908.
36. Klochikhin, V.L. and Trakhtenberg, L.I. (1983) *Elektrokhimiya*, **19**, 1212.
37. Klochikhin, V.L. and Trakhtenberg, L.I. (1983) *Khim. Fiz.*, **1**, 810.
38. Ovchinnikova, M.Ya. (1979) *Chem. Phys.*, **36**, 85.
39. Lines, M. and Glass, A. (1977) *Principles and Applications of Ferroelectrics and Related Materials*. Clarendon Press, Oxford, 1977.
40. Nikolskii, V.G. and Buben, N.Ya. (1962) *Dokl. AN SSSR*, **134**, 134.
41. Klochikhin, V.L., Pshezhetskii, S.Ya., Slavinskaya, N.A. and Trakhtenberg, L.I. (1981) *Vysokomol. Soedin.*, **23B**, 855.
42. Dakhnovskii, Yu. I., Ovchinnikov, A.A. and Benderskii, V.A. (1982) *Chem. Phys.*, **66**, 93.
43. Bell, R.P. (1980) *The Proton in Chemistry. Chapman and Hall*, London.
44. Zamaraev, K.I. and Khairutdinov, R.F. (1978) *Usp. Khim.*, **47**, 992.
45. Dogonadze, R.R., Kuznetsov, A.M. and Marsagishvili, T.A. (1980) *Electrochim. Acta*, **25**, 1.
46. Klochikhin, V.L., Pshezhetskii, S.Ya. and Trakhtenberg, L.I. (1978) *Dokl. AN SSSR*, **239**, 879.
47. Ansel'm, A.I. and Porfir'eva, N.N. (1949) *Zh. Eksp. Teor. Fiz.*, **19**, 438.
48. Aleksandrov, I.V. and Zhidomirov, G.M. (1961) *Zh. Eksp. Teor. Fiz.*, **41**, 127.
49. Trakhtenberg, L.I. (1982) *Khim. Fiz.*, **1**, 53.
50. Zamaraev, K.I., Khairutdinov, R.F., Mikhailov, A.I. and Goldanskii, V.I. (1971) *Dokl. AN SSSR*, **199**, 640.

51. Parmon, V.N., Khairutdinov, R.F. and Zamaraev, K.I. (1974) *Fiz. Tverd. Tela*, **16**, 2572.
52. Marcus, R.A. (1956) *J. Chem. Phys.*, **24**, 966.
53. Levich, V.G. and Dogonadze, R.R. (1959) *Dokl. AN SSSR*, **124**, 123.
54. De Vault, D. (1980) *Q. Rev. Biophys.*, **13**, 387.

CHAPTER 4
Influence of Defects on the Rate of Solid-Phase Tunneling Reactions

The influence of environment on the chemical reaction rate as presented in the previous chapter has demonstrated that the value and temperature dependence of the rate constant is largely determined by intermolecular vibrations. The rate constants of solid-phase reactions obtained for different models differ markedly, especially at low temperatures. Thus the Debye model for an ideal lattice results in a power-like temperature dependence of the logarithm of the rate constant, whereas the Einstein model and the local mode yield at low temperatures an exponential temperature dependence of the logarithm of the rate constant.

At the same time in fact the reaction rate constant is affected by various types of vibrations. In low-temperature solid-phase chemical reactions one of the reactants, as a rule, differs markedly from the host molecules in its mass and its interaction with the environment. Hence the active particle participating in the reaction is a (point) defect of the crystal lattice. Such a situation arises, for example, in irradiated molecular crystals in which radiation-stimulated radical formation takes place, and these are defects due to their differences from the host molecules.

The defects change the spectrum and dynamics of the crystal-lattice vibrations, which affect the reaction rate constant. Since these defects are reactants themselves and therefore a distortion of the regular lattice occurs within the reaction zone, solid-state defects should be taken into account even when the total abundance of the radiation (or other) damage is not large.

Below we shall dwell upon the general issues of how the point defects influence the amplitudes and spectrum of the intermolecular vibrations. The presentation assumes that the reader is familiar with the phonon spectrum of a crystal with point defects (see e.g. [1–3]). Most attention is paid to the calculation of the normalized solutions and

Green functions of the equations of motion for a crystal lattice with defects, which are necessary to find the rate constants of solid-phase reactions.

The solutions obtained are used to reveal the relative role of the different types of vibrations in the elementary processes of tunneling transitions.

4.1. VIBRATIONAL SPECTRUM AND DYNAMICS OF A LINEAR LATTICE WITH DEFECTS

Point defects are known to influence the spectrum and dynamics of the solid-lattice vibrations. Frequencies within the allowed bands are shifted a little, while one or several frequencies that were near the edges of the allowed bands before the defect insertion might be expelled into the forbidden bands. Such frequencies are specific local vibrations, and the amplitudes of the atomic displacements decrease on going away from the defect, i.e. only a small number of neighbouring lattice atoms vibrate together with the defect.

A local mode arises in several cases, for example when the defect mass M′ is smaller than the lattice-atom mass M and/or when the defect-environment force constant g′ is larger than the force constant g between the atoms constituting the solid ($\delta M = M' - M < 0$, $\Delta g = g' - g > 0$). If these inequalities are reversed, then there is no local mode but rather a so-called resonance quasilocal mode arises within the allowed band. Near the frequency corresponding to the resonance mode the spectral density increases, resulting in strong phonon–impurity scattering.

The frequencies of normal vibrations in the defect crystal lattice obey laws that are formulated as theorems known as the "Rayleigh theorems" (e.g. [1]). Since the thermodynamical properties as well as the scattering in solids have been determined mainly by the spectral density of the lattice vibrations it is the phonon spectrum of the defect solid that has been most studied, whereas less attention has been paid to the dynamics of the lattice vibrations.

The results of Chapter 3 have shown that the relative reactant vibrations influence the chemical reaction rate, which makes it very important to study the equations of motion of the defect crystal lattice. The necessity of accounting for the role of defects in vibration dynamics and spectrum is demonstrated by the example of atomic tunneling between molecules located at sites 0 and 1 of a one-dimensional lattice. Use is made of the fact that intramolecular frequencies are much higher than intermolecular ones and a temperature range is studied where the intramolecular vibrations are still not excited. This allows the analysis to be restricted to the low-frequency acoustic phonon modes — the internal degrees of freedom of the molecules can be ignored. It is also assumed that there is only one molecule in an elementary cell, i.e. the lattice is monoatomic.

The equations of motion for the one dimensional monoatomic crystal lattice containing N-1 atoms (molecules) with mass M and one atom (in the site 0) with mass M′ are considered; the 0–1 force constant is g′ and differs from all other force constants, which are equal to g. These equations are

$$\left.\begin{aligned} &M(1 + \epsilon\,\delta_{j0})\frac{\partial^2 X_j}{\partial t^2} = g(X_{j-1} + X_{j+1} - 2X_j) + \Delta g(X_1 - X_0)(\delta_{j0} - \delta_{j1}),\\ &\epsilon = \delta M/M, \quad \Delta g = g' - g, \quad X_j = X_{j+N}, \quad -\tfrac{1}{2}N < j \leqslant \tfrac{1}{2}N. \end{aligned}\right\} \tag{1.1}$$

Here $\delta_{jj'}$ is the Kronecker symbol and X_j is the jth atom displacement.

It is convenient to seek solutions of (1.1) in the form

$$X_j = a_j \exp[-i\omega t], \tag{1.2}$$

where a_j does not depend on time and specifies the displacement of the jth particle. Then substitution of (1.2) into (1.1) gives

$$-\omega^2 a_j = g(a_{j-1} + a_{j+1} - 2a_j)/M + \omega^2\epsilon a_0\,\delta_{j0} + \Delta g(a_1 - a_0)(\delta_{j0} - \delta_{j1})/M, \tag{1.3}$$

which can be solved by Fourier analysis. Equation (1.3) is multiplied by $\exp[-ika_j]$, summed over all j and the Fourier transform is found:

$$b(k) = \frac{\omega^2\epsilon a_0 + \Delta g(a_1 - a_0)(1 - \exp[-ika])/M}{4g\sin^2(\tfrac{1}{2}ka)/M - \omega^2}, \tag{1.4}$$

where k is the wavenumber of the ideal lattice ($M = M'$, $g = g'$); a is the lattice spacing.

A quantum number g (generally complex) for the defect lattice is defined in the same fashion as in the ideal lattice, i.e. by using the equation

$$\omega^2 = 4g\sin^2(qa/2)/M, \tag{1.5}$$

which may be treated as a dispersion relation for the phonon frequencies in the linear defect lattice ($\omega = \omega_q$).

Then inversion of the Fourier transform gives [4]

$$a_j(q) = \frac{1}{N}\sum_k b(k)\exp[ikaj]$$

$$= \frac{1}{N} \sum_{k} \{M\omega_q^2 \epsilon a_0 + \Delta g(a_1 - a_0)(1 - \exp[-ika])\}$$

$$\times \frac{\exp[ikaj]}{2g[\cos(qa) - \cos(ka)]}. \tag{1.6}$$

This set of solutions depends on the sums

$$G_j^0 = \frac{1}{N} \sum_{k} \frac{\exp[ikaj]}{2g[\cos(qa) - \cos(ka)]}, \tag{1.7}$$

which may be interpreted as the Green functions of the ideal lattice. These are well known for the one-dimensional case [5]:

$$G_j^0(q) = -\frac{\cos(jqa)}{2g \sin(qa)} \cot \frac{Nqa}{2} - \frac{\sin(|j|qa)}{2g \sin(qa)}. \tag{1.8}$$

Thus (1.7) for a_j results in

$$a_j = M\omega_q^2 \epsilon a_0 G_j^0(q) + \Delta g(a_1 - a_0) [G_j^0(q) - G_{j-1}^0(q)]. \tag{1.9}$$

where the system

$$\left.\begin{aligned} a_0 &= M\omega_q^2 \epsilon a_0 G_0^0(q) + \Delta g(a_1 - a_0) [G_0^0(q) - G_1^0(q)], \\ a_1 &= M\omega_q^2 \epsilon a_0 G_1^0(q) + \Delta g(a_1 - a_0) [G_1^0(q) - G_0^0(q)] \end{aligned}\right\} \tag{1.10}$$

has solutions for the allowed values of q. This system has been obtained by inserting $j = 0$ and 1 into (1.9) and taking account of $G_j^0(q) = G_{-j}^0(q)$.

Equating the determinant of the system (1.10) to zero gives an equation for the allowed values of the wavenumber q:

$$1 + 2\Delta g(G_0^0 - G_1^0) - M\omega_q^2 \epsilon\, G_0^0 - M\omega_q^2 \epsilon\, \Delta g(G_0^{0^2} - G_1^{0^2}) = 0. \tag{1.11}$$

This equation also yields a condition that must be satisfied by the force constants and atomic (molecular) masses in the lattice in order to give rise to a local mode:

$$\Delta M/M' < \Delta g/g.$$

Now (1.11) and the solutions of the system (1.10) are used and (1.9) is rewritten in the form

$$\left.\begin{aligned} a_j(q) &= a_0(q) [p_0(q) G_j^0(q) - p_1(q) G_{j-1}^0(q)], \\ p_{0,1} &= (1 - M\omega_q^2 \epsilon\, G_{0,1}^0)/(G_0^0 - G_1^0). \end{aligned}\right\} \tag{1.12}$$

It can easily be seen that (1.11) and (1.12) incorporate the particular cases studied in [1].

(i) $\epsilon = 0$, $\Delta g \neq 0$. Substituting $\epsilon = 0$ into (1.11) and (1.12) gives solutions (antisymmetric and symmetric with respect to the point equally dividing the segment connecting the site 0 and 1):

$$\left.\begin{aligned} a_j'(q') &= -2\Delta g\, a_0(q')\,(G_j^0(q') - G_{j-1}^0(q')), \\ a_j''(q'') &= A(q'')\cos[q''a(j - \tfrac{1}{2})], \end{aligned}\right\} \tag{1.13}$$

with an additional condition for the first solution:

$$1 + 2\Delta g(G_0^0 - G_1^0) = 0. \tag{1.14}$$

(ii) $\epsilon \neq 0$, $\Delta g = 0$. Then the solutions of the equations of motion for symmetric and antisymmetric vibrations with respect to the site 0 are

$$\left.\begin{aligned} a_j'(q') &= M\omega_{q'}^2\epsilon a_0(q')\, G_j^0(q'), \\ a_j''(q'') &= A'(q'')\sin(q''aj), \end{aligned}\right\} \tag{1.15}$$

and the allowed wavenumbers for the symmetric vibrations are determined by

$$1 - M\omega_{q'}^2\epsilon\, G_0^0(q') = 0. \tag{1.16}$$

The wavenumber q'' for the solutions a_j'' in (1.13) and (1.15) corresponds to the unperturbed part of the phonon spectrum, while the solutions $a_j'(q')$ contain the local mode.

Generally, $\epsilon_1\Delta g \neq 0$, and the allowed q values are determined by the transcendental equation (1.11), which, using (1.15), leads to a frequency shift in the allowed band of the phonon spectrum and the local-mode frequency. For this the most general form

$$q = q_1 + iq_2 \tag{1.17'}$$

should be substituted into (1.5), which results in the equation for the frequency:

$$\omega^2 = \frac{2g}{M}\{1 - \cos(q_1a)\cosh(q_2a) - i\sin(q_1a)\sinh(q_2a)\}. \tag{1.17''}$$

Here ω^2 must be real and positive for the allowed q values. Then, as follows from (1.17″), this condition holds only in two cases.

$$\text{(a)}\ q_2 = 0, \quad \omega^2 = \frac{4g}{M}\sin^2(\tfrac{1}{2}q_1a); \tag{1.18'}$$

the frequencies lie in the dense part of the phonon spectrum.

$$\text{(b)}\ q_1 = \pi(2n + 1)/a, \quad \omega^2 = 4g\,\mathrm{ch}^2(q_2a/2)/M; \tag{1.18''}$$

The frequency found is split off from the top of the allowed band and is associated with the local mode. It is clear that the periodicity of the frequencies and Green functions results in $n = 0$, and q_2 is determined by (1.11). Substituting (1.17′) into (1.11) and making use of (1.18″) gives the frequency of the local mode:

$$\left.\begin{aligned} \omega_0 &= \omega_D/(1 - \xi_0^2)^{1/2}, \\ \xi_0 &= \tfrac{1}{2}\left\{\left[\left(\frac{g}{g'} - \epsilon\right)^2 + \frac{4M'\Delta g}{Mg'}\right]^{1/2} - \left(\frac{g}{g'} + \epsilon\right)\right\} \end{aligned}\right\} \tag{1.19}$$

It can easily be seen that the particular local-mode frequencies corresponding to the cases (i) and (ii) can be obtained from (1.19).

4.2. RATE CONSTANT OF ATOMIC TUNNELING IN A LINEAR LATTICE WITH DEFECTS

In studying the influence of defects on the rate of tunneling reactions, we start from (7.2) of Chapter 3. Equation (7.1) of Chapter 3 is used and the fact is taken into consideration that the coefficients of the expansion of the molecular displacements in normal coordinates are solutions of the equations of motions for the lattice (e.g. (7.4) of Chapter 3); then (7.2) of Chapter 3 can be rewritten as

$$K = \frac{1}{\hbar}\,|V(\{q_\mu^0\})|^2 \exp\left[\frac{\Delta E}{2k_BT} + \sum_\mu |c_\mu|^2 \coth\left(\frac{\hbar\omega_\mu}{2k_BT}\right)\right]$$
$$\times \int_{-\infty}^{\infty} ds \exp\left[-\,i\Delta Es + \sum_\mu \frac{|c_\mu|^2 \cos(\hbar\omega_\mu s)}{\sinh(\hbar\omega_\mu/2k_BT)}\right], \tag{2.1}$$
$$c_\mu = J'(\hbar/8M\omega_\mu)^{1/2}\,[a_\mu - a'_\mu],$$

where a_μ, a'_μ are the solutions of the equations of motion for the molecules between which the transition occurs.

As follows from (2.1), the tunneling rate constant depends exponentially on the following type of expression:

$$S(\epsilon,\Delta g) = \sum_\mu |c_\mu|^2 f(\omega_\mu, T). \tag{2.2}$$

All the types of vibrations associated with both the split-off mode and the dense part of the perturbed and unperturbed spectrum are included in the sum in (2.2).

To calculate the sums (2.2) the solutions of the equations of motion should be normalized; the normalization condition is known to be

$$\sum_j |a_j|^2 M_j/\overline{M} = 1, \tag{2.3}$$

where M_j is the mass of the jth site molecule and $\overline{M} = (1/N) \sum_j M_j$ is the mean molecular mass in the chain. The normalized solutions for arbitrary values of Δg and ϵ turn out to be rather complicated. We therefore consider the sums (2.2) for just three cases: (i) $\epsilon = 0, \Delta g \neq 0$; (ii) $\epsilon \neq 0, \Delta g = 0$; (iii) $\epsilon \neq 0, \Delta g \gg g$. It is also assumed that at least one of the inequalities $\epsilon < 0$ or $\Delta g > 0$ holds.

Then (1.13) and (1.14) together with the normalization yield in the first case [4] the equation

$$S(0,\Delta g) = \frac{\hbar (J')^2}{M\omega_0} \frac{\Delta g g'}{(\Delta g + g')^2} f(\omega_0,T) + \frac{\hbar (J')^2}{2MN} \sum_{q'>0} \frac{1}{\omega_{q'}} \frac{g^2 \sin^2(q'a)}{(\Delta g^2 + (g')^2) + g(g' + \Delta g)\cos(q'a)}, \tag{2.4}$$

from which it follows that the vibrations associated with the unperturbed part of the spectrum (the solutions $a_j''(q'')$ correspond to them) does not affect the tunneling rate constant. There is a contribution only due to the vibrations associated with the dense part of the perturbed spectrum (solutions $a_j'(q')$) and the split-off mode. If $g' \gg g$ then the local mode mainly contributes to the relative reactant vibrations, as if the nonlocal vibrations had been forced out of the reaction zone. Therefore the rate constant of the solid-phase tunneling reactions is described by (5.26).

In the second case (2.2) becomes [4]

$$S(\epsilon,0) = \frac{\hbar (J')^2}{2MN} \sum_{q'>0} \frac{1}{\omega_{q'}} \frac{(1+\epsilon)^2 \sin^2(q'a)\sin^2(\frac{1}{2}q'a)}{(1+\epsilon^2) + (1-\epsilon^2)\cos(q'a)} f(\omega_{q'},T) + \frac{\hbar (J')^2}{4MN} \sum_{q''>0} \frac{1}{\omega_{q''}} \sin^2(q'a) f(\omega_{q''},T) + \frac{\hbar (J')^2}{M\omega_0} \frac{|\epsilon|}{(1-\epsilon^2)(1-\epsilon)^2} f(\omega_0,T). \tag{2.5}$$

It can be seen that all three types of vibrations influence the tunneling rate constant in solids. If the condition $M' \ll M$ holds for the system studied then the contribution of the dense part of the unperturbed spectrum to the relative reagent displacement is not large and the importance of the local mode is increased. As for the vibrations associated with the unperturbed part of the phonon spectrum, their contribution does not depend on the defect mass, and the rate constant of solid-phase chemical reactions is determined by the vibrations of the lattice as a whole.

In the third case the normalized solutions are sought by means of (1.12) and an additional condition,

$$2(G_0^0 - G_1^0) - \epsilon M\omega_q^2[(G_0^0)^2 - (G_1^0)^2] = 0, \tag{2.6}$$

which arises from (2.11) on neglecting the terms $\sim \Delta g/g$. If $\epsilon \neq 0$ then (2.6) yields two conditions for two types of vibrations,

$$\left.\begin{aligned} G_0^0 - G_1^0 &= 0, \\ \epsilon M\omega_q^2(G_0^0 + G_1^0) &= 2. \end{aligned}\right\} \tag{2.7}$$

Equation (1.12) and the additional conditions (2.7) then allow (2.2) to be rewritten in the form [6]

$$\begin{aligned} S(\epsilon,\Delta g) = {} & \frac{\hbar(J')^2(\epsilon + 2)}{8M(\epsilon + 1)\,\omega_0} f(\omega_0,T) \\ & + \frac{\hbar(J')^2}{4MN} \sum_{q'>0} \frac{g^2 \sin^2(q'a)}{\Delta g^2} f(\omega_{q'},T) \\ & + \frac{\hbar(J')^2}{MN} \sum_{q''>0} \frac{(g/\Delta g)^2 \sin^2(\frac{1}{2}q''a)}{1 + \tan^2\{(\frac{1}{2}q''a) + [\epsilon \sin^2(\frac{1}{2}q'a)]^{-1}\}^2} f(\omega_{q''},T). \end{aligned} \tag{2.8}$$

Now it is easy to see that it is only the local mode that contributes to the relative reactant displacements, whereas the contribution of the dense part of the phonon spectrum is $\sim (g/\Delta g)^2$. In this case a complex of reactants (radical and molecule) forms, and the tunneling rate constant is determined by the internal vibrations of the complex [7,8], which corresponds to an oscillator with force constant g' and reduced mass

$$MM'/(M + M') = M[(\epsilon + 1)(\epsilon + 2)].$$

At low temperatures, $T < \hbar\omega_D/2k_B$, the Debye model is valid and the one-dimensional summation in (2.4) and (2.5) can be carried out analytically, so the transition rate constant becomes

$$K = K_0 \exp\left[\frac{\Delta E - |\Delta E|}{2k_BT} + \alpha T^2\right], \tag{2.9}$$

where in the first and second cases the αs have the forms:

$$\alpha_1 = \frac{\pi M\omega_D\, k_B^2(J')^2}{48\hbar(g')^2}, \qquad \alpha_2 = \frac{\pi M\omega_D\, k_B^2(J')^2}{48\hbar g^2} \tag{2.10}$$

Here K_0 is the preexponential factor, which depends weakly on the temperature. The first term in the exponent has the same meaning as in (7.3) of Chapter 3. The coefficient of T^2 (here the exponent is $\ell + 1$, where ℓ is the dimensionality) in the first case, $\epsilon = 0$, $\Delta g \neq 0$, is due to the vibrations associated with the dense part of the unperturbed part of the spectrum, and the crystal defects do not change the temperature dependence within this temperature range. An essential influence on the tunneling rate constant in both cases results from the zero-point vibrations.

At high temperatures ($T > \hbar\omega_D/2k_B$) the tunneling rate constant in both cases is

$$K = K_0' \, I_r \left\{ \frac{(J')^2 \hbar b}{M\omega_0 \sinh(\hbar\omega_0/2k_B T)} \right\}$$
$$\times \exp\left[\frac{\Delta E}{2k_B T} + \tilde{\beta} T - \frac{\Delta^2}{\gamma' T} + \frac{\hbar J' b}{M\omega^2} \coth\left(\frac{\hbar\omega_0}{2k_B T}\right)\right], \qquad (2.11)$$
$$\tilde{\beta}_1 = \frac{(J')^2 g \, k_B}{2g'(g' + \Delta g)}, \qquad \gamma_1' = \frac{(J')^2 g^2 \hbar^2 \, k_B}{M(g' + \Delta g)^2},$$
$$\tilde{\beta}_2 = \frac{(J')^2 (1 + \epsilon^2) \, k_B}{2g(1 - \epsilon)^2}, \qquad \gamma_2' = \frac{(J')^2 \hbar^2 \, k_B}{M(1 - \epsilon)},$$
$$\Delta = |\Delta E| - r\hbar\omega_0, \quad r = [|\Delta E|/\hbar\omega_0]$$

Here [. . .] means integer part; K_0' is the preexponential factor, which is proportional to $(\gamma' T)^{1/2}$. This equation holds if $\hbar\omega_D \Delta \ll \gamma' T$, the relation between the temperature and local mode frequency ω_0 being arbitrary. The Bessel-function argument depends on the local mode: it contains $b_1 = g'\Delta g/(g' + \Delta g)^2$ in the first case and $b_2 = |\epsilon|/(1 - \epsilon^2)(1 - \epsilon)^2$ in the second. The coefficients β and γ in the exponent are expressed in terms of the system parameters: in the first case they are determined by the vibrations associated with the dense part of the perturbed spectrum, and tend to zero for $\Delta g \gg g$; in the second case they are due to the vibrations associated with the whole dense part of the spectrum, and do not tend to zero as $M'/M \to 0$.

References

1. Maradudin, A.A., Montroll, E.W. and Weiss, G.H. (1963) *Theory of Lattice Dynamics in the Harmonic Approximation*. Academic Press, New York, London.
2. Maradudin, A.A. (1966) *Theoretical and Experimental Aspects of the Effects of Point Defects of Crystals*. Academic Press, New York.
3. Stoneham, A.M. (1975) *Theory of Defects in Solids*. Clarendon Press, Oxford.
4. Trakhtenberg, L.I. (1982) *Khim. Fiz.*, **1**, 53.
5. Ludwig, W. (1966) *Dynamics of a Crystal Lattice with Defects* (Proc. of the Summer School, 1964 Hrazany). Academia, Prague.

6. Trakhtenberg, L.I., Klochikhin, V.L. and Pshezhetskii, S.Ya. (1982) *Chem. Phys.*, **69**, 121.
7. Klochikhin, V.L., Pshezhetskii, S.Ya. and Trakhtenberg, L.I. (1978) *Dokl. AN SSSR*, **239**, 879.
8. Klochikhin, V.L., Pshezhetskii, S.Ya. and Trakhtenberg, L.I. (1980) *Zh. Fiz. Khim.*, **54**, 1324.

CHAPTER 5
Rate Constant of Solid-Phase Chemical Reactions. Comparison of Theory and Experiment

In this chapter the theory of radiationless transitions in solids is applied to bimolecular radical reactions. Fluctuations of the potential barrier are shown to result in a specific temperature dependence of the rate constant of solid-phase chemical reactions, which is indeed observed experimentally.

We start with calculation of the transition matrix element over the intramolecular wave functions, assuming the states between which the transition occurs to be parabolic in form. This allows determination of how the heat of reaction is distributed among the internal molecular degrees of freedom.

Then the transition matrix element, and hence J' and J'' (see Chapter 3), are calculated for some realistic potential surfaces. In particular, J' and J'', which determine the dependence of the potential barrier on the intermolecular coordinates, are found analytically. It is shown that at low temperatures the quadratic term in the expansion of the logarithm of the matrix element may be discarded.

Theory and experiment are compared for five low-temperature radical reactions with hydrogen-atom transfer. Assuming reasonable parameters of the systems studied, i.e. tunneling length and intermolecular frequency, good agreement between theoretical predictions and experimental data is obtained. The parameters yielding the best fit of the theory to the data are close to those found from independent experiments. The isotope effect is calculated, and satisfies an empirical inequality.

To conclude the Chapter, the kinetics of bimolecular (quasimonomolecular) reactions is studied, with account being taken of their polychronic character. The spread of the reactant-to-reactant distances is shown to result in nonexponential kinetics and explains the observed time dependence of the "rate constant" of low-temperature solid-phase reactions.

5.1. TRANSITION MATRIX ELEMENT. DISTRIBUTION OF HEAT OF REACTION OVER INTERNAL MOLECULAR DEGREES OF FREEDOM

Chemical reactions, generally yielding a change of the numbers of atoms in each of the reactants, changes the system of normal coordinates of the reaction participants. It is clear that under such conditions the rearrangements of the reacting molecules cannot be described in terms of varying equilibrium coordinates and frequencies, since the number of oscillators specifying the intramolecular subsystem of each reactant changes. The intramolecular wave functions of the initial and final states therefore depend on different sets of normal coordinates, which makes calculation of the transition matrix element more complicated.

However, if the reaction results in bond rupture and formation of new bonds associated with some characteristic molecular vibrations then the reactant wave function may be presented as a product of the wave function describing the ruptured (or newly formed) bond and the wave function corresponding to the remaining part of the molecule (i.e. its "core"), which does not change essentially during the transition (to be more precise the system of normal coordinates corresponding to the internal core vibrations does not change). For example, when B-atom transfer is studied,

$$\mathrm{A{-}B} + \mathrm{D} \longrightarrow \mathrm{A} + \mathrm{B{-}D} \tag{1.1}$$

where A—B and B—D are chemical bonds corresponding to some characteristic vibrations, then the intramolecular wave functions are

$$\Psi_{\mathrm{i}} = \varphi_{\mathrm{i}}(\mathrm{A{-}B})\,\varphi_{\mathrm{i}}(\mathrm{A})\,\varphi_{\mathrm{i}}(\mathrm{D}) \tag{1.2'}$$

for the initial state and

$$\Psi_{\mathrm{f}} = \varphi_{\mathrm{f}}(\mathrm{B{-}D})\,\varphi_{\mathrm{f}}(\mathrm{A})\,\varphi_{\mathrm{f}}(\mathrm{D}) \tag{1.2'}$$

for the final state. The wave functions $\varphi_{\mathrm{i,f}}(\mathrm{A})$ and $\varphi_{\mathrm{i,f}}(\mathrm{D})$ of the initial and final intramolecular states for the cores may differ in the frequencies of the normal vibrations and the equilibrium normal coordinates.

If a *molecule* B is transferred then the wave function (1.2) should be multiplied by the wave function of the intramolecular system of the B molecule in the initial and final states.

The vibrations of the R—H bonds (C—H, O—H, N—H) are known to be the characteristic ones, so for hydrogen-atom transfer the intramolecular wave functions have the form (1.2). Such chemical reactions (in particular, hydrogen abstraction from organic molecules by radicals) have been extensively studied over a wide temperature range, and hydrogen tunneling at low temperatures has been experimentally verified (see Chapter 2). Besides, the double adiabatic approximation holds especially well for hydrogen-atom transfer (i.e. B = H). These are the reasons why most attention is paid here to the comparison of the concepts developed in the above chapters with the experimental data for the reactions

$$\mathrm{R{-}H + R' \longrightarrow R + R'{-}H} \tag{1.1'}$$

These reactions proceed, as a rule, with considerable heat release, which is distributed over the internal degrees of freedom of the products, and hence determines the final intramolecular state. The total transition probability per unit time is obtained by summing all partial probabilities of all possible final states. However, a particular excitation of one or several vibrations might appear to be more favourable than the others, in which case there is no need to sum over the final intramolecular states, and only transitions into that single final state (discrete intramolecular states of the product are implied) need be considered.

To elucidate the distribution of the heat of chemical reaction over the intramolecular degrees of freedom, the matrix element (to be more specific, a nonadiabatic transition is considered; (2.12) of Chapter 3) for the states, described by the wave functions (1.2), is written as

$$\langle \mathrm{f}|\mathrm{U}|\mathrm{i}\rangle = \frac{\delta \mathrm{U}'}{2} \langle \varphi_\mathrm{f}(\mathrm{D{-}H})|\varphi_\mathrm{i}(\mathrm{A{-}H})\rangle \langle \varphi_\mathrm{f}(\mathrm{A})|\varphi_\mathrm{i}(\mathrm{A})\rangle \langle \varphi_\mathrm{f}(\mathrm{D})|\varphi_\mathrm{i}(\mathrm{D})\rangle . \tag{1.3}$$

As pointed out in Chapter 2, the transition from the lowest intramolecular level makes the principal contribution to the rate constant at low-enough temperatures. Assuming the intramolecular vibrations to be harmonic, the squared matrix element (1.1) becomes

$$|\langle \mathrm{f}|\mathrm{U}|\mathrm{i}\rangle|^2 = (\tfrac{1}{2}\delta \mathrm{U}')^2 \exp\left\{-\tfrac{1}{2}(\Delta\xi_\mathrm{H})^2 - \sum_\mathrm{j} \tfrac{1}{2}(\Delta\xi_\mathrm{j})^2\right\}$$

$$\times \frac{(\Delta\xi_H)^{2n_H}}{n_H!} \prod_j \frac{(\Delta\xi_j)^{2n_j}}{n_j!}, \tag{1.4}$$

where j stands for the normal modes of the cores A and D; n_H and n_j denote the quantum numbers of the D—H bond and of the jth mode in the final state; $\Delta\xi_H$ and $\Delta\xi_j$ are the dimensionless displacements of the final terms with respect to the initial ones.

As follows from (1.4), the most probable excitation is that of a vibration with maximum distance between the initial and final components. It is obvious that $\Delta\xi_H$ greatly exceeds $\Delta\xi_j$. In fact, hydrogen-atom transfer in low-temperature solid-phase reactions [1–5] takes place over a distance ~ 1.5 Å [6], whereas the variations of the normal equilibrium coordinates in cores A and D barely exceed 0.1 Å.

Therefore it is the excitation of the valence vibrations of the newly formed bond that is most probable, i.e. n_H is at maximum, while $n_j = 0$. If n_H decreases by one, while n_j, according to energy conservation, increases by one (or maybe by a larger integer if the frequency of the jth mode is much smaller than that of the newly formed bond), then (1.4) is multiplied by the factor

$$n_H(\Delta\xi_i/\Delta\xi_H)^2. \tag{1.5}$$

Evaluation of the ratio $(\Delta\xi_j/\Delta\xi_H)^2$ yields $\sim \frac{1}{200}$, so the coefficient (1.5) is much smaller than one.

Therefore the chemical heat of reaction is almost completely transferred to the vibrations of the newly formed bond, and only the energy defect is compensated by the excitation of some other, lower frequency, vibrational modes and phonons. It should be emphasized that the above conclusion relates only to the elementary process of atomic transfer; later, the excitation energy may pass to other vibration of modes of the reaction products.

5.2. TRANSITION MATRIX ELEMENT FOR NONADIABATIC AND ADIABATIC REACTIONS

Chapter 3 has shown that the temperature dependence of the rate constant of solid-phase radical reactions is governed by potential-barrier fluctuations due to the interreactant vibrations and depends on the values J′ and J″ (see (5.19) of Chapter 3). Thus these values J′ and J″ can be found if the transition matrix element is calculated as a function of the reactant-to-reactant spacing. It is clear that for both adiabatic and nonadiabatic transitions it is the overlap of the wave functions associated with the ruptured and newly formed bonds that mainly depends on the spacing. The overlap of the intramolecular wave functions of the initial and final core states, A and D, depends more weakly on the intermolecular coordinates. This may be only due to the effect of the environment on the internal potential energy of molecules, which is known to be rather small in molecular crystals. Therefore in studying bimolecular radical reactions we shall assume that the A and D cores (R and R′ radicals) are solid structureless particles and the total dependence of the matrix element on the phonon coordinates and the energy transferred into the phonon subsystem is determined by the overlap of the ruptured and newly formed wave functions.

We now turn to the calculation of the nonadiabatic matrix element. Since particle transfer in these reactions proceeds over a considerable distance that substantially exceeds the localization radius of the particle both in the ruptured and newly formed bonds, in order to calculate the matrix element, we must know the corresponding wave functions, not only in the vicinity of the equilibrium intramolecular coordinates where the harmonic approximation holds, but also far from equilibrium. Therefore the wave function should be used that corresponds to realistic potentials that correctly describe the asymptotics of the molecular term at large distances. The Morse potentials are known to be of this type:

$$U(r) = D_0(e^{-2\alpha r} - e^{-\alpha r}), \tag{2.1}$$

where D_0 is the dissociation energy, r is the coordinate associated with the bond under consideration, and α is the parameter specifying the potential localization.

The Schrödinger equation with potential (2.1) has solution (see e.g. [7])

$$\varphi_n(r) = \left[\frac{\alpha(2s_0 - n)!}{n!(2s_0 - 2n)!(2s_0 - 2n - 1)!}\right]^{1/2} y^{s_0 - n} F(-n, 2s_0 - 2n + 1; y),$$

$$y = (2s_0 + 1)\exp[-\alpha r], \tag{2.2}$$

where n is the number of the energy level, s_0 determines the number of the vibrational levels in the term (2.1) ($s_0 = (2MD_0)^{1/2}/\alpha\hbar$), M is the reduced mass associated with the bond, F(n,n′;y) is a confluent hypergeometric function. The eigenenergies are

$$E_n = -[(s_0 - n)/\alpha\hbar]^2/2M. \tag{2.3}$$

When studying the intramolecular radiationless transitions, the matrix elements were calculated for the Morse potentials in the same

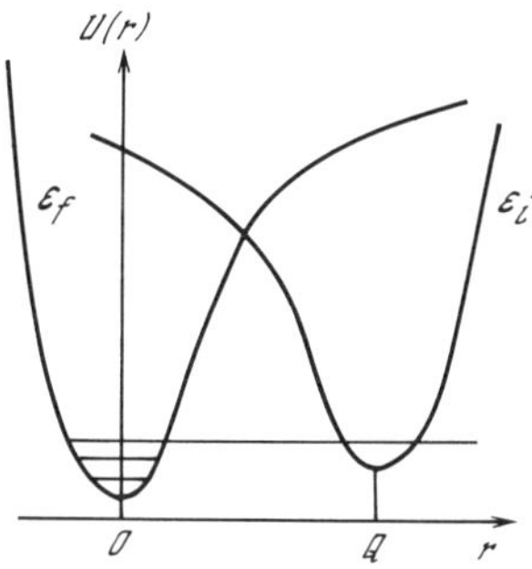

Figure 5.1 Counter-directed Morse potentials describing the initial (ϵ_i) and final (ϵ_f) states in the low temperature reactions of the hydrogen atom transfer.

direction (see e.g. [8]). But, as pointed out in [9], it is necessary to consider the oppositely directed Morse potentials (Fig. 5.1).

If the wave function associated with the newly formed bond has the form (2.2) then the minimum of the corresponding term is at $r = 0$. The minimum of the ruptured term is at $r = Q$, so the initial wave function may be written as

$$\varphi_0(Q - r) = \alpha (y')^{s_0^i} \exp[-y'/2]/(2s_0^i - 1)!, \qquad (2.4)$$
$$y' = (2s_0^i + 1)\exp[\alpha r - \alpha Q].$$

Using the (2.2) and (2.4), the wave function overlap is given by

$$\langle n|0\rangle \sim \int dr \exp\left\{[-\alpha r(s_0^f - n - s_0^i) - \alpha Q s_0^i] - (2s_0^f + 1)\exp\left(-\frac{\alpha r}{2}\right)\right.$$
$$\left. -(2s_0^i+1) \quad \exp[-\tfrac{1}{2}\alpha(Q-r)]\right\} F\{-n, 2s_0^f - 2n + 1, (2s_0^f + 1)\exp(-\alpha r)\}. \qquad (2.5)$$

The exponent in the integrand contains a term $\alpha(s_0^f - n - s_0^i)\, r$, which can be transformed by means of (2.3) to

$$\alpha(s_0^f - n - s_0^i)\, r = [(-E_{h_f})^{1/2} - (-E_{n_i})^{1/2}](2M)^{1/2}\, r/\hbar. \qquad (2.6)$$

As is emphasized above, nearly all of the heat of reaction is spent in exciting the newly formed bond, so it may be assumed that $E_{n_f} = E_{n_i}$, and hence that the value of (2.6) is close to zero.

Now it should be kept in mind that the confluent hypergeometric function in the integrand varies much more slowly than the exponential, so the integral may be calculated by means of steepest descent to give

$$\langle n|0\rangle \sim \exp[\tfrac{1}{4}\alpha Q - \alpha Q s_0^i + 2(1 - e^{-\frac{1}{2}\alpha Q})\,(s_0^f + \tfrac{1}{2})^{1/2}\,(s_0^i + \tfrac{1}{2})^{1/2}]$$

$$\times\ F\{-n,\ 2s_0^f - 2n + 1;\ [(2s_0^i + 1)(2s_0^f + 1)]^{1/2} \exp(-\tfrac{1}{2}\alpha Q)\}. \tag{2.7}$$

Then, substituting the values $\alpha = 2\ \text{Å}^{-1}$ [10], $Q = 1.5$ Å, $n = 0, \ldots, 2$, $s_0 \approx 20$ (the n and s_0 values are for heat of reaction not higher than 1 eV and dissociation energy of 4–5 eV) for the C—H bond into (2.7), it is easy to see that the confluent hypergeometric function does indeed vary slowly near the saddle point in comparison with the exponential, and additionally, $F(n,n';y) \approx 1$. Therefore, finally,

$$\langle n|0\rangle \approx \exp\{-\alpha Q(s_0^i - \tfrac{1}{4}) + 2(1 - \exp[-\tfrac{1}{2}\alpha Q])\,[(s_0^f + \tfrac{1}{2})(s_0^i + \tfrac{1}{2})]^{1/2}\}. \tag{2.8}$$

Proceeding from (2.8), the parameters J′ and J″ are found that govern the temperature dependence of the rate constant of solid-phase chemical reactions. The distance between the minima of the terms is given as $Q = Q_0 + \xi$ (Q_0 is the equilibrium distance) and the exponent is expanded in a series in ξ, retaining the quadratic addends. Then at $s_0^i \approx s_0^f \approx s_0$ it becomes

$$-2s_0[\tfrac{1}{2}\alpha Q_0 + \tfrac{1}{2}\alpha(1 - \exp[\tfrac{1}{2}\alpha Q_0])\xi + \tfrac{1}{8}\exp[-\alpha Q_0 + \alpha^2 \exp[-\tfrac{1}{2}\alpha Q_0]\xi^2], \tag{2.9}$$

from which the equations for J′ and J″ follow (cf. (8.10) in Chapter 3):

$$J' = \tfrac{1}{2}\alpha\,(1 - e^{-\alpha Q_0/2}),\quad J'' = \tfrac{1}{8}\alpha^2 e^{-\alpha Q_0/2}. \tag{2.10}$$

Thus the parameters found depend on the equilibrium distance between the initial- and final-state minima. At small distances ($\alpha Q_0 < 1$) (8.10) of Chapter 3 arises from (2.10) if it assumes that the reaction is symmetric, i.e. for a short range transfer the harmonic approximation holds for the ruptured and the newly formed terms. This is easy to understand, since in this case the overlap of the initial and final wave functions is due to the region where the bond potential energies are well approximated by parabolas.

If a long-range particle transfer takes place ($\alpha Q_0 > 1$), as for example in the experiments reported in [1–5], then it follows from (2.9) and (2.10) that the transition-matrix element depends on the intermolecular coordinates only owing to the linear term in the exponent (2.9).

The quadratic term is now considered in greater detail. At high temperatures, when the intermolecular vibrations are classical (this occurs in the same region where the quadratic term might be of importance) the rate constant is obtained by averaging the transition probability per unit time over the intermolecular vibrations (see (4.5) in Chapter 2). For this the squared matrix element (2.9) is multiplied by the factor $\exp[\gamma_R \xi^2/2k_B T]$ and integrated over ξ.

The terms in the exponent in the integrand are now grouped as follows:

$$- 4s_0 \left\{ \tfrac{1}{2}\alpha Q_0 + \tfrac{1}{2}\alpha\,[1 - \exp(-\tfrac{1}{2}\alpha Q_0)]\,\xi + \exp(-\tfrac{1}{2}\alpha Q_0) \right.$$
$$\left. + \left[\frac{\gamma_R}{8k_B\, Ts_0} + \tfrac{1}{8}\alpha^2 \exp(-\alpha Q_0/2) \right] \xi^2 \right\}. \tag{2.11}$$

Thus the quadratic term in the expansion of the logarithm of the matrix element may be discarded if the second term in the coefficient of ξ^2 is small compared with the first one. A criterion for the validity of doing this follows immediately from considering when the quadratic term in the expansion (4.4) of Chapter 2 may be neglected:

$$T < \hbar\gamma_R \exp[\tfrac{1}{2}\alpha Q_0]/M_H \omega k_B. \tag{2.12}$$

Substituting the values $\gamma_R = 10^4$ dyn cm^{-1}, $\omega = 6 \times 10^{14}$ s^{-1} and $Q_0 = 1.5$ Å yields $T < 300$ K for hydrogen-atom transfer. Therefore, comparing the theory with the experimental data on low-temperature radical solid-phase reactions [1–5], we may confine attention to the linear term, i.e. it may be assumed that $J'' = 0$.

If adiabatic low-temperature radical reactions are considered then the matrix element (3.14) of Chapter 3 should be used, and the parameter J' is then

$$J' = \frac{\partial}{\partial R}\, \frac{2}{\hbar} \int_{a(R)}^{b(R)} dr\, \{2M_H[\epsilon_\ell(r,R) - E(R)]\}^{1/2}, \tag{2.13}$$

where a(R) and b(R) are the potential-barrier boundaries for the reactant-to-reactant spacing R.

The comparison of theory and experimental data to be presented below is carried out under the assumption of either nonadiabatic or adiabatic low-temperature solid-phase reactions [1–5].

5.3. COMPARISON OF THEORETICAL AND EXPERIMENTAL DATA – THE ISOTOPE EFFECT

Five bimolecular reactions of the type (1.1′) [1–5] have so far been investigated over a wide temperature range:

$$CH_3OH + \dot{C}H_3 \longrightarrow \dot{C}H_2OH + CH_4, \tag{I}$$

$$C_2H_5OH + \dot{C}H_3 \longrightarrow CH_3\dot{C}HOH + CH_4, \tag{II}$$

$$CH_3CN + \dot{C}H_3 \longrightarrow \dot{C}H_2CN + CH_4, \tag{III}$$

$$CH_3NC + \dot{C}H_3 \longrightarrow \dot{C}H_2NC + CH_4, \tag{IV}$$

$$HONC(CH_3)C(CH_3)N\dot{O} + HONC(CH_3)C(CH_3)NOH \longrightarrow HONC(CH_3)C(CH_3)NOH + \dot{O}NC(CH_3)C(CH_3)NOH. \tag{V}$$

A non-Arrhenius temperature dependence of the rate constant is characteristic of these reactions, and it is of interest to compare the theory presented above with the experimental data on the reactions (I)–(V).

The following model is used when making numerical calculations of the rate constants of solid-phase radical reactions.

(i) Two types of vibration influence the reaction rate: the intramolecular vibrations of the ruptured and newly formed bonds and relative (intermolecular) vibrations of the reactants.

(ii) The particles A, D and H in their initial and final states lie along one line (the colinear model of chemical reaction). The potential surface is determined by two coordinates: the hydrogen-atom-to-radical A distance and the A-to-B distance.

(iii) Initially, the intramolecular subsystem (ruptured bond) is in its ground state. For the symmetric reactions (radical pair conversions in dimethylglyoxime) the intramolecular subsystem is in its ground level in the final state too. In the case of exothermic reactions (methyl radical with host molecules) the final state of the intramolecular subsystem coincides with that of the ground level of the initial state (see Fig. 5.1).

(iv) The environmental-rearrangement energy due to the reaction is zero.

(v) The harmonic approximation is used for intermolecular vibrations.

(vi) For the reaction of the methyl radical with the host molecules (I)–(IV) the main contribution to potential-barrier fluctuations is due to a local mode. This follows from the facts that the CH_3 mass is much smaller than the mass of the host molecules and that the radical-to-molecule force constant is assumed to exceed the molecule-to-molecule force constant. The rate constant of the reactions (I)–(IV) is calculated by means of (5.26) of Chapter 3 at $\Delta E = 0$ and $\rho_f = 1/\hbar\Omega_0$. Thus

$$K = \frac{2\pi}{\hbar^2\Omega_0} |V(R_0)|^2 \exp\left[\frac{\hbar(J')^2 \coth(\hbar\Omega_0/2k_BT)}{8M\Omega_0}\right] \times I_0\left\{\frac{\hbar(J')^2}{8M\Omega_0 \sinh(\hbar\Omega_0/2k_BT)}\right\}. \tag{3.1}$$

When calculating the rate constant of radical-pair conversions in dimethylglyoxime the relative reactant fluctuations are assumed to be due to ideal lattice vibrations with the dispersion law, $\omega(K) = \omega_D \sin(sK/\omega_D)$ (s is the velocity of sound). In contrast with the reaction of the methyl radical with a host molecule, in this case the mass of the radical (M = 115 daltons) nearly coincides with the mass of the host molecule (M = 116 daltons). The intermolecule force constant in dimethylglyoxime is determined by a strong hydrogen bond, and hence the radical-to-molecule force constant is not much greater. Therefore, when processing the experimental data on radical-pair conversions in dimethylglyoxime, (7.5) of Chapter 3 with $\Delta E = 0$ should be used:

$$K = \frac{1}{\hbar} |V(R_0)|^2 \int_{-\infty}^{\infty} ds \exp\left\{ \frac{3\hbar(J')^2}{4M} \int_0^1 dx \frac{x^2(1 - \sin(\pi x)/\pi x}{\omega(k_D x)} \right.$$
$$\left. \times \left[\coth\left(\frac{\hbar\omega(k_D x)}{2k_B T}\right) + \frac{\cos(s\hbar\omega_D(k_D x))}{\sinh(\hbar\omega(k_D x)/2k_B T)} \right] \right\}, \qquad (3.2)$$

$$\omega(k_D x) = \omega_D \sin(\pi x/2).$$

Here k_D is the Debye wavenumber. The above model is based on the theory of Chapters 3 and 4 as well as Section 5.1 and 5.2.

To calculate the chemical-reaction rate constants, it is necessary to know the relevant potential surfaces, which can be computed by means of the semiempirical LEPS technique [11], which is widely used in calculating the potential surfaces for bimolecular reactions. As pointed out above, the ruptured and newly formed bond terms, necessary for the LEPS technique, are described well by the Morse potential (2.1) and depend on two parameters D_0 and α. The dissociation energy D_0 varies for different substances, even for the same type of bonds, α is a geometric factor determining the Morse potential behavior at large distances and is closely connected to the length of the chemical bond. Since the bond lengths associated with the characteristic vibrations do not in fact vary from molecule to molecule, the parameter α is assumed to be the same for all C—H and O—H bonds [11]. Table 5.1 shows the D_0 and α values used in the calculations.

According to the LEPS technique, the potential energy of a three-particle system A—H- - - - D is written in the form [11]

$$U_0 = \frac{1}{1+p} \{Q_1 + Q_2 + Q_3 - [\tfrac{1}{2}((I_1 - I_2)^2 + (I_1 - I_3)^2 + (I_2 - I_3)^2)]^{1/2}\}, \qquad (3.3)$$

where Q_1, Q_2 and Q_3 are the Coulomb integrals, and I_1, I_2 and I_3 are the

Table 5.1 Morse-potential parameters used in the calculation

Bond	$\alpha\,(\mathring{A}^{-1})$	$D_0 \times 10^{12}$ *(erg)*	*Bond length* ($\mathring{A}$)	$S_0 + \frac{1}{2}$
$H—CH_2OH$	1.78	6.75	1.1	25.43
$H—CH_2CH_2OH$				
$H—CH_2CN$(I,II)	1.78	5.64	1.1	23.24
$H—CH_2NC$				
H—O in DMG	2.22	5.24	0.96	17.96
$H—CH_2$	1.78	7.49	1.1	26.78

exchange integrals for the A—H, D—H and A—D bonds respectively. The trial value P is found by fitting the potential-barrier height along the reaction path to the gas-phase activation energy that is obtained experimentally. The Coulomb and exchange integrals are connected with the Morse, U, and anti-Morse, U′, terms:

$$\left.\begin{aligned} Q + I &= (1 + p)U, \\ Q - I &= (1 - p)U'. \end{aligned}\right\} \tag{3.4}$$

The Morse term is associated with a bound state, and has the form (2.1), whereas the anti-Morse term is associated with a repulsive state, and can be approximated as [11]

$$U'(r) = \frac{1}{2} D_0[\exp(-2\alpha r) + 2\exp(-\alpha r)]. \tag{3.5}$$

Calculations of the potential-energy surfaces lead to schemes similar to those depicted in Fig. 5.2. Here the distances between the hydrogen

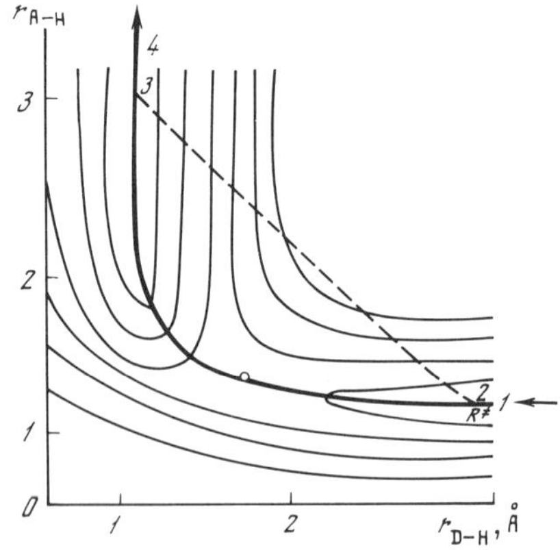

Figure 5.2 Two-dimensional potential surface corresponding to the reaction of the hydrogen atom abstraction by the methyl radical from the acetonitril molecule.

atom and the radicals A and D respectively are presented on the axes. The cross-section of the potential surface by a plane parallel to the energy axis (i.e. perpendicular to the plane of the figure) and cutting segments of length R on the axes $r_{A—H}$ and $r_{D—H}$ corresponds to a fixed reactant-to-reactant spacing, $R = r_{A—H} - r_{D—H}$. The initial state is in the lower right-hand corner of the scheme (reactant valley), whereas the final state is the upper left-hand corner (product valley). The reaction coordinate is the steepest descent from the reactant valley to the product valley (the solid curve 1–4).

In gas-phase reactions at high temperatures the system proceeds mainly along the reaction coordinate; at low temperatures the probability to reach the saddle point is small. First the reactant approaches along the reaction coordinate to a point with the distance $R^{\neq}$ (segment 1–2); then the hydrogen atom tunnels from A to D at a fixed reactant-to-reactant distance (dashed line 2–3), and finally the products are removed (segment 3–4).

A cut of the potential-energy surface for the reaction of the methyl radical with the acetonitrile molecule with distance 4 Å between them (A–D) is exhibited in Fig. 5.3. The lower double-well potential is calculated according to (3.3), but the upper term corresponds to an equation that differs from (3.3) by the sign in front of the square root. The

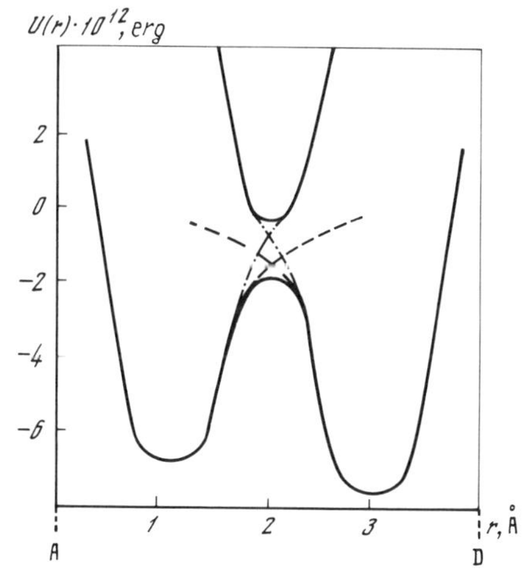

Figure 5.3 Potential energy of the hydrogen atom in the reaction $CH_3 + CH_3CN \rightarrow CH_4 + CH_2CN$ in case of the distance between the carbon atoms in the molecule and the radical being equal to 4 Å.

Solid lines correspond to the adiabatic potential surface (see fig. 5.2); dashed lines correspond to the Morse potentials of the unperturbed chemical bonds C—H in the radical and the molecule; dashed-point lines correspond the potential energy of the chemical bonds accounting for the diagonal correction to the interaction of terms.

dashed lines show the terms (Morse potentials) of the A—H and D—H chemical bonds, disregarding the molecule–radical interaction. When only the diagonal correction is taken into account, the potentials that arise are depicted by the dash–dotted line, they are differ (at the intersection point) from the lower double-well term.

It is interesting that, ignoring the interaction terms (dashed lines), gives a potential barrier very close to the adiabatic one. This justifies the assumption that, despite of the large splitting between the adiabatic terms ($\approx$ 1 eV), it is possible to use the theory of nonadiabatic transitions if the unperturbed (Morse) potentials of A—H and D—H are used for the ruptured and newly formed bond terms. The calculations of the potential surface yield the parameters specifying the transparency of the potential barrier versus the reactant-to-reactant spacing [6,12,13].

For the adiabatic transition the exponent

$$\tilde{J}(R) = \frac{2}{\hbar} \int_{a(R)}^{b(R)} dr \, \{2M_H[\epsilon_\ell(r,R) - E(R)]\}^{1/2}, \tag{3.6}$$

by virtue of the numerical integrations in the matrix element (3.15) of Chapter 3. Calculations of the functions $\tilde{J}(R)$ are presented in Fig. 5.4 for reactions (I)–(V). If the intermolecular amplitude does not exceed several tenths of an Ångström then $\tilde{J}(R)$ at $R > 3.5$ Å is described well by a linear function of the reactant displacements from the equilibrium positions, and hence the quadratic term in the exponent expansion may be disregarded.

It is convenient to perform the computations by interpolating the function J(R) by a second-order polynomial

$$\tilde{J}(R) = A_0 + A_1R + A_2R^2, \tag{3.7}$$

which readily yields the derivative J$'$. The values A_0, A_1 and A_2 for reactions (I)–(V) are summarized in Table 5.2.

The potenital-barrier transparency, assuming the transition to be nonadiabatic, is determined by the matrix element (2.8). The calculations of the exponent are shown in Fig. 5.4 (dashed line). It can easily be seen that both ways of calculating (adiabatic and nonadiabatic) lead to quite close results. The somewhat overestimated values of J(R) used in the calculation under the nonadiabatic approximation should have resulted in somewhat lower equilibrium reactant-to-reactant distances as compared with those obtained in the adiabatic calculations.

The formulae used to calculate the rate constant of the reactions (I)–(V) depend on two independent parameters: (3.1) depends on the values R_0 and Ω_0, (3.2) depends on R_0 and ω_D. To compare the theory

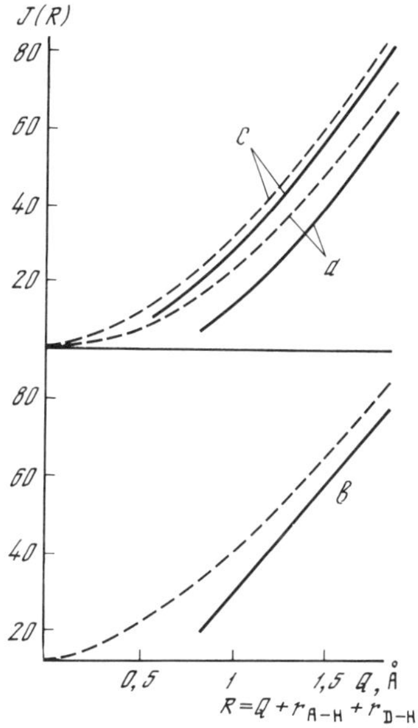

Figure 5.4 A calculation of the function J(R) for the reactions I–V. (a) — for the reactions I and II; (b) — for the reactions III and IV; (c) — for the reaction V. Solid line depicts the results of the calculation for the adiabatic reactions, dashed line is for the nonadiabatic reactions.

Table 5.2 Parameters for interpolating the function J(R) (adiabatic calculation)

Reaction	A_0	$A_1(\text{Å}^{-1})$	$A_2(\text{Å}^{-2})$
CH_3 + CH_3OH, CH_3CH_2OH	−91.183	9.344	8.140
CH_3 + CH_3CN, CH_3NC	−80.234	4.383	7.802
DMG	−79.598	15.316	7.504

with experimental data, the following procedure is adopted [6,12,13]. The independent parameters R_0 and Ω_0 (or ω_D) are chosen so as to achieve the best fit of the curves drawn using (3.1), (3.2) and the experimental temperature dependences of the rate constants. The r.m.s. technique is applied in fitting the optimal parameters by minimizing the functional

$$\bar{\phi} = \sum_i [y_i(T_i) - K(R_0, \Omega_0, T_i)]^2, \qquad (3.8)$$

where y_i are the experimental rate constants at temperatures T_i. All the experimental points are incorporated in the sum. The equilibrium

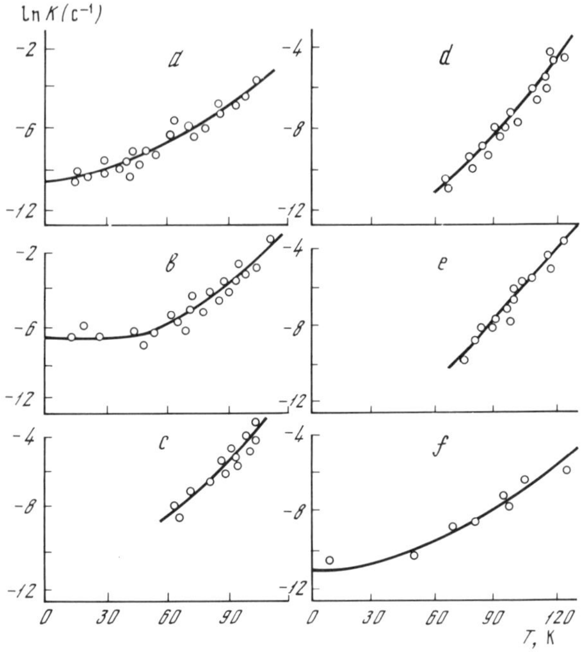

Figure 5.5 Temperature dependent rate constants for the reactions of the hydrogen atom abstraction by methyl radical from the molecules of methanol (a), ethanol (b), acetonitril I (c), acetonitril II (d), and methylisocyanide (e), the conversion of the radical pairs in dimethylglyoxime are shown in the plot (f).

Open circles are for the experimental values, solid lines for the calculation.

Table 5.3 Optimal parameters R_0 and Ω_0 obtained by comparing the theory with experimental data

Matrix	$R_0(\mathring{A})$	$\Omega_0 \times 10^{-13}(s^{-1})$	$\sigma_{R_0}(\mathring{A})$	$\sigma_{\Omega_0} \times 10^{-13}(s^{-1})$
	Adiabatic calculation technique			
CH_3OH	3.6625	3.5608	0.0120	0.1742
CH_3CH_2OH	3.6446	3.5078	0.0125	0.1445
$CH_3CN(I)$	3.8224	3.2860	0.0132	0.0982
$CH_3CN(II)$	3.8967	3.1049	0.0132	0.0693
CH_3NC	3.8595	3.2616	0.0133	0.0819
DMG	3.1701	2.9196	0.0141	0.3354
	Nonadiabatic calculation technique			
CH_3OH	3.7140	3.6601	0.0110	0.1340
CH_3CH_2OH	3.7082	3.6904	0.0110	0.1448
$CH_3CN(I)$	3.9340	3.4770	0.0119	0.1026
$CH_3CN(II)$	4.0079	3.2426	0.0120	0.0701
CH_3NC	3.9709	3.4276	0.0120	0.0844
DMG	3.1854	3.0657	0.0114	0.3571

radical-to-molecule distances and intermolecular frequencies found using this procedure are compared with those measured in independent experiments.

If these values appear to be close then the agreement of the theory and experiment may be assumed to be satisfactory. The fits are presented in Fig. 5.5; the optimum parameters are in Table 5.3.

At present there are no crystallographic studies available of the structures corresponding to the reactions (I)–(IV), which is why it is not possible to directly check the values R_0 found for the reactions of the methyl radical with molecules. However, there are data in the literature [14] on the structure of the dimethylglyoxime crystal, which, as has been pointed out above, changes only slightly during radical formation, the difference between the equilibrium distances being 0.5–0.8 Å. This difference is evaluated using the van der Waals radii [15]:

$$R_0(\text{C—H----C}) - R_0(\text{O—H----O}) = \\ = (\rho^{\text{C—H}} + \rho^{\text{H}} + \rho^{\text{C}}) - (\rho^{\text{O—H}} + \rho^{\text{H}} + \rho^{\text{O}}) \approx 0.5\ \text{Å} \qquad (3.9)$$

which correspond to the values presented in Table 5.3.

The optimal local mode frequencies of reactions of methyl radicals with molecules are (3–3.5) $\times\ 10^{13}\ \text{s}^{-1}$. As is well known, the local mode frequency is than the upper boundary of the allowed spectrum (i.e. the Debye frequency). Thus the theory agrees with experiment if the host Debye frequencies in acetonitrile and methyl isocyanide satisfy the condition $\omega_D < \Omega_0 \approx 3 \times 10^{13}\ \text{s}^{-1} \approx 150\ \text{cm}^{-1}$. The Debye frequencies of the molecular crystals lie within the range 50–200 cm^{-1}, which overlaps the range where the fitted frequencies lie.

The rate constant of solid-phase chemical reactions according to (3.1) and (3.2) turns out to be very sensitive to the deviations of the parameters $R_0(Q_0)$ and $\Omega_0(\omega_D)$ from their optimal values. If these deviations do not exceed the standard deviations $\sigma_{R_0,\,Q_0}$ and $\sigma_{\Omega_0,\,\omega_D}$ (see Table 5.3) then the divergence of theory and experiment is smaller than the statistical measurement error. Therefore both the equilibrium distances and frequencies obtained from comparison of theory with experimental data are reasonable. This means that the physical pattern of the chemical reactions in solids has been constructed correctly.

Attention should be drawn to the fact that the rate constant of radical-pair conversions in dimethylglyoxime increases smoothly, passing from the low-temperature limit to the region of rapid temperature rise. This is explained by the fact that potential-barrier fluctuations are determined by ideal-crystal dynamics, and even at liquid-helium temperatures the logarithm of the rate constant is proportional to T^4 (see (3.2) and (7.6) of Chapter 3). At the same time, the rate constant (3.1) is virtually constant at $T \leqslant 40$–50 K in the reactions of methyl radicals with molecules where the reactant distances vary mainly as a result of the local mode. Then with increasing temperature there is a swift rise in the radical reaction rate constant. A similar behaviour of the rate constant also appears when, instead of the translational, the orientational molecule vibrations play the principal role in the creation of the temperature dependence.

It should be kept in mind that, in contrast with normal quasimonomolecular processes (where the concentration of one reactant, the host molecule, is much larger than the concentration of the other reactant, the radical), the kinetics of the reactions (I)–(IV) deviates markedly from the first-order law. This is why the authors of [2,3,5] measured the rate constant only by processing the initial parts of the kinetic curves of the loss of the radical CH_3. Some features of the kinetics of this reactions and of complications relevant to the experimental data processing are discussed in Section 5.4.

The above parameters of the potential barrier are now used to study the kinetic isotope effect in solid-phase reactions with hydrogen transfer. At low temperatures the atomic transfer proceeds via tunneling, and since the tunneling probability is sensitive to the particle mass, hydrogen-to-deuterium exchange leads to a rather large isotope effect. In fact, assuming the characteristic parameter of the ruptured and newly formed bonds to be (as in Table 5.1 and the distance between the term minima to be $Q \approx 1.5$ Å, the Franck–Condon factor (see (2.8)) for the hydrogen atom is approximately 10^{10}–10^{12} times larger than for deuterium. This value would have determined the kinetic isotope effect if there was no need to account for potential-barrier fluctuations caused by intermolecular vibrations. In fact, the zero-point vibrations at the characteristic potential barrier parameters on their own make the isotope effect nearly 10^3 times weaker. In the temperature range in which the intermolecular vibrations may be assumed to be classical, the logarithm of the kinetics isotope effect decreases linearly with increasing temperature (i.e. $\ln [K(H)/K(D)] \sim -T$). At the same time, the quasiclassical isotope effect caused by the difference of the zero-point frequencies discussed in Chapter 2 leads to the dependence $\ln [K(H)/K(D)] \sim 1/T$.

The isotope-effect decrease caused by intermolecular vibrations is easy to interpret: the potential-barrier fluctuations exert a greater influence on the tunneling of the heavier deuterium atom. When the reactants are at the same distances the deuterium tunneling transfer increases more rapidly than that of the hydrogen atom. This is the mechanism that governs the temperature-dependent isotope effect in low-temperature bimolecular reactions.

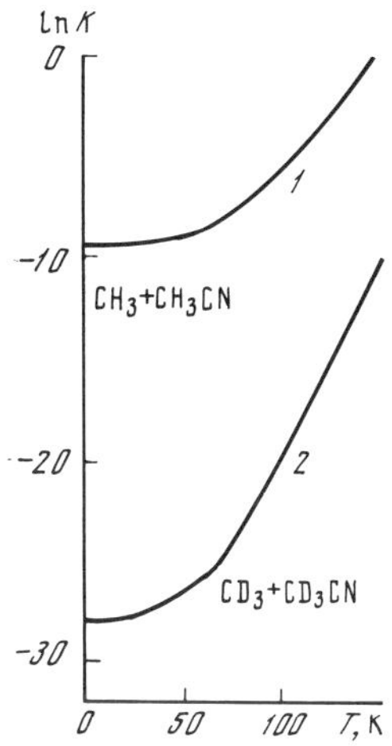

Figure 5.6 A calculated temperature dependent rate constant of the hydrogen atom abstraction by the methyl radical from the molecule of ordinary (1) and deuterated (2) acetonitril.

Figure 5.6 shows calculated temperature dependences of the rate constants of hydrogen and deuterium abstraction by methyl radicals from the molecules of both normal and deuteriated acetonitrile. The deuterium-transfer curve is lower than the protium one, but in the temperature range where intermolecular vibration may be treated as classical it grows twice as rapidly with the temperature. This results in the curves approaching one another, and hence in a decrease of the isotope effect. The calculated temperature dependence of the kinetic isotope effect for hydrogen (deuterium) abstraction by methyl radical from acetonitrile is shown in Fig. 5.7 [16]. At a temperature of 77 K the theoretical value of the isotope effect is 5×10^5, which is in satisfactory agreement with the experimental result $\eta \geqslant 28000$ [3].

The potential barriers found above are used to compare the transition rate constants for the various excitations of the newly formed bonds. The rate constant is calculated by means of (5.26) of Chapter 3 at $\rho_f = 1/\hbar\Omega_0$; $\Delta\nu$ is determined by the energy-conservation law. The calculations of the reaction III are presented in Fig. 5.8. It can readily be seen that the transition rate constant is at a maximum when the newly formed bond is excited to the level $n_f = 3$, which is energetically the closest (from below) to the initial one, whereas the excitation to the fourth level has no low-temperature rate limit at all, since the energy-conservation law requires in this case an absorption of phonon energy. The dependences exhibited in Fig. 5.8 qualitatively illustrate our discussion of (5.26) and (5.26′) of Chapter 3.

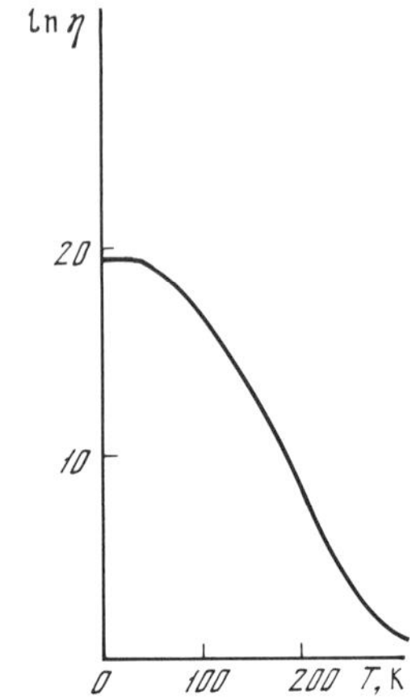

Figure 5.7 The kinetic isotope effect calculated for the reaction of the methyl radical with the acetonitril molecule.

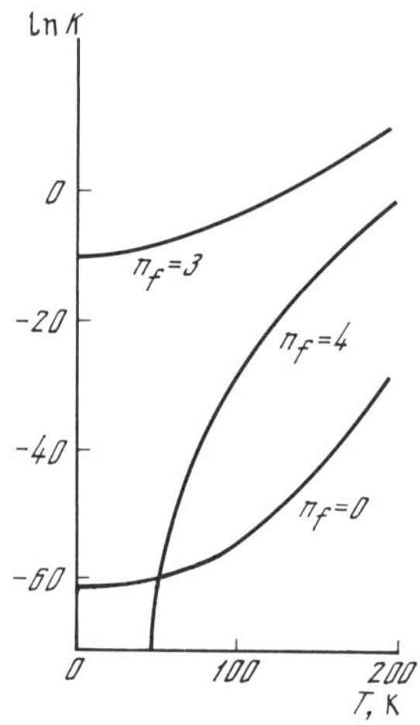

Figure 5.8 Temperature dependent rate constant calculated* for the reaction $CH_3 + CH_3CN \rightarrow CH_4 + CH_2CN$ for different states of the C—H bond in the methan molecule.

*Calculations were carried out at $R_0 = 4$ Å, $\Omega_0 = 3.10^{13}$ s^{-1}.

5.4. POLYCHRONIC KINETICS OF TUNNELING RADICAL CAGE REACTIONS

Nonstationary time development is well known to be a common feature of low-temperature solid-phase reactions — the kinetics of such chemical reactions observes neither a first- nor a second-order law, and has as a rule a rather complicated form. The free radical recombination [17,18] (see also [19]), the alkyl-radical reaction with glassy alcohol matrices [20,21], oxidation of the alkyl radicals by the molecular oxygen dissolved in a solid host [21] and other reactions may be attributed to this type of process. In many cases the "rate constant" of these reactions decreases with time according to a square-root law: $k(t) \sim t^{-1/2}$ [22]. That is why there has been a tendency — to treat these reactions as nonstationary stages of diffusion-controlled reactions [23,24]. However, the conversion depth at such a stage is not large for three-dimensional diffusion. Therefore, in order to explain the nonstationary development of the process within a two-order range of the concentrations of reacting particles, it was necessary either to artificially overestimate their reaction radius [23,24] or to assume the relative motion in the geminal pairs to be of a one-dimensional character. Both explanations do not suffice for the cage model when the particle reacts with its nearest-neighbour molecules constituting the cage. These reactions are in principle not diffusion-controlled. It is natural to assume that the kinetically controlled reactions are nonstationary owing to the rate dispersion caused by energetic and spatial nonuniformity. This kind of kinetics is called polychronic: first, the particles with highest rate constant react, next those with a lower rate, and so on. A rate-constant spread can be due to a wide spectrum of potential barriers with different heights (energetic nonuniformity) and widths (spatial nonuniformity) [26–30]. Meanwhile, a permanent conversion of one type of reactant into another one might occur in the system (the reagent distribution over the potential barriers varies with time); then if the rate of such conversions does exceed the chemical reaction rate the polychronity no longer results in a deviation from the first or second-order kinetic law [31].

In exploring the processes that are non-stationary in time, the problem arises of how to interprete the experimental data, i.e. to reveal the nonstationary mechanism. A detailed discussion of the papers devoted to this problem (based, in many cases on polychronic kinetics) may be found in review papers (see e.g. [22,32,33]). Only some aspects of the issues relevant to polychronic kinetics of solid-phase cage reactions are touched upon here. Of principal interest is the explanation of the square-root law of the tunneling rate constant decay revealed in [20,21] for the reactions (I)–(IV) in the context of polychronic kinetics.

One possibility of interpreting the square-root law in terms of polychronic kinetics was first demonstrated in [34], which considers atomic transfer as a quasiresonant process, with the rate dispersion associated with the energetically nonuniform system. The square-root nonstationary radiationless conversion of electronic excitation into vibrational excitation has been explained in [35] in such a manner. At high temperatures, when particle transfer is overbarrier with a small activation energy spread but not a resonant one, its polychronic kinetics has been shown [36] to be close to the square-root law

$$c(t) = c(0) \exp[-\tilde{k}t^{1/2}],$$

where c(t) is the concentration of the reacting particles at time t.

Cage reactions may become nonstationary not only owing to the energetic but also to spatial nonuniformity.† This is true for all kinetically controlled tunneling reactions, of both resonant and activation-like type. In particular, these include low-temperature reactions (I)–(IV) [2,3,5], where the tunneling rate depends on the reactant-to-reactant spacing so steeply that its 0.03–0.04 Å decrease induces an order-of-magnitude increase in the rate [6]. Because of this, the modulation of intermolecular spacing by the thermal and even the zero-point vibrations of the reactants is seen to influence decisively the scale and temperature dependence of the tunneling probability. The transfer mainly occurs when the barrier is narrowed by fluctuations, i.e. for spacings that are smaller than the mean intermolecular ones. At low temperatures the thermal fluctuations are frozen in, but a nonuniform distribution of the equilibrium intermolecular spacings in the cage remains, which leads inevitably to a rate dispersion.

Strong nonstationarity of the cage reactions and the square-root law of the rate decay can both be interpreted as polychronic tunneling effects, with reasonable assumptions concerning the scale of the intermolecular spacing dispersion [37] being made. There are also some kinetic peculiarities in reactions in mixed proton- and deuterium-containing matrices, which can be also explained in terms of the polychronic model.

5.4.1. Tunneling Square-Root Law

Quenching of an excited molecule by randomly distributed surrounding energy acceptors is well known to proceed according to a square-root law if the energy transfer is due to dipole–dipole interaction, i.e. $K \sim R_0^{-6}$ [38–40]. With a transferring electron and atom the rate constant depends exponentially on the distance:

† A study of how spatial nonuniformity affects the kinetics of electron tunneling transfer was first carried out in [30], where an algorithm for experimental data processing in the polychronic tunneling kinetics was presented.

$$K(R_0) = \bar{K} \exp[-J'(R_0 - \bar{R}_0)], \tag{4.1}$$

where $\bar{R}_0$ is the mean equilibrium reactant-to-reactant spacing (the cage radius). The parameter decrease J' for an electron is nearly forty times smaller than for the hydrogen atom. Therefore the electronic excitation may be transferred to rather distant impurities, but nevertheless because of the exponential dependence of $K(R_0)$ the kinetics of this process is completely different from that of the Förster square-root law [30,41]. However, when a heavy particle (atom) tunnels it is transferred only to the nearest neighbors separated by distances deviating only slightly from a mean value. But since the tunneling probability is extremely sensitive to such deviations, the averaging results in a kinetics indistinguishable from the square-root law over two orders of magnitude of the initial radical conversion depth [37]. For the sake of simplicity, the deviation distribution is assumed to be uniform within the range ΔR:

$$\varphi(R_0) = \begin{cases} 1/\Delta R & (\bar{R}_0 - \tfrac{1}{2}\Delta R \leqslant R_0 \leqslant \bar{R}_0 + \tfrac{1}{2}\Delta R), \\ 0 & (R_0 < \bar{R}_0 - \tfrac{1}{2}\Delta R, \quad R_0 > \bar{R}_0 + \tfrac{1}{2}\Delta R). \end{cases} \tag{4.2}$$

The distribution (4.2) can be used to average the radical concentration; thus the kinetics of the quasimonomolecular process is given by

$$\frac{c(t)}{c(0)} = \frac{1}{\Delta R} \int_{\bar{R}_0 - \Delta R/2}^{\bar{R}_0 + \Delta R/2} dR_0 \exp[-\bar{K}t \exp[-J'(R_0 - \bar{R}_0)]]$$

$$= \frac{1}{J'\Delta R} \int_{-\ell n\bar{K}t - J'\Delta R/2}^{-\ell n\bar{K}t + J'\Delta R/2} dx \exp[-\exp(-x)]. \tag{4.3}$$

Although the integrand in (4.3) is the same as that given in [36], the principal difference between the over- and underbarrier reactions is in the limits of integration. Here they depend on the scale of the spatial nonuniformity, ΔR, and, via J', on the mass of the tunneling particle. They are not arbitrary. Not only is it important that the dependence calculated from (4.3) is linear in the coordinates $\ell g\ (c/c(0))$ and $(\bar{K}t)^{1/2}$, but also this occurs with quite reasonable parametrizations. Calculations for different products $J'\Delta R$ are shown in Fig. 5.9, from which it can easily be seen that (4.3) is approximated well by the dependence $\ell g\ (c/c(0)) - a - bt^{1/2}$ and for values of $J'\Delta R$ between 4.5 and 7 it may be assumed that $a = 0$ to within experimental error [20,21]. Thus the relationship $\ell g\ (c(t)/c(0)) \sim t^{1/2}$ holds until the concentrations are reached that are two and a half orders of magnitude smaller than the

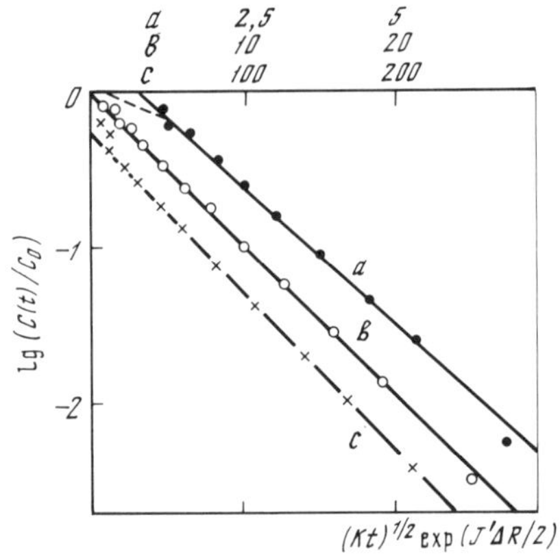

Figure 5.9 Kinetics of a tunnel reaction calculated with due account for the dispersion of the equilibrium distances between the reagents.
Values of J' ΔR are: a — 3, b — 5.5, c — 10.

initial one. The relationship is most closely observed for $J'\Delta R = 5.5$. The width of the ΔR spread is easily estimated if it is recalled that $J' = 7 \times 10^9$ sm^{-1} for reaction of the methyl radical with a hydrogen-containing alcohol molecule [6], and hence $\Delta R = 0.06$–0.1 Å.

Figure 5.10 compares the experimental data from [20] and the kinetics of the solid-phase reaction

$$\dot{C}H_3 + CH_3OD \longrightarrow CH_4 + \dot{C}H_2OD$$

calculated by means of (4.3). The measurements were carried out

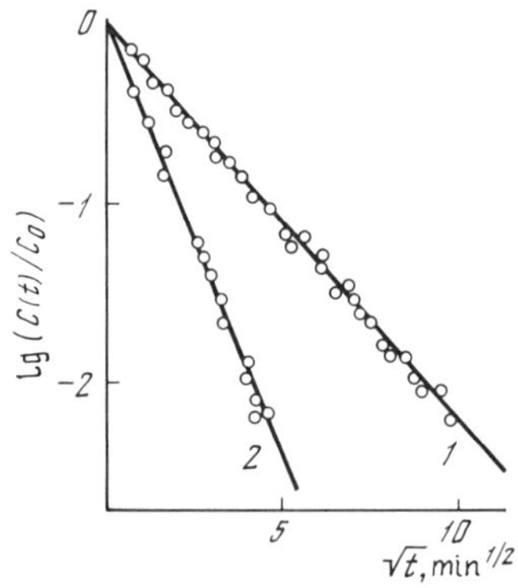

Figure 5.10 Kinetics of the reaction $CH_3 + CH_3OD \rightarrow CH_4 + CH_2OD$ at 77K (curve 1) and at 87K (curve 2). Open circles stand for the experimental values [20]; solid lines show the calculation at R = 3.69, = $2.75 \cdot 10^{13}$ s^{-1}, ΔR = 0.08 Å.

in a glassy matrix of CH_3OD at 77 and 87 K, and the values $K = 5.3 \times 10^{-3}\ s^{-1}$ were found, assuming $J'\Delta R = 5.5$. Two K values yield, according the procedure in [6], two parameters: the mean equilibrium spacing $\overline{R}_0$ and the intermolecular frequency Ω_0. Since these are determined using only two experimental points, the error is rather large, so the values found, $\overline{R}_0 = 3.69$ Å and $\Omega_0 = 2.73 \times 10^{13}\ s^{-1}$, give only an estimate. Nevertheless, they agree with the corresponding values found in [6] by treating the experimental data from [5] (see Table 5.3). These data on the temperature-dependent rate constant for the reaction of the radical $\dot{C}H_3$ with the molecules CH_3OH and CH_3CH_2OH are obtained from the initial segment of the kinetic curve of the radical decay; therefore the R_0 values found by such a treatment of the experimental data in [6] are effective values close to $\overline{R}_0 - \frac{1}{2}\Delta R$. This treatment yields R_0 values that are somewhat smaller than the mean values, the deviation of R_0 from its mean being not larger than 0.05 Å.

Therefore a square-root nonstationary process observed in a single-component medium can be explained if the spread of the equilibrium reactant-to-reactant spacings in the matrix is assumed to be of the order of 0.1 Å. The situation is more complicated in mixed matrices such as proton- or deuterium-containing alcohols, which are examined in the next section.

5.4.2. Concentration Dependence of the Isotope Effect

A free radical is assumed to be in a cage containing $\tilde{N}$ alcohol molecules; then the probability that i nondeuteriated molecules are in the cage is $\binom{\tilde{N}}{i}\alpha_0^i(1 - \alpha_0)^{\tilde{N}-i}$, where α_0 is the proportion of the nondeuteriated molecules in the matrix containing a mixture of normal and deuteriated alcohols. $\alpha_0 = 1$ is for a completely nondeuteriated host, whereas $\alpha_0 = 0$ is for a completely deuteriated one. Now the index $\ell = 1, 2, \ldots, i$ enumerates i nondeuteriated cage molecules, while $m = i + 1, \ldots, \tilde{N}$ enumerates $\tilde{N} - i$ deuteriated molecules. The reorientation rate of small free radicals in the cage is known to be much more rapid than their reaction with the matrix ($10^{-2}\ s^{-1}$ at liquid-nitrogen temperature), so we believe that at each moment a statistical equilibrium exists with respect to different intracage orientations. If K_H^ℓ and K_D^m are the rate constants of the radical reactions with normal and deuteriated cage molecules then the total radical rate constant is $\overline{K}_0 = \sum_\ell K_H^\ell + \sum_m K_D^m$ and the initial radical loss is governed by the law $\exp[-\bar{K}_0 t]$, which is to be averaged over the distances to all cage particles.

The averaging is denoted

$$\langle \ldots \rangle = \int \ldots \int dR_0^1\, dR_0^2 \ldots dR_0^{\tilde{N}} \ldots \prod_{i=1}^{\tilde{N}} \varphi(R_0^i).$$

Distances between the free radicals and the neighboring cage molecules are assumed to be uncorrelated, i.e. all R_0^i are independent random variables with the same distributions. So the averaging yields

$$\begin{aligned}\langle \exp[-\bar{K}_0 t] \rangle &= \prod_{\ell=1}^{i} \int dR_0^\ell \exp[-K_H^\ell t]\, \varphi(R_0^\ell) \prod_{m=i+1}^{\tilde{N}} \int dR_0^m \exp[-K_D^m t]\, \varphi(R_0^m) \\ &= [\int dR_0 \exp[-K_H(R_0)t]\, \varphi(R_0)]^i\, [\int dR_0 \exp[-K_D(R_0)t]\, \varphi(R_0)]^{N-i}.\end{aligned} \tag{4.4}$$

Then the total kinetics of radical loss is given by the equation

$$\begin{aligned}\frac{c(t)}{c(0)} = \sum_{i=0}^{\tilde{N}} \binom{\tilde{N}}{i} \alpha_0^i (1 - \alpha_0)^{\tilde{N}-i}\, [\int dR_0 \exp[-K_H(R_0)t]\, \varphi(R_0)]^i \\ \times [\int dR_0 \exp[-K_D(R_0)t]\, \varphi(R_0)]^{\tilde{N}-i}.\end{aligned} \tag{4.5}$$

It is experimentally known [20,21] that the laws:

$$\left.\begin{aligned} \left.\frac{c(t)}{c(0)}\right|_{\alpha_0=1} &= \exp[-\bar{K}_H t^{1/2}], \\ \left.\frac{c(t)}{c(0)}\right|_{\alpha_0=0} &= \exp[-\bar{K}_D t^{1/2}], \end{aligned}\right\} \tag{4.6}$$

are observed in both the nondeuteriated and completely deuteriated matrices in the temperature range under study. Assuming now $\alpha_0 = 0$ and $\alpha_0 = 1$ in (4.5) and comparing with (4.6) gives

$$\left.\begin{aligned} n_H &= \int dR_0 \exp[-K_H(R_0)t]\, \varphi(R_0) = \exp\left[-\frac{\bar{K}_H}{\tilde{N}}\, t^{1/2}\right], \\ n_D &= \int dR_0 \exp[-K_D(R_0)t]\, \varphi(R_0) = \exp\left[-\frac{\bar{K}_D}{\tilde{N}}\, t^{1/2}\right]. \end{aligned}\right\} \tag{4.7}$$

Then on substituting (4.7) into (4.5) an equation is found that describes the radical-loss kinetics [37]:

$$\frac{c(t)}{c(0)} = \sum_{i=0}^{\tilde{N}} \binom{\tilde{N}}{i} \alpha_0^i (1 - \alpha_0)^{\tilde{N}-i} \exp\left[-\frac{i}{\tilde{N}}\, \bar{K}_H\, t^{1/2} - \frac{\tilde{N} - i}{\tilde{N}}\, \bar{K}_D\, t^{1/2}\right]$$

$$= \left\{ \alpha_0 \exp\left[-\frac{\overline{K}_H}{\widetilde{N}} t^{1/2} \right] + (1 - \alpha_0) \exp\left[-\frac{\overline{K}_D}{\widetilde{N}} t^{1/2} \right] \right\}^{\widetilde{N}}. \tag{4.8}$$

From here it is easy to calculate the derivative of the radical concentration at the initial instant:

$$\left. \frac{\partial \ln\{c(t)/c(0)\}}{\partial t^{1/2}} \right|_{t=0} = \alpha_0 K_H + (1 - \alpha_0) \overline{K}_D \tag{4.9}$$

This law (a linearly α_0 dependent derivative at the initial instant) is indeed observed experimentally [20,21].

5.4.3. Migration of Active Particles in Radical Cage Reactions

Reactions in mixed alcohol matrices (C_2H_5OD, C_2D_5OD) have been carried out for three types of free radicals: CH_3, C_2H_5 and C_4H_9 [20]. Since the dimensions of these radicals markedly differ, it may be assumed that the number of molecules in a cage, $\widetilde{N}$, increases with radical size. This assumption is readily verified, since the kinetics of radical loss in mixed alcohol matrices (see (4.8)) depends strongly on $\widetilde{N}$, and, comparing the concentration time dependence (4.8) with the experimental curves the values of $\widetilde{N}$ are found for various radicals.

Figure 5.11 and 5.12 show experimental data from [20] and the curves calculated according to (4.8). Only the data for matrices containing small concentrations of the nondeuteriated molecules ($\alpha_0 = 0.05 \div 0.08$) are processed, since the reactions not only with normal but also with the deuteriated molecules contribute the curves.

The parameters used in the calculations are listed in Table 5.4 [42]. In

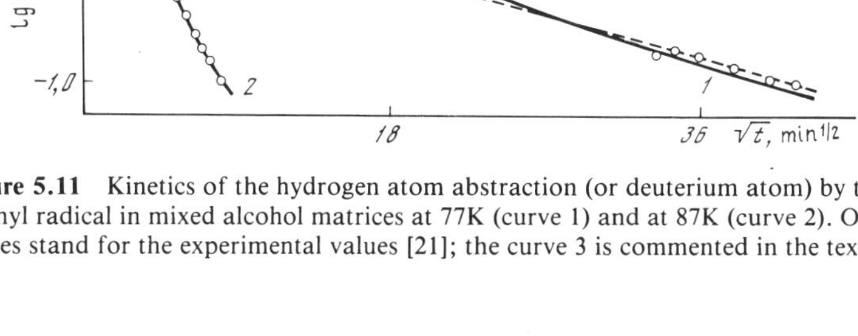

Figure 5.11 Kinetics of the hydrogen atom abstraction (or deuterium atom) by the methyl radical in mixed alcohol matrices at 77K (curve 1) and at 87K (curve 2). Open circles stand for the experimental values [21]; the curve 3 is commented in the text.

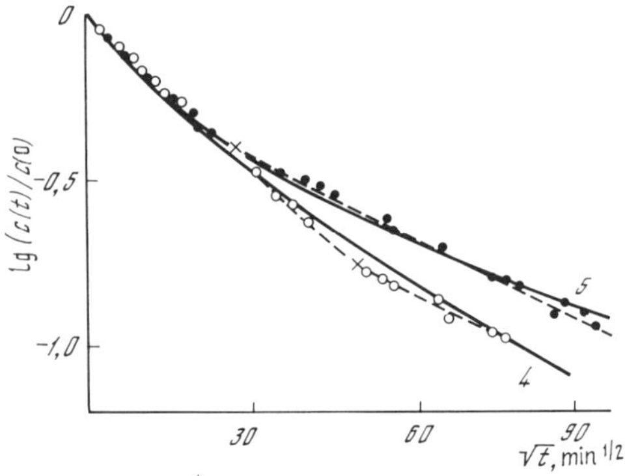

Figure 5.12 Kinetics of the hydrogen atom abstraction (or deuterium atom) by the C_2H_5 (curve 4) and C_4H_9 (curve 5) radicals in mixed alcohol matrices at 87K. Open circles stand for the experimental values [21].

a mixed matrix the large difference in the rate constants allows a deuteriated molecule to participate in the reaction only when there are no nondeuteriated molecules among the nearest neighbours of the radicals. Addition of a small proportion of hydrogen-containing molecules cannot change the $\widetilde{K}_D$ value; therefore in performing the calculations $\widetilde{K}_D$ is taken from independent experiments in completely deuteriated matrices. The constant $\widetilde{K}_H$ for reactions of the free radicals with normal molecules inserted into a nondeuteriated matrix is not known, and hence it is fitted in line with $\widetilde{N}$ by comparing the experimental data with (4.8). The parameters. $\widetilde{K}_H$ and $\widetilde{N}$ are sought that give the best fit of theory and experiment. The constants $\widetilde{K}_H$ for the matrices studied (α_0 = 0.05–0.08) appear to be larger than for the hydrogen-containing matrices, and they vary versus the radical dimensions.

What was quite unexpected, was the behaviour of $\widetilde{N}$. Instead of the expected increase with growing radical dimensions, $\widetilde{N}$ rapidly decreases. Also, the $\widetilde{N}$ values found in [42] are considerably larger than the expected number of free radical neighbors in a cage, which should have corresponded to the first coordination sphere ($\leqslant$ 10).

These facts lead us to think that the reaction in deuteriated matrices weakly doped by hydrogen-containing molecules is not a cage reaction, i.e. a motion of the free radical through the sample (or some other kind of free valence migration) takes place, until it meets a hydrogen-containing molecule. This might be interpreted as an increase of the cage dimensions, the "cage" being larger for smaller radicals owing to their higher mobility. The time available for the radical to find a hydrogen-containing partner is determined by the rate constant of the radical–deuteriated-molecule reaction. Another argument in favour of the above pattern is that the cage size decreases with falling temperature, which is accompanied by a decrease of active-particle mobility.

Thus a doubt arises as to whether the cage model holds for the reactions (I)–(IV). Presumably, the "cage" concept for two-component matrices may be applied only during the initial time period until the free valence has had time to displace. Figure 5.11 shows a curve 3 constructed using the same $\widetilde{K}_H$ and $\widetilde{K}_D$ values as for the curve 1, but with $\widetilde{N}$ equal to the number of radical nearest neighbours in the cage ($\widetilde{N} = 8$). After a certain time, the curves 1 and 3 diverge more than the experimental error and the cell model can no longer be applied to the reaction. It should be emphasized that the most important and experimentally verifiable result of Section 5.4.2 — a linear dependence of the radical-concentration derivative on α_0 — is still valid, since (4.9) is for the initial period, when the cage model still holds.

For a one-component matrix the concept of the cage reaction is quite justified, since the isotope content of a new cage visited by the radical does not differ from the preceeding one; moreover, the rate constant of the elementary chemical reaction in hydrogen-containing matrices greatly exceeds the rate of the radical "hopping" over the matrix, so the radical has time to react within the initial cell. It should be kept in mind that the application of (4.8) in developing the above model of radical reactions is not completely justified, since the (4.8) is deduced assuming a cage character for the reactions (I)–(IV) whereas the conclusions go beyond the scope of the cage model. However, the number of the molecules in a cage corresponding to the experimental data may be estimated without applying (4.8). If it is assumed that a considerable change in slope of the experimental curve corresponds to the termination of the "burning down" of the hydrogen-containing molecules, then only

Table 5.4 Optimal parameters of the kinetic curves (see (4.12)) obtained by comparing the theory with experimental data

Figure	*Curve*	*Temperature, (K)*	*Radical*	*Isotope content of the matrix (%)*	$\widetilde{K}_H$ *(min*$^{-1/2}$*)*	$\widetilde{K}_D$ *(min*$^{-1/2}$*)*	*Number of neighbours in a cage N*
5.11	1	77	CH_3	0.08	0.422	0.0078	24
5.11	2	87	CH_3	0.08	1.75	0.0114	62
5.11	3	77	CH_3	0.08	0.422	0.0078	8
5.12	4	87	C_2H_5	0.05	0.239	0.00625	30
5.12	5	87	C_2H_5	0.055	0.265	0.0056	16

those radicals do not react by this time that have been in the cage with exceptionally deuteriated molecules. The proportion of these radicals is $c(t)/c(0) = (1 - \alpha_0)^{\tilde{N}}$, which yields $\tilde{N}$: $\tilde{N}_{CH_3(77K)} = 19$, $\tilde{N}_{C_2H_5} = 34$, $\tilde{N}_{C_4H_9} = 15$. These values are in accord with those presented in Table 5.4.

References

1. Yakimchenko, O.E. and Lebedev, Ya.S. (1971) *Int. J. Rad. Phys. Chem.*, **3**, 17.
2. Hudson, R.L., Shiotany, M. and Williams, F. (1977) *Chem. Phys. Lett.*, **48**, 193.
3. Sprague, E.D. (1977) *J. Phys. Chem.*, **81**, 516.
4. Toriyama, K., Nunome, K. and Iwasaski, M. (1977) *J. Am. Chem. Soc.*, **99**, 5823.
5. Le Roy, R.J., Murai, H. and Williams, F. (1980) *J. Am. Chem. Soc.*, **102**, 2325.
6. Trakhtenberg, L.I., Klochikhin, V.L. and Pshezhetskii, S.Ya. (1981) *Chem. Phys.*, **59**, 191; Ibid., **69**, 121.
7. Landau, L.D. and Lifshitz, E.M. Quantum Mechanics. Nonrelativistic Theory. Pergamon, Oxford.
8. Plotnikov, V.G. and Ovchinnikov, A.A. (1978) *Usp. Khim.*, **47**, 44.
9. Goldanskii, V.I. (1959) *Dokl. AN SSSR*, **127**, 1037.
10. Energy of Chemical-Bond Rupture. Ionization Potentials and Electron Affinity (ed. V.N. Kondrat'ev). (1974) (In Russian.) Nauka, Moscow.
11. Kondrat'ev, V.N., Nikitin, E.E., Reznikov, A.I. and Umanskii, S.Ya. (1976) *Thermal Bimolecular Reactions in Gases*. (In Russian.) Nauka, Moscow.
12. Klochikhin, V.L. and Trakhtenberg, L.I. (1981) *Dokl. AN SSSR*, **259**, 399.
13. Trakhtenberg, L.I. (1982) *Khim. Fiz.*, **1**, 53.
14. Merit L., Lauterman E. (1952) *Acta Crystallogr.*, **5**, 811.
15. Kitaigorodskii, A.I. (1971) *Molecular Crystals*. (In Russian.) Nauka, Moscow.
16. Klochikhin, V.L. and Trakhtenberg, L.I. (1983) *Khim. Fiz.*, **2**, 810.
17. Mikhailov, A.I., Lebedev, Ya.S. and Buben N.Ya. (1965) *Kinetika i kataliz*, **6**, 48.
18. Ermolaev, V.K., Molin, Yu.N. and Buben, N.Ya. (1962) *Kinetika i kataliz*, **3**, 314.
19. Pshezhetskii, S.Ya., Kotov, A.G., Milinchuk, V.K. *et al.* (1972) *ESR of Free Radicals in Radiation Chemistry*. (In Russian.) Khimiya, Moscow.
20. Bol'shakov, B.V., Stepanov, A.A. and Tolkachev, V.A. (1980) *Int. J. Chem. Kinet.*, **12**, 271.
21. Zaskul'nokov, V.M., Vyazovkin, V.L., Bol'shakov, B.V. and Tolkachev, V.A. (1981) *Int. J. Chem. Kinet.*, **13**, 707.
22. Burshtein, A.I. (1978) *Usp. Khim.*, **42**, 212.
23. Burshtein, A.I. and Tsvetkov, Yu.D. (1974) *Dokl. AN SSSR*, **214**, 369.
24. Butyagia, P.Yu. (1974) *Vysokomol. Soedin.*, **16**, 63.
25. Kofman, A.G. and Burshtein, A.I. (1978) *Chem. Phys.*, **27**, 217.
26. Primack, W. (1955) *Phys. Rev.*, **100**, 1677.
27. Primack, W. (1960) *J. Appl. Phys.*, **31**, 1524.
28. Mikhailov, A.I., Kuzina, S.I., Lukovnikov, A.F. and Goldanskii, V.I. *Dokl. AN SSSR*, **204**, 383.
29. Mikhailov, A.I., Bolshakov, B.I., Lebedev, Ya.S. and Goldanskii, V.I. (1972) *Fiz. Tverd. Tela*, **14**, 1172.
30. Zamaraev, K.I., Khairutdinov, R.F., Mikhailov, A.I. and Goldanskii, V.I. (1971) *Dokl. AN SSSR*, **199**, 640.
31. Goldanskii, V.I. and Kozhushner, M.A. (1984) *Dokl. AN SSSR*, **274**, 632.
32. Yakimchenko, O.E. and Lebedev, Ya.S. (1978) *Usp. Khim.*, **47**, 1018.
33. Karpukhin, O.N. (1978) *Usp. Khim.*, **47**, 1119.
34. Bol'shakov, B.V., Fuks, M.P., Tolkatchev, B.A. and Burshtein, A.I. (1977) In: *Proc. 4th Symp. on Radiation Chemistry*. Acad. Kiado, Budapest.

35. Avanesov, A.G., Burshtein, A.I., Depker, B.I., Osiko, V.V., Pirumov S.S. and Shchebakov I.A. (1980) *Dokl. AN SSSR*, **254**, 593.
36. Yakimchenko, O.E., Degtyarev, E.I., Prusakov, V.N. and Lebedev, Ya.S. (1980) *Teor. Eksp. Khim.*, **16**, 75.
37. Burshtein, A.I., Klochikhin, V.L. and Trakhtenberg, L.I. (1984) *Khim. Fiz.*, **3**, 155.
38. Förster, Th. (1949) *Z. Naturforsch*, **4a**, 321.
39. Förster, Th. (1959) *Discuss. Faraday Soc.*, **27**, 7.
40. Agranovich, V.M., Galanin, I.D. (1978) *Transfer of Energy of Electronic Excitations in Condensed Media*, p. 383. (In Russian.) Nauka, Moscow, 1978.
41. Scher, H. and Montroll, E.W. (1975) *Phys. Rev.*, **B12**, 2455.
42. Klochikhin, V.L. and Trakhtenberg, L.I. (1984) *Zh. Fiz. Khim.*, **58**, 2877.

CHAPTER 6
Tunneling Phenomena in Amorphous Solids

The basics of the modern theory of kinetic processes in amorphous solids (glasses) are presented. Section 1 gives a brief historical introduction and points out the main physical effects observed in glasses that are assumed to be relevant to heavy-particle tunneling.

The double-well potentials (DWP) and the relevant two-level systems (TLS) are treated from the point of view of the general principles of statistical physics as being a consequence of the nonequilibrium amorphous state. A simple model of the DWP is discussed. The study of the distributions of the DWP parameters indicates the possibility of two types of DWP — soft and rigid.

The TLS contribution to the specific heat, depending linearly on temperature, is considered. A possible dependence of the specific heat on time and the connection of this effect with the distribution of the TLS parameters is also discussed.

Within the framework of single-phonon interaction with the soft DWPs, the relaxation time, ultrasonic attenuation, renormalization of the velocity of sound, and the thermal conductivity are studied. This approach holds for temperatures that are not too high, when the relaxation broadening of the levels is still less than their energy difference.

The main experimental data observed in the high-temperature range (10–100 K and higher) are accounted for the structural-defect model and its relation to the TLS model are discussed.

A multiphonon theory of kinetic processes in glasses is presented that uses barrier fluctuations when considering heavy-particle tunneling in DWP. The double adiabatic approximation is shown to hold for the rigid DWPs but to be less reliable for soft DWPs. The temperature-dependent amplitude of coherent tunneling transitions is calculated.

The multiphonon TLS relaxation is investigated. The temperature dependence of the relaxation time and possible deviations from the Arrhenius law are studied.

The multiphonon sound–TLS interaction is considered, and the inverse length of ultrasonic attenuation and renormalization of the velocity of sound are calculated. Conditions under which the multiphonon mechanism dominates over the relaxation mechanism in the ultrasonic attenuation are found. The contribution of the multiphonon sound–TLS interaction to the thermal conductivity is calculated and its relation to the experimentally observed temperature plateau is discussed.

6.1. INTRODUCTION

The modern development of physical concepts concerning the kinetic phenomena in amorphous dielectrics is usually assumed to have started in 1971 with the experiments of Zeller and Pohl [1], who demonstrated that the low-temperature behaviour of the specific heat and thermal conductivity of glasses strongly deviated from that expected on the basis of the then current theory, which predicted that the specific heat should be proportional to the third power of temperature, $C \sim T^3$, at liquid-helium temperatures. It had been supposed that under these conditions the specific heat would have been due to the long-wave phonons, which are not affected by the detailed structure of the substance on the atomic scale. Because of the supposition the impression arose that no qualitative differences between the crystalline and amorphous solids were to be expected. The Debye model (see e.g. [2]), which accounted well for the long-wave phonons, would have yielded the above result, $C \sim T^3$, for the specific heat of the glass. The behaviour of the thermal conductivity, should have depended on the prevailing scattering mechanism. At low temperatures the scattering of phonons by the sample boundaries would be the most likely determining factor, in which case the phonon free path would be independent of frequency, and the conductivity would then be proportional to the third power of temperature as well, $\kappa \sim T^3$. The fact that these ideas are so simple and seemingly reliable explains the sensational reaction to the results of Zeller and Pohl [1], who observed a linear law for the specific heat ($C \sim T$) and a quadratic law for the thermal conductivity ($\kappa \sim T^2$) in various glasses at liquid-helium temperatures.

It soon became clear that anomalies occurred not only at very low temperatures and not only in these two quantities. Further measurements showed that in amorphous solids nearly all kinetic processes with the participation of phonons or electromagnetic waves within a temperature range from the lowest currently accessible to several hundred degrees proceed in ways completely different from the same processes in crystalline systems. Simultaneously it was realized that many effects had in fact been observed long before the Zeller and Pohl experiments. Mason and McScimin [3] investigated scattering of relatively short-wave acoustic phonons. These phonons play an essential part in thermal conductivity in the temperature-plateau region at temperatures above 10 K (see [4–6]). Anomalies in the specific heat and thermal conductivity were observed by Anderson [7] and Hornung *et al.* [8]. The quadratic law for the conductivity was also reported by O. Anderson *et al.* [9]. An unusual temperature behaviour of the inverse length of ultrasonic attenuation was studied by A. Anderson and Bömmel [11] and Krause and Kurkjan [11]. Nevertheless, it was Zeller and Pohl's results that initiated the recent development of the studies in this field. At present the total number of papers on this subject is so large that we are not able to present within this introduction an exhaustive review and we recommend to the interested reader a collection of very detailed reviews edited by Phillips [12] and written by those authors who have contributed appreciably to the development of this field (see also [13]).

Several theoretical models were proposed to explain the observed experimental facts. We shall dwell here upon only one of them, which is the most widely accepted at present. It was formulated simultaneously and independently by Phillips [14] and P.W. Anderson *et al.* [15], who introduced the hypothesis that there are entities in glasses (atoms, molecules or, maybe, more complicated formations) that may occur in two energetically and spatially close states divided by an energy barrier. Transitions between these two states at low temperatures proceed mainly via tunneling. There are a large number of such TLSs in the glass, they are characterized by an energy difference ΔE between two states and by the tunneling amplitude $\nu e^{-J/2}$. Both values are statistically spread over certain ranges.

When the glass is heated, these TLSs are excited and hence contribute to the heat capacity. They usually scatter phonons more strongly than other entities (sample boundaries, anharmonicity, etc.), and so determine the thermal conductivity and other kinetic coefficients. On the basis of these concepts, it became possible to explain a wide range of kinetic phenomena in glasses at low temperatures. These are the specific heat, thermal conductivity, various acoustic phenomena, relaxation processes that take place when the amorphous substance interacts with electromagnetic radiation, collective phenomena relevant to the TLS–TLS interaction, etc.

However, in spite of the outstanding success of the TLS model, it

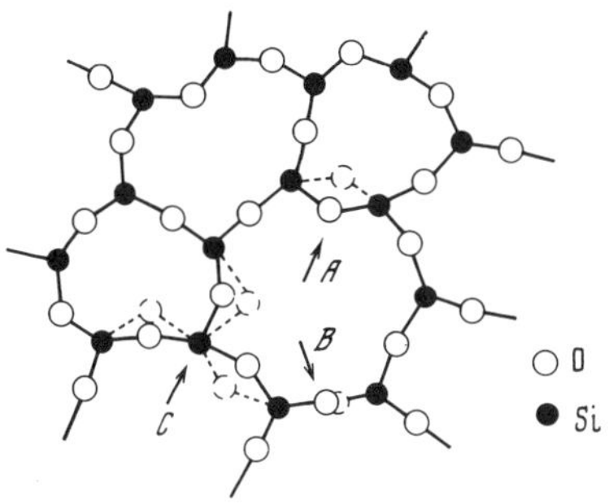

Figure 6.1 A fragment of the SiO_2 amorphous structure. Three ways (A,B,C) of how the double-well potentials might appear are shown.

cannot be claimed that all the problems have been solved. First, the microscopic structure of TLSs is still unclear. It is not yet possible to tell exactly how these TLSs are constructed and what entity it is that tunnels in them: atom, molecule or maybe a more complicated particle. Some of the simplest ways of constructing the TLSs can be visualized in Fig. 6.1 (taken from [16]), which shows how, after the amorphization of crystabalite, the possibility arises that some atoms can occupy different positions that were absent in the crystalline state. Phillips [77] has presented reasons why TLSs appear between the boundaries of clusters in which the atoms are grouped in the glass. An interesting hypothesis was put forward by Rivier and Duffy [18]. Starting from the assertion that on a scale larger than the interatomic distance an amorphous substance is homogeneous and invariant with respect to the local rotation group SO(3), they formulate a gauge theory in which the minimization the free-energy density yields two classical solutions. Tunneling transitions are possible between these two states. The TLS in their model is a rather complicated linear construction containing a large number of atoms. Another model that also yields two solutions, but on the atomic scale, was proposed by Klinger and Karpov [19] (see also [20]).

The fact that a variety of TLS formation mechanisms are available, of which only some have been mentioned here, does not make our task easier. As a matter of fact, we cannot at present prove convincingly on the basis of the experimental data which of the abovementioned mechanisms really does apply to glasses. Therefore our understanding of the processes in amorphous substances remains, to this extent, phenomenological.

The TLS concept gives theorists a possibility to interpret processes in glasses below liquid-helium temperatures. Serious difficulties arise, however, when trying to extend this approach to higher temperatures (a detailed discussion of this issue can be found in A.C. Anderson's review published in [12]).

A possible way of overcoming these difficulties was proposed by Trakhtenberg and Fleurov [21–25]. The reader who has read the preceding chapters of this book will easily see that the fluctuations of the barrier that play such an important role in chemical tunneling reactions should also be taken into account when analysing tunneling processes in TLSs. As will be shown below, the TLS–phonon coupling is very strong because of this mechanism, and a tunneling transition at high temperatures might be a multiphonon process. This permits the explanation of a variety of kinetic phenomena in glasses in the whole temperature range.

There are still many unclear issues in this rapidly developing branch of physics. Many problems are still under discussion and there are a number of different opinions on some of them. So we should point out that this chapter deals with such debatable issues from the present authors' point of view.

6.2. GENERAL IDEAS ON TWO-LEVEL SYSTEMS (TLS)

A substance in an amorphous state is not in equilibrium. The fact that we can observe it and even make experiments is due to the large relaxation times, which exceed not only the duration of the experiment but sometimes that of the human life! Correspondingly, the entropy of an amorphous substance does not become zero even at zero temperature. According to the definition of the entropy (e.g. [26]) the inequality $S \neq 0$ is due to the fact that a given macroscopic state can be the result of a number of microscopic states (structures) that do not differ significantly in energy from each other. These microscopic states are specified by different spatial arrangements of atoms — and here lies the principal difference between a glass and an ideal crystal. The latter at zero temperature is in the single ground state with the minimum energy, which is characterized by a definite spatial arrangement of atoms, the entropy in this case being zero.

It is natural to assume that transitions are possible between the various microscopic structures of a given macroscopic state of glass, which at low temperatures might proceed via tunneling. This changes the spatial arrangement of atoms in the glass, and shifts of small groups or individual atoms not exceeding the interatomic spacings appear to be elementary acts. This consideration leads us directly to the hypothesis of the tunneling modes and TLSs related to them, which were first put forward in [14, 15].

As has been mentioned above there are at present a number of TLS models, although a choice among them is still difficult. Here one of the possible versions is discussed that seems to be the most probable or at least the simplest.

To be more precise tunneling of an individual atom is considered (e.g. of oxygen in glassy SiO_2) and the simplest ideas on TLS formation mechanism are assumed (see Fig. 6.1). Let a TLS in SiO_2 be formed as the result of an extension of the chain $>$Si — O — Si$<$. If the Si–Si distance coincides with twice the length of the chemical bond then the oxygen atom is in the potential well as shown in Fig. 6.2(a), in which the lowest level has energy $\frac{1}{2}\hbar\omega$ (ω is the zero-point frequency, $\hbar\omega/k_B \sim 1000$ K)†. When the Si–Si distance increases, the potential well for the O atom broadens (Fig. 6.2b); then a small potential hump appears in it (Fig. 6.2c), which later becomes higher than $\frac{1}{2}\hbar\omega$ (Fig. 6.2d). Here, however, it should be kept in mind that the energy $\hbar\omega$ of the zero-point vibrations may decrease owing to distortion of the potential curve near the bottom of the well. It is possible to talk about TLS formation only in the latter case, when there are two deep enough wells, each of which is capable of holding the particle, i.e. the condition $V_{tot} > \frac{1}{2}\hbar\omega$ holds. In

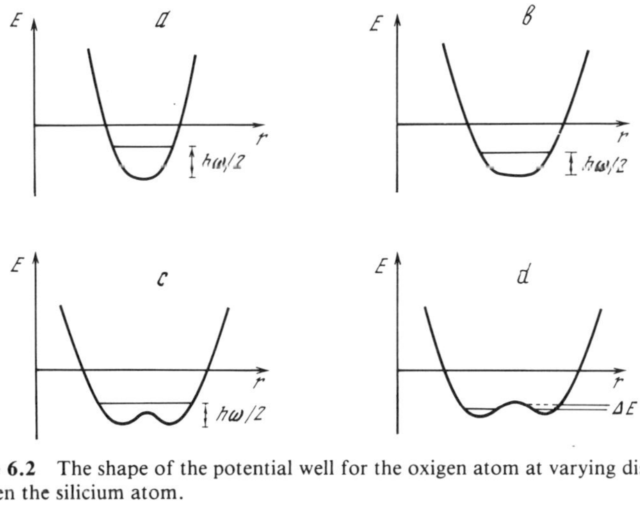

Figure 6.2 The shape of the potential well for the oxigen atom at varying distance between the silicium atom.

† Here only the order of magnitude is given, since not only SiO_2 but also other glasses are being considered.

the case depicted in Fig. 6.2(b) the TLS has not yet formed, since the energy of the particle zero-point vibrations is larger than the barrier height, i.e.

It is important to distinguish between the total height of the barrier, V_{tot}, and the barrier height above the level of the zero-point vibrations, $V = V_{tot} - \frac{1}{2}\hbar\omega$. At this level the barrier width Δd is measured. The distribution of the V and Δd values are important characteristics of glasses. Unfortunately, we know no direct technique allowing the determination of the actual form of these distributions. It is only possible to say that there are some empirical indications that they are not uniform. Probably the distribution of the parameter

$$J \sim (2m\,V)^{1/2}\,\Delta\,d/\hbar,$$

where m is the tunneling mass, has the form shown in Fig. 6.3. The following mechanism of formation of such a distribution can be proposed.

The TLSs relate to the frozen-in fluctuations of the spatial atomic arrangement that took place in the liquid phase. The probability of such flucutations decreases exponentially with the relevant energy i.e.

$$W \sim \exp\,[-\,R_{min}/k_B T_g],$$

where R_{min} is the minimum work performed to form such a fluctuation (see e.g. [26]); T_g is the melting temperature of the glass. In the present case the scale of the fluctuation might be specified by the same linear size as that of the potential barrier:

$$R_{min} = \tfrac{1}{2}\gamma\Delta d^2$$

Hence it can be inferred that the TLSs with large barrier widths, and consequently with large parameters J, are less probable, since more work corresponds to these TLSs. Therefore Cohen and Grest [27], who

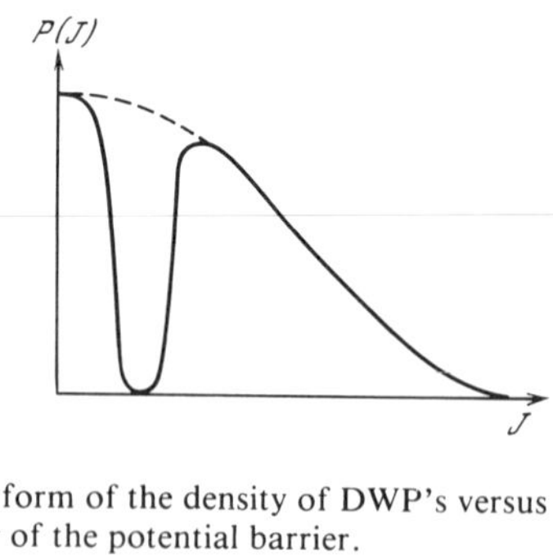

Figure 6.3 A possible form of the density of DWP's versus the magnitude of the transparency parameter of the potential barrier.

have related the tunneling states to the free volumes of the atomic scale in glasses, believe that the tunneling transition probability amplitude $\nu\, e^{-J/2}$ does not exceed $k_B T_g$.

On the other hand, the TLSs with relatively small potential barriers appear to be unstable [24]. For systems with small potential barrier V and asymmetry parameter ϵ_a (i.e. the level-to-level energy difference) a deformation might turn out to be energetically favourable at which the potential barrier nearly vanishes, i.e. V_{tot} becomes nearly equal to $\frac{1}{2}\hbar\omega$. It is energetically favourable for the particle to occupy the lowest level, whose energy decreases when the barrier width decreases. The corresponding gain in energy compensates for the elastic energy spent on the local deformation of the atomic structure.

For the TLSs with extremely small values of V (<100 K, soft double-well potentials) the above mechanism [24] is weak. In this case the potential curves both near the maximum and near the minima differ considerably from the two original potentials describing the interaction of the silicon and hydrogen atoms. That is why not only do the height and width of the potential barrier vary when the Si–Si distance varies, but also the frequency of the zero-point vibrations in each well changes significantly, i.e. the TLS levels shift relatively to each other.† The properties of such systems were investigated in [19, 20], where double-well potentials (DWP) were discussed in which the frequency of the zero-point vibrations in each well and the potential-barrier height do not exceed about 30 K. It is clear that they may play an essential role in low-temperature processes, whereas at higher temperatures (~ 50–100 K) the DWPs with higher characteristic energies might be of importance.

As we shall see below, there are a number of experimental results favouring the existence of two groups of TLSs with different values of principal parameters. This point of view was first expressed in [28, 29]. Nevertheless, this issue as well as many other issues concerning the nature of the TLS remain debatable and require a further study.

To conclude this section, a simple mathematical problem is discussed that illustrates a mechanism for the origin of the DWP (see [25, 30] and references therein). Let the interaction of the oxygen and silicon atoms be described by an anharmonic Morse oscillator. Then the oxygen atom in the chain $\rangle$Si — 0 — Si$\langle$ is subject to the action of the sum of the potentials of two Si atoms, which can be written as

† The double adiabatic approximation appears to be violated under these conditions, so the fluctuating barrier used in [24] cannot be effectively applied.

$$U_{(x)} = U_M(x + \delta a) + U_M(-x + \delta a), \tag{2.1}$$

where

$$U_M(x) = D(1 - e^{-\alpha x})^2.$$

Here D ~ 2–3 eV is the chemical-bond energy. The zero-point frequency of a single potential $U_M(x)$ can be estimated as

$$\omega_0 \sim (2D\,\alpha^2/m)^{1/2} \tag{2.2}$$

The quantity δa gives the variation of the Si–Si spacing with respect to its crystallographic value. Extrema of the potential (2.1) are determined from the equation

$$\sinh\alpha x - 2e^{-\alpha\delta a} \sinh\alpha x \cosh\alpha = 0, \tag{2.3}$$

which always has one solution at $x = 0$. But if the condition

$$e^{\alpha\delta a} > 2 \tag{2.4}$$

holds then there are two more solutions

$$x_0 = \pm\alpha^{-1} \operatorname{arcosh}\left(\tfrac{1}{2}e^{\alpha\delta a}\right). \tag{2.5}$$

The condition (2.4) therefore determines the value δa at which the single-well potential becomes a double-well potential. The variable

$$\xi = \tfrac{1}{2}e^{\alpha\delta a} - 1 \tag{2.6}$$

is now introduced and the region $\xi \ll 1$ is considered. Then the potential-barrier height is

$$V_{tot} = V(0) - V(x_0) \approx D\xi^2, \tag{2.7}$$

and the characteristic width of each well is $x_0 \sim \alpha^{-1}(2\xi)^{1/2}$. Therefore the zero-point energy in such a well is

$$\hbar\omega \sim \frac{(\hbar\omega_0)^2}{D}\,\frac{1}{8\xi}. \tag{2.8}$$

A TLS appears if $\frac{1}{2}\hbar\omega < V_{tot}$; hence

$$\xi \geqslant \xi_c = (\hbar\omega_0/4D)^{2/3} \tag{2.9}$$

Thus TLSs first appear with rather small frequencies

$$\hbar\omega_c \approx \frac{\hbar\omega_0}{2}\left(\frac{\hbar\omega_0}{4D}\right)^{1/3}$$

and barrier heights $V = V_{tot} - \frac{1}{2}\hbar\omega$. The corresponding shifts δa are several times larger than the zero-point amplitude

$$u_0 \approx \alpha^{-1} (\hbar\omega_0/D)^{1/2}$$

in the potential $V_M(x)$. Increasing δa further, the parameter ξ grows rapidly, and TLSs with $\hbar\omega \approx \hbar\omega_0$ and large potential barriers appear.

The above model for small ξ reproduces a simplified version of the theory of critical soft DWPs proposed in [19,20]. One of the most interesting predictions of this theory seems to be an explanation of the anomalously large deformation potential observed experimentally. Differentiation of (2.8) with respect to ξ gives, in the critical region ($\xi \approx \xi_c$),

$$B_\epsilon \approx \hbar\omega_0 (4D/\hbar\omega_0)^{1/3}$$

which may be as large as several tenths of an eV. However, for larger values lying outside the critical region the deformation potential decreases proportionally to ξ^{-2}. Thus the system swiftly enters the region of rigid DWPs (second peak in Fig. 6.3) with small deformation potentials, which must be treated within the framework of multi-phonon theory [21–25].

6.3. SPECIFIC HEAT OF AMORPHOUS DIELECTRICS

The experimentally observed linear temperature dependence of the specific heat in the low-temperature limit ($T < 1$–2 K) indicates the existence of elementary excitations whose density of states in the corresponding energy range is virtually constant. The TLS model due to Phillips [14] and Anderson *et al.* [15] assumes these excitations to be transitions from lower to upper states of TLSs. There are a large number of TLSs in glass with various energy differences ΔE, and it is natural to assure that the density of TLSs, $n(\Delta E)$ with green ΔE depends only weakly on ΔE — at any rate over a rather wide energy range. The density distribution $n(\Delta E)$ seems to be bounded from above by an energy $\Delta E_{max} \sim kT_g$. There is probably also a lower bound ΔE_{min}. Judging from the large number of experimental data, whatever the lower boundary is, it corresponds to extremely low temperatures ($\Delta E_{min/k} < 0.1$ K).

Using standard formulae of statistical physics (see e.g. [26]), the internal energy of a glass due to TLS excitations is

$$E_{in} = \int_{\Delta E_{min}}^{\Delta E_{max}} d\Delta E \, \Delta E n(\Delta E) \frac{\exp[-\beta\Delta E]}{1 + \exp[-\beta\Delta E]}, \beta = 1/k_B T. \tag{3.1}$$

If $n(\Delta E)$ is assumed to be weakly dependent on the energy difference ΔE and this function is taken out of the integration (3.1) then

$$E_{in} = \frac{1}{12}\pi^2 n(0)(k_B T)^2 \tag{3.2}$$

where $n(0)$ is an average value of the function $n(\Delta E)$. Differentiating (3.2) with respect to temperature, the specific heat is

$$C(T) = \frac{1}{6}\pi^2 n(0)\, k_B^2 T. \tag{3.3}$$

The above considerations lead not only to an understanding of the nature of the linear temperature law for the specific heat of glasses, but also to a quantitative evaluation of the TLS density. For example, Stephens [31] finds for silicate glasses

$$n(0) \sim 10^{32}\text{–}10^{33} \text{erg}^{-1}\text{cm}^{-3}.$$

The linear law for the specific heat holds in the temperature range $\Delta E_{min} < k_B T$. At present many measurements of specific heat at temperature down to $T \sim 0.1$ K have been made, and no essential deviations from the dependence (3.3) have been found (see e.g. [12]). This circumstance allows the above estimate for ΔE_{min} to be obtained.

It is not possible to evaluate the upper boundary ΔE_{max} of the TLS density distribution directly from experiment, since, from liquid-helium temperature upwards, it is not TLSs that mainly contribute to the specific heat, but rather phonons, and the linear law is replaced by a cubic one. As we pointed out above, the TLSs are "frozen-in" fluctuations that occurred in the liquid phase. Hence the melting temperature of glass appears as a natural scale of the ΔE distribution. It is relevant both to the quantity $\Delta E_{max} \sim kT_g$ [15, 27] and to the density of states [32]

$$n(0) \sim (k_B T_g)^{-1}.$$

Such a law for the density of states has been observed experimentally in specific-heat [33] and phonon-free-path [34, 35] measurements. On making further measurements for a wider variety of glasses, a behaviour

$$n(0) \sim \exp[-\alpha T_g]$$

was found in [36]

The specific heat is a thermodynamical quantity, and strictly speaking, it should be measured in a system that has reached an equilibrium state. However, a peculiar property of disordered systems is that they do not come to equilibrium. Each individual TLS is specified not only by the energy difference, but also by the potential-barrier parameters. Then the amplitude of tunneling probability

$$\frac{\Delta}{\hbar} \approx \nu \exp[-\tfrac{1}{2}J] \tag{3.4}$$

may vary owing to variations of the transparency parameter J. If the J values are distributed over a wide enough range then for each observation time t, TLS can be found with such large J values that they have not relaxed till that time. Their number will decrease with time, and the energy collected by them at a given temperature will increase. Correspondingly, the measured value of the specific heat will also increase.

The simplest way of finding the specific-heat time dependence is to assume the parameter J to be homogeneously distributed within a range (J_{min}, J_{max}). At large enough times the TLSs with large J values remain out of equilibrium, for which the inequality

$$\Delta \ll \Delta E \sim k_B T$$

holds. Then the characteristic relaxation time of these TLSs is given by

$$\frac{1}{\tau} \sim \hbar \nu^2 \rho_{ph} \exp[-J], \tag{3.5}$$

where ρ_{ph} is the phonon density of states. By time t the only TLSs that are in equilibrium with phonon are those for which the relaxation times satisfy the condition $\tau < t$ or $J > J_b(t) \sim \ell n(t/\hbar\nu^2\rho_{ph})$. The specific heat is proportional to the number of TLSs at equilibrium by the time of observation; hence it depends on time as

$$\left.\begin{aligned} C(T, t) &= C(T)\,\frac{J_b(t) - J_{min}}{J_{max} - J_{min}} = \frac{C(T)}{J_{max} - J_{min}}\,\ell n\left(\frac{t}{\tau_{min}}\right), \\ \frac{1}{\tau_{min}} &\sim \hbar\nu^2\rho_{ph}\exp[-J_{min}]. \end{aligned}\right\} \tag{3.6}$$

This logarithmic time dependence of the specific heat is one of the most important inferences of the TLS model.† Its experimental observation (or, in contrast its absence) would allow judgement to be made on how the potential-barrier parameters are distributed. For a long time no time dependence of the specific heat was seen. This led to the

† Here a deep similarity between this approach and polychronic kinetics is emphasized. Some applications of the latter to the description of chemical kinetics is described in Section 5.4, and applications to electron tunneling scavenging are described in Chapter 9.

conclusion that the distribution of the J parameter is narrow and rather close to a δ-like one. Only recently have experiments conducted over a small time range ($10\ \mu s < t < 100$ ms) revealed a dependence of the specific heat on time that is close to the logarithmic one [37]. However, a doubt has been expressed [38] about the validity of these results, since the phonon system might not have reached equilibrium during so short a period.

We have mentioned in Section 6.2 the finding by Black and Halperin [28, 29] that the measurement of specific heat, on the one hand, and that of the thermal conductivity and phonon echo, on the other, yield different values of the TLS spectral density n(0). To explain this circumstance, they proposed two groups of TLSs with different relaxation times (see Fig. 6.2). Then only the rapidly relaxing TLSs contribute to the kinetic effects, while both groups contribute to the specific heat. The time dependence of the specific heat would thus have a behaviour more complicated than logarithmic, (3.6), and to calculate it the distribution P(J) must be known in detail.

Even when no attention is paid to the difficulties connected with the time-dependent specific heat, some more problems remain if a detailed comparison of the theory is made with the experimental data on the temperature-dependent specific heat. As a rule, experiments are interpreted by means of the equation

$$C = C_1 T^n + C_3 T^3. \tag{3.7}$$

Sometimes an additional term

$$C_E \left(\frac{T_E}{T}\right)^2 \frac{\exp\,[T_E/T]}{[\exp[T_E/T] - 1]^2}$$

is introduced; this is due the Einstein-like transverse vibrations with characteristic temperature $T_E \sim 50$–100 K. Comparing (3.7) with the experimental data, it is found that the exponent $n \sim 1.1$–1.2 is practically never equal to unity. The value of C_3, is, as a rule, more than twice as large as the value that follows from the Debye model with known velocity of sound.

This latter circumstance is often ascribed to an energy dependence of the TLS density, which is assumed to be

$$n(\Delta E) = a + b\Delta E^2. \tag{3.8}$$

Then the characteristic scale of this dependence is ~ 1 K. This is introduced purely phenomenologically and has no microscopical grounds. A formula

$$C(T) = C_1 T + C_2 T^2 + C_{Deb} T^3 \tag{3.9}$$

has also been proposed to describe the specific heat. Here C_{Deb} is the value of C_3 from the Debye model. The quadratic temperature term is assumed [39] to be connected with the creation of new TLSs on heating. Another motivation is pointed out in [40], where the quadratic term is due to the renomralization of the density of state caused by thermal fluctuations of the energy differences ΔE of TLSs.

6.4. RELAXATION OF TWO-LEVEL SYSTEMS AT LOW TEMPERATURES

When calculating the energy difference between the TLS states it is convenient to make use of the so-called nondiagonal or local representation. Let the particle be in one of the two wells of the potential curve depicted in Fig. 6.2(d). The wave functions Ψ_1 and Ψ_2 correspond to these states, each of which is localized within the corresponding well. Transitions due to subbarrier tunneling are possible between these two states; hence the total TLS Hamiltonian in matrix form may be written as

$$H = \frac{1}{2}\begin{pmatrix} \epsilon_a & \Delta_0 \\ \Delta_0 & -\epsilon_a \end{pmatrix}, \tag{4.1}$$

where

$$\epsilon_a = \langle \Psi_1 | \mathscr{H} | \Psi_1 \rangle - \langle \Psi_2 | \mathscr{H} | \Psi_2 \rangle,$$

$$\frac{1}{2}\Delta_0 = \langle \Psi_1 | \mathscr{H} | \Psi_2 \rangle,$$

$$\mathscr{H} = -\frac{\hbar}{2m}{}^2 \nabla^2 + V(\mathbf{R});$$

V(R) is the double-well potential (see Fig. 6.2d). This Hamiltonian is diagonalized as

$$H' = T^{-1}HT = \frac{1}{2}\begin{pmatrix} \Delta E & 0 \\ 0 & -\Delta E \end{pmatrix} \tag{4.2}$$

(where $\Delta E = (\epsilon_a^2 + \Delta_0^2)^{1/2}$) by means of the transformation

$$T = \frac{1}{(2\Delta E)^{1/2}}\begin{pmatrix} (\Delta E + \epsilon_a)^{1/2} & (\Delta E - \epsilon_a)^{1/2} \\ -(\Delta E - \epsilon_a)^{1/2} & (\Delta E + \epsilon_a)^{1/2} \end{pmatrix} \tag{4.3}$$

Thus this transformation describes a transition to the diagonal representation, i.e. to the eigenstates of the Hamiltonian (4.1), which may be written as superpositions of the states Ψ_1 and Ψ_2. Here the energy difference ϵ_a, which is often called the asymmetry parameter, is renormalized so that the level-to-level difference becomes equal to ΔE.

Transitions between these diagonal states are also possible in real systems. They are caused by a coupling between the TLSs and the atomic vibrations in a glass, so they are accompanied by phonon emission or absorption. When a phonon wave passes by, the atoms causing the double-well potential curve of the TLS displace. This yields a distortion of the curve and a displacement of the prime energy levels, on one hand, and a variation of the potential-barrier form resulting in a variation of the tunneling transition probability amplitude. These variations may be written as

$$\delta\epsilon_a = B_{ij}\,\zeta_{ij}, \tag{4.4}$$

$$\delta\Delta_0 = D_{ij}\,\zeta_{ij}, \tag{4.5}$$

where B_{ij} and D_{ij} are the deformation-potential tensors, which, generally speaking, depend on the local glass structure at the place where the given TLS is located. The deformation tensor

$$\zeta_{ij} = \frac{1}{2}\left(\frac{\partial u_i}{\partial x_j} + \frac{\partial u_j}{\partial x_i}\right)$$

may be expressed in terms of the displacement vector, which is expanded over the phonon normal coordinates:

$$u = \frac{1}{(2V\rho)^{1/2}} \sum_{\mathbf{k}\lambda} \frac{\mathbf{e}_{\mathbf{k}\lambda}}{\omega_{\mathbf{k}\lambda}} \left\{ b_{\mathbf{k}\lambda} e^{i(\mathbf{kr} - \omega_{\mathbf{k}\lambda} t)} + b^{+}_{\mathbf{k}\lambda} e^{i(\mathbf{kr} - \omega_{\mathbf{k}\lambda} t)} \right\},$$

where $b^{+}_{\mathbf{k}\lambda}$ and $b_{\mathbf{k}\lambda}$ are the creation and annihilation operators of the phonons with wave vector k, polarization vector $\mathbf{e}_{\mathbf{k}\lambda}$ and frequency $\omega_{\mathbf{k}\lambda}$, V is the sample volume, and ρ is the mass density of the glass.

Here we retain only linear terms in the expansions of the quantities ϵ_a and Δ_0 over deformations that were studied by Sussman [41, 42]. The applicability of such an approximation will be discussed in the following sections of this chapter, where it will be shown that it holds only at sufficiently low temperatures.

The transformation (4.3) allows the Hamiltonian of the TLS–phonon interaction to be presented in the form

$$H_{int} = \frac{1}{2}\begin{pmatrix} M_{d,ij} & M_{n,ij} \\ M_{n,ij} & -M_{d,ij} \end{pmatrix} \zeta_{ij}, \tag{4.6}$$

where

$$\left.\begin{aligned} M_{d,\,ij} &= \frac{\epsilon_a}{\Delta E}\,B_{ij} + \frac{\Delta_0}{\Delta E}\,D_{ij}, \\ M_{n,\,ij} &= -\,\frac{\Delta_0}{\Delta E}\,B_{ij} + \frac{\epsilon_a}{\Delta E}\,D_{ij}. \end{aligned}\right\} \tag{4.7}$$

The diagonal matrix element of interaction $M_{d,\,ij}$ corresponds to variation of the TLS energy difference ΔE; the off-diagonal matrix element $M_{u,\,ij}$ corresponds to the transitions between TLS states in the diagonal representation (4.2). This alows determination of the relaxation time of these states in the TLS with energy difference ΔE:

$$\frac{1}{\tau} = \frac{\pi}{\hbar V \rho} \sum_{k\lambda} \frac{(\mathbf{M}_{n\lambda}\mathbf{k})^2}{\omega_{\mathbf{k}\lambda}} (2N_{\mathbf{k}\lambda} + 1)\delta(\Delta E - \hbar\omega_{\mathbf{k}\lambda}), \tag{4.8}$$

where

$$M_{n\lambda,\,i} = \frac{1}{2}\,(M_{n,\,ij} + M_{n,\,ji})\, e_{k\lambda j}.$$

At low temperatures it is sufficient to take account only of the contribution of the acoustic phonons, whose spectrum is described well by the Debye approximation

$$\omega_{\mathbf{k}\lambda} = s_\lambda\, k,$$

where s_λ is the velocity of sound for the polarization λ. Then (4.8) becomes

$$\frac{1}{\tau} = \sum_{\lambda} \frac{M_{n\lambda}^2}{s_\lambda^5}\, \frac{\Delta E^3}{2\pi\rho\hbar^4}\, \coth\left(\frac{\beta\Delta E}{2}\right). \tag{4.9}$$

It describes the relaxation of the TLS with energy difference ΔE, which proceeds via state-to-state transitions and is accompanied by emmission or absorption of a single phonon with energy

$$\hbar\omega_{\mathbf{k}\lambda} = \Delta E.$$

Usually TLSs with $\Delta E \leqslant kT$, for which $\Delta_0 < kT$, contribute to the observed quantities. The deformation potential D at low temperature is therefore small because of the exponential smallness of Δ_0. At the same time, the deformation potential B is large for soft DWPs (see Section 6.2), which seem to contribute mainly to the kinetic effects at low temperatures. For these, the deformation potential is large ($B \geqslant 0.5$ eV), which allows neglect of the second term in the expression (4.7) for the off-diagonal matrix element $M_{n\lambda}$; (4.9) for the inversed relaxation time can then be rewritten in the form

$$\frac{1}{\tau} = \sum_{\lambda} \frac{B_{\lambda}^2}{s_{\lambda}^5} \frac{\Delta E \Delta_0^2}{2\pi\rho\hbar^4} \coth\left(\frac{B\Delta E}{2}\right), \tag{4.10}$$

which is most frequently presented in the literature.

6.5. SOUND ATTENUATION AND THERMAL CONDUCTIVITY

Sound attenuation in glasses as well as the thermal conductivity is determined by TLS–phonon scattering. To describe this it is convenient to use a standard analogy between a TLS and a spin-$\frac{1}{2}$ particle, which also has two states [42]. The TLS Hamiltonian has the form

$$H = H_0 + H_{int}, \tag{5.1}$$

where

$$H_0 = \frac{1}{2} \Delta E \, \sigma_z, \tag{5.2}$$

$$H_{int} = \frac{1}{2} M_d \, \sigma_z \, \zeta + \frac{1}{2} M_n \, \sigma_x \, \zeta; \tag{5.3}$$

σ_i are the Pauli matrices. (Hereinafter, for clarity of presentation, the Cartesian indices of the deformation and deformation-potential tensors are omitted.) This Hamiltonian describes the dynamics of a spin-$\frac{1}{2}$ particle in a constant external magnetic field

$$\mathbf{B}_0 = (0, 0, -\Delta E) \tag{5.4}$$

directed along the z-axis and in a variable magnetic field

$$\mathbf{B}' = (-M_n, 0, -M_d)\, \zeta \tag{5.6}$$

Using this notation the Hamiltonian (5.1) is rewritten in the form

$$H = -(\mathbf{B}_0 + \mathbf{B}')\mathbf{S},$$

which makes the above analogy obvious. Identifying the TLS with a spin-$\frac{1}{2}$ particle in an external magnetic field is consistent only if the tensor structure of the deformation potentials is neglected. In the general case this analogy is rather conditional.

The dynamics of our system is described by the Bloch equation

$$\frac{\partial\langle \mathbf{S}(t)\rangle}{\partial t} = \langle \mathbf{S}(t)\rangle \, [\mathbf{B}_0 + \mathbf{B}'(t)] + \left[\frac{\partial\langle \mathbf{S}(t)\rangle}{\partial t}\right]_{rel}, \tag{5.6}$$

where $\langle \mathbf{S}(t)\rangle$ is the quantum-mechanical average of the spin operator

$$\mathbf{S} = \frac{1}{2}\,\sigma.$$

Two timescales are introduced, longitudinal τ_1 and transverse τ_2, which correspond to the relaxation of the longitudinal

$$\left[\frac{\partial\langle S_z(t)\rangle}{\partial t}\right]_{rel} = -\frac{1}{\tau_1}\left[\langle S_z(t)\rangle - Z_{in}^{0}(t)\right] \tag{5.7}$$

and transverse

$$\left[\frac{\partial\langle S_{x,y}\rangle}{\partial t}\right]_{rel} = -\frac{1}{\tau_2}\,\langle S_{x,y}(t)\rangle \tag{5.8}$$

spin components respectively. The longitudinal component relaxes to the equilibrium value corresponding to the instantaneous value of the z-component of the magnetic field, i.e.

$$Z_{in}^{0}(t) = \frac{1}{2}\tanh\left[\tfrac{1}{2}\beta(B_{0z} + B_z'(t))\right]. \tag{5.9}$$

The longitudinal time scale is relevant to spin-flip, i.e. to transition of the TLS from one state to the other. Therefore it coincides with the relaxation time found in the previous section:

$$\frac{1}{\tau_1} = \frac{1}{\tau}$$

The transverse time scale determines the time during which the phase memory is lost, which occurs owing to the usual relaxation and to dissipationless spin–spin interaction between different TLSs:

$$\frac{1}{\tau_1} = \frac{1}{2\tau} + \frac{1}{\tau_\varphi}.$$

When the usual spin ensembles are studied, the spin–spin interaction is caused by the magnetic dipole interaction. In the present case, the terms "spin" and "magnetic field" are only analogies. It is clear that the TLSs do not give rise to any long-range magnetic field. Nevertheless, an indirect spin–spin interaction due to resonant phonon exchange is possible between two TLSs with close energy differences ΔE. However, the characteristic density of TLSs that are in resonance with each other is too small ($\sim 10^{12}\mathrm{cm}^{-3}$) for this process to contribute significantly to the transverse time scale τ_φ. A larger contribution might be due to nonresonant processes caused by the diagonal interaction $M_d\,\zeta$ in (5.3). It is interesting to note that the energy of such an indirect spin–spin interaction in this case decreases as the third power of the distance between the TLSs [44, 28], thus having the same dependence as the usual dipole–dipole interaction. These issues are of great importance in the analysis of various collective phenomena in glasses but do not play an essential role in the behaviour of the single TLS. That is why we shall not dwell upon them in detail here.

We now turn to the analysis of (5.6). In Cartesian coordinates it is represented as a system of three equations for the projections of the spin S:

$$\left.\begin{aligned} \dot{X} &= -\frac{X}{\tau_2} - \frac{\Delta E + M_d\zeta}{\hbar}\,Y, \\ \dot{Y} &= -\frac{Y}{\tau_2} + \frac{\Delta E\, M_d\zeta}{\hbar}\,X - \frac{M_n\zeta}{\hbar}\,Z, \\ \dot{Z} &= \frac{M_n\zeta}{\hbar}\,Y - \frac{1}{\tau_1}\,(Z - Z^0_{in}). \end{aligned}\right\} \qquad (5.10)$$

Let a monochromatic sound wave be excited in the glass, with the deformation

$$\zeta = \zeta_0 \cos\omega_\lambda t$$

(ω_λ is the sound frequency) corresponding to it.

The solution of (5.10) is sought in an oscillatory form, for example for the x-component

$$X = X_a \cos(\omega t + \varphi),$$

where

$$X_a = \frac{M_n\zeta}{\hbar}\,(Z_0 + \bar{Z})\left[\frac{1}{\omega_0 - \omega_\lambda - i/\tau_2} + \frac{1}{\omega_0 + \omega_\lambda - i/\tau_2}\right], \qquad (5.11)$$

$$\omega_0 = \Delta E/\hbar.$$

$Z_0 = \tanh(\frac{1}{2}\beta\Delta E)$ is obtained from (5.9), neglecting the diagonal interaction $B_z'(t)$:

$$\bar{Z} = -\frac{1}{4\hbar^2}\,\frac{M_n^2\zeta_0\tau_1\tau_2\,Z_0}{1 + (\omega_0 - \omega_\lambda)^2\tau_1\tau_2 + M_n^2\zeta_0\tau_1\tau_2/\hbar^2}\,. \qquad (5.12)$$

Correspondingly, the amplitudes of the transverse vibrations can be found in the form

$$Z_a = -M_d\zeta\,\frac{dZ_0}{d\Delta E}\,\frac{1 + i\omega_\lambda\tau_1}{1 + \omega_\lambda^2\tau_1^2}\,. \qquad (5.13)$$

These formulae allow calculation of the ultrasonic attenuation in glasses. For this is necessary to find a correction to the total internal energy of the glass relevant to the TLS–sound interaction. Owing to the spin analogy used here, this correction has the form of the energy of spins interacting with the external variable magnetic field:

$$\delta U = \Sigma\, \mathbf{B}'\langle \mathbf{S} \rangle. \tag{5.14}$$

The summation sign Σ indicates that the contributions of all TLSs should be taken into account. Consequently the correction to the elastic modulus is

$$\delta C = \frac{1}{\zeta}\frac{\partial \delta U}{\partial \zeta} = -\frac{1}{\zeta}\sum (M_n X + M_d Z), \tag{5.15}$$

which allows calculation of both the renormalization of the velocity of sound

$$\delta s = \frac{1}{2\rho s}\,\mathrm{Re}\,(\delta C(\omega_\lambda)) \tag{5.16}$$

and its attenuation coefficient (i.e. inverse attenuation length)

$$\ell^{-1} = -\frac{\omega_\lambda}{\rho s^3}\,\mathrm{Im}\,(\delta C(\omega_\lambda)). \tag{5.17}$$

The energy (5.14) as well as the correction to the dynamical elastic modulus (5.15) contain two terms proportional to the squared matrix elements of the nondiagonal and diagonal interactions respectively, which are called the resonant and relaxation contributions [43].

First we consider the resonant mechanism due to the nondiagonal interaction M_n. In this case the attenuation coefficient is given by the expression

$$\ell^{-1}(\omega_\lambda) = \ell^{-1}_{res}(\omega_\lambda)\left(1 + \frac{I}{I_c}\right)^{-1/2}, \tag{5.18}$$

where I is the sound intensity, $I_c = 2\hbar^2\rho^2 s^3/M_n^2\tau_1\tau_2$ is the critical intensity,

$$\ell^{-1}_{res}(\omega_\lambda) = \frac{n(0)\, M_n^2\omega_\lambda}{4\rho s^3}\tanh\left(\tfrac{1}{2}\beta\hbar\omega_\lambda\right) \tag{5.19}$$

is the sound attenuation coefficient at low intensities. Such an attenuation is caused by the TLSs whose energy difference is equal to the energy of the sound quantum $\Delta E = \hbar\omega_\lambda$. Here only a single phonon $\hbar\omega_\lambda$ is absorbed or emitted, and a TLS increases or decreases its energy by the value ΔE.

As a rule, experiments are performed at frequencies and temperatures for which the condition $\beta\hbar\omega_\lambda \ll 1$ holds. So the sound attenuation

$$\ell_{\text{res}}^{-1}(\omega_\lambda) \sim \omega_\lambda^2/T \tag{5.20}$$

must increase with a fall in temperature. However, such a dependence is far from being observed in all experiments, since, additionally, the sound intensity must be small ($I \ll I_c$). The critical intensity in glasses is extremely small (e.g., $I_c \sim 10^{-3}$W cm^{-2} at T ~ 1.5 K), and special precautions (such as lowering the sound intensity) must be taken in order to measure in linear regime. Only then is a dependence of the form (5.20) seen.

At large sound intensities ($I \gg I_c$) there is an induced transparency, since the populations of the TLS states that are in resonance with the sound wave are nearly equal and these TLSs cease to interact with the sound. A similar effect is known in nonlinear optics. There, however, it became observable only after powerful laser light sources had been constructed.

Since the sound intensities in the majority of the experiments are large enough ($I \gg I_c$), the resonant mechanism does not contribute noticeably to the attenuation, which is now determined mainly by the relaxation mechanism (the second term in (5.15)). In this case the sound vibrations do not by themselves induce transitions between the TLS states, but only modulate the initial state, which then relaxes by means of the single-phonon processes described in the previous section. It is a double-phonon process in this sense. Therefore there is no direct relation between the energy of the sound quantum and the energy difference ΔE, so the whole TLS spectrum contributes to the attenuation. It is found from (5.15) and (5.17) that

$$\ell_{\text{rel}}^{-1}(\omega_\lambda) = \int d\Delta E\, N(\Delta E)\, \frac{M_d^2}{4\rho s^3}\, \frac{\beta \exp[\beta\Delta E]}{\{1+\exp[\beta\Delta E]\}^2}\, \frac{\omega_\lambda^2\tau(\Delta E)}{1+\omega_\lambda^2\tau^2(\Delta E)}\,. \tag{5.21}$$

Here we have assumed all TLSs to be specified by a single value of the deformation potential. This integral over ΔE cannot be found in the general case, but it is, however, clear that only TLSs with $\Delta E \sim k_B T$ contribute to it. Then at relatively low temperatures, when the condition $\omega_\lambda\tau(k_BT) \gg 1$ holds, the attenuation increases proportionally to the third power of temperature and does not depend on the sound frequency

$$\ell_{\text{res}}^{-1} \sim 1/\tau(k_BT) \sim T^3. \tag{5.22}$$

This growth continues until the opposite inequality $\omega_\lambda \tau)k_B T) \gg 1$ becomes valid. Then the main contribution to the integral (5.21) is due to those TLSs for which the equality $\omega\tau(\Delta E) \approx 1$ holds; hence

$$\ell_{rel}^{-1} \sim \omega_\lambda. \tag{5.23}$$

Similarly, renormalization of the velocity of sound due to TLS–sound interaction can be found. In this case the resonance mechanism is the main contribution at practically all sound intensities. The induced transparency has only a weak influence, since significant overoccupation only takes place for the small proportion of the TLSs for which $\Delta E \approx \hbar\omega_\lambda$, whereas the velocity of sound is renormalized by all TLSs.

Retaining only the nondiagonal processes in (5.15) and (5.16) gives

$$\delta s = \frac{M_n^2}{8\rho s} \int d\Delta E \, n(\Delta E) \left[\frac{1}{\hbar\omega_\lambda - \Delta E} - \frac{1}{\hbar\omega_\lambda + \Delta E} \right] \tanh\left(\frac{\beta \Delta E}{2} \right). \tag{5.24}$$

In the limit $\beta\hbar\omega_\lambda \ll 1$ the integral (5.24) gives

$$\delta s = \frac{M_n^2}{4\rho s} \ell n \left(\frac{k_B T}{\Delta E_{max}} \right) \tag{5.25}$$

(A more exact calculation of this integral is presented in [12]). In the experiment the velocity of sound at a given temperature T is usually compared with the velocity measured at a reference temperature T_0, i.e. the difference

$$s(T) - s(T_0) \sim \ell n(T/T_0) \tag{5.26}$$

is measured. In this form this result was first obtained both experimentally and theoretically in [45].

Therefore the velocity of sound in glass increases with increasing temperature. This can be understood as follows. All the TLSs with $\Delta E > k_B T$ contribute to the correction to the velocity of sound. The TLSs with smaller ΔE values scatter the sound more weakly, since their states are nearly equally populated. Any scattering results in a decrease of the velocity of sound, but when the temperature rises the number of the effectively scattering centers decreases and the velocity correspondingly increases.

We now calculate the thermal conductivity of glass. Here again, the principal contribution is from resonant sound scattering, since the intensities of thermal currents are so low that the condition $I \ll I_c$ always holds very well. That is why the thermal conductivity is determined by the expression

$$\kappa = \sum_{\mathbf{k}\lambda} \frac{s\hbar^2\omega_\lambda^2(\mathbf{k})}{4V\,k_BT^2}\,[\cosh(\tfrac{1}{2}\beta\hbar\omega_\lambda(\mathbf{k}))]^{-2}\,\ell_{res}(\omega_\lambda(\mathbf{k})), \tag{5.27}$$

in which (5.19) should be used when estimating the phonon free path. Carrying out all the necessary integrations, the thermal conductivity is found in the form

$$\kappa = \frac{8\rho\,k_B}{\pi^2 n(0)\hbar^2}\left(\sum_\lambda \frac{s_\lambda}{M_{n\lambda}}\right)^2 T^2. \tag{5.28}$$

The quadratic temperature dependence (5.28) can be interpreted as follows. The principal contribution to the heat transfer is from phonons with energies $\hbar\omega_\lambda(\mathbf{k}) \lessdot k_BT$, whose energy increases with temperature as T^3. At the same time, the thermal phonon free path ($\hbar\omega_\lambda(\mathbf{k}) \approx k_BT$) decreases with increasing temperature as T^{-1}. The desired law $\kappa \sim T^2$ thereby appears.

All the results of this section agree with a large number of experimental data obtained for different glasses (see e.g. [12]). Their applicability is confined to the region of low temperatures and frequencies ($\hbar\omega < k_BT$). This was analysed by Gurevich and Parshin [46, 47], who showed that the perturbation theory used above is valid only at temperatures satisfying the condition

$$T < T_c = \frac{(ms^2\,k_B^3\,T_D^3)^{1/2}}{k_B\,B_\epsilon} \sim 10\text{–}20\ \text{K}. \tag{5.29}$$

For $T \gtrsim T_c$ the width of the TLS energy levels becomes equal to the TLS energy difference

$$\hbar/\tau(k_B\,T_c) \sim k_B\,T_c \tag{5.30}$$

and there is no sense in talking about the TLSs. Gurevich and Parshin also studied relaxation sound absorption at $\hbar\omega_\lambda \gg k_BT$, and found that the dependence (5.22) is replaced by a law $\ell_{rel}^{-1} \sim \omega^3$.

6.6. KINETIC PROCESSES IN GLASSES AT HIGH TEMPERATURES

The theory presented in Sections 6.3–6.5 gives a satisfactory enough qualitative and quantitative explanation of a variety of phenomena observed in glasses at low temperatures (as a rule, below liquid-helium temperature). Apart from the acoustic phenomena mentioned above, the effects that are observed when high-frequency electromagnetic radiation passes through amorphous substances should be pointed out.

Here again the TLSs play an important role. The tunneling particle or the surrounding atoms that form the DWP have a dipole moment. The electromagnetic radiation therefore interacts with the TLS in practically the same way as the sound does. Then all the results of the previous sections may be used if M_d and M_n are interpreted not as the deformation potentials but rather as permanent and induced dipole moments of the TLS, the substitution $\rho s^3 \rightarrow c\,\epsilon^{-1/2}/4\pi$ is made when studying the absorption and $\rho V^2 \rightarrow -8\pi$ when studying the propagation velocity of the electromagnetic wave. (Here c is the velocity of light and ϵ' is the real part of the dielectric constant).

Difficulties arise in trying to expand these theoretical results over the high-temperature region (from liquid-helium to the vitrification temperature). All the quantities studied in the preceeding sections behave completely differently in the high-temperature experiment from the predictions of the TLS theory as presented above. The simplest case is that of the specific heat, where the linear temperature law, $C \sim T$, is replaced by a cubic one, $C \sim T^3$, for $T > 1$–2 K, since the phonon contribution to the specific heat grows with temperature more rapidly than that of the TLSs, and the latter become negligible. However, here a question concerning the magnitude of the coefficient in the cubic law arises that has already been mentioned in Section 6.2. The situation is more complicated for sound attenuation and the renormalization of the velocity of sound, as well as the thermal conductivity. Comparison of the data obtained for crystals and glasses shows that the scattering of phonons by the sample boundaries or the phonon–phonon iteration due to the anharmonicity that is characteristic of crystals give contributions that are two or three orders of magnitude lower than the value of the thermal conductivity actually observed in glasses. On the other hand, the TLS model as presented above cannot adequately explain the observed effects.

These effects are discussed below in more detail using the data presented in [12]. As above, the most important experimentally measured quantities are discussed: thermal conductivity, sound attenuation and renormalization of the velocity of sound. The temperature dependence of the thermal conductivity characteristic of practically all amorphous substances is plotted in Fig. 6.4. A curve of the temperature-dependent thermal conductivity of crystalline dielectric is shown in the same figure for purposes of comparison. At low temperatures there is quadratic growth of the thermal conductivity of glass as discussed above; near 5–10 K a temperature plateau begins, which is up to 20 K wide; after this, the thermal conductivity starts to grow again. It should also be noted that the fall of the thermal

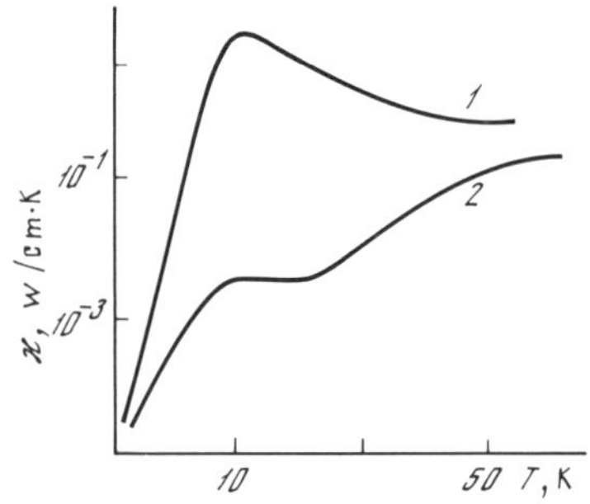

Figure 6.4 Dependence of the thermoconductivity of the crystalline (1) and amorphous (2) solid on temperature.

conductivity in this region is reported in [48, 49]. Such a pattern is completely different from the temperature behaviour of the thermal conductivity of crystalline dielectric, which are characterized by a monotonic decrease in the high-temperature region that is caused by the growth of the intensity of anharmonic processes. At low temperatures the thermal conductivity of crystals determined by the scattering of phonons from the sample boundaries. It should be emphasized again that the thermal conductivity of glasses is two or three orders of magnitude lower than that of crystals.

Figure 6.5 shows the temperature dependence of the inverse length of sound attenuation (i.e. the absorption coefficient). The existence of two maxima at 5–10 K and at 50–100 K is characteristic of practically all amorphous solids, with only the heights and positions of these maxima varying (e.g. amorphous GeO_2 has its second maximum near room temperature). The first maximum behaves in different ways in different glasses. Only a weakly pronounced plateau is observed in such

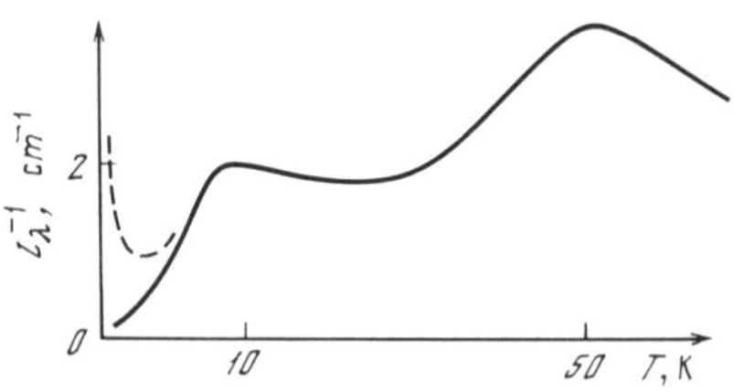

Figure 6.5 Temperature dependence of the inverse length of the ultrasonic attenuation.

Dashed line shows this dependence accounting for the contribution of the resonant TLS-phonon interaction which is not seen in many cases because of the induced transparency effect.

glasses as $Zn(PO_3)_2$ and B_2O_3, whereas in other glasses (SiO_2, BeF_2, GeO_2, etc.) the first maximum and the subsequent minimum are distinctly pronounced.

Finally, the temperature-dependent correction δs to the velocity of sound behaves nonmonotonically. At low temperatures (Fig. 6.6a) δs increases as the logarithm of temperature, goes through a maximum at 5–10 K, and then decreases. This decrease continues (Fig. 6.6b) up to 50–100 K, where δs go through a minimum and then starts to increase. This growth in some glasses (BeF_2, SiO_2) proceeds over rather a long temperature range and is approximately linear in the temperature. It is interesting that the velocity of sound at high-enough temperatures in such glasses may become even larger than its value at liquid-helium temperature.

Studies of high-temperature anomalies in the kinetic-coefficient behaviour had started well before the TLS hypothesis was formulated. To explain these, Anderson and Bömmel [10] proposed a structural-defect (SD) model that assumes that there are local structural defects in glasses that may occupy at least two configurations. Some specific ways of constructing such defects were considered in [10, 50, 51] (see Fig. 6.1). The difference between the SD and TLS models is mainly due to the transition mechanism between two wells of the DWP. The SD model assumes that the transitions occur by means of thermally activated processes.

It seems quite natural to treat both models from a common point of view and to connect two absorption peaks with two types of structural defects having different parameters. Section 6.2 considers a mechanism by which DWPs arise, and the origin of two DWP groups is seen. On the basis of this mechanism, ultrasonic attenuation can be considered as follows.

The ultrasonic attenuation over the whole temperature range is due to a relaxation mechanism

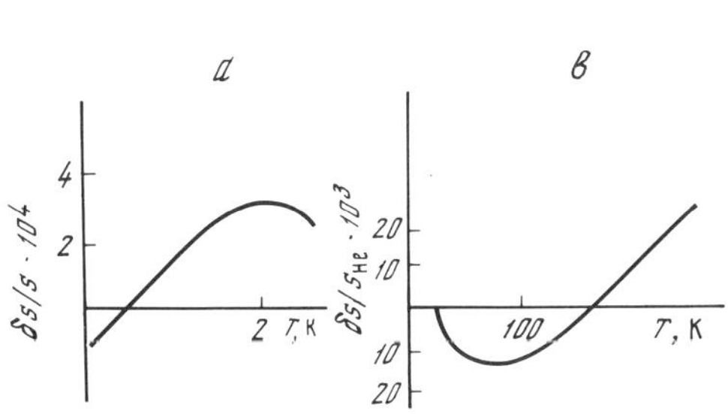

Figure 6.6 Temperature dependence of the correction to the sound velocity caused by its interaction with the TLS's.

$$\ell_{rel}^{-1}(\omega_\lambda) \sim \frac{\tau(T)\omega_\lambda^2}{1 + \tau^2(T)\omega_\lambda^2}, \tag{6.1}$$

where $\tau(T)$ is the relaxation time at a given temperature. The resonant absorption is weak at low temperatures, either because of the induced transparency or because of small frequencies of the sound waves. In addition, the resonant absorption decreases rapidly with increasing temperature, and can almost always be neglected at high enough temperatures. The relaxation absorption (6.1) goes through a maximum at temperatures satisfying the condition

$$\tau(T)\omega_\lambda \sim 1. \tag{6.2}$$

Stronger attenuation is caused by the soft DWPs at low temperatures than by the rigid ones, since the deformation potential for the latter is small. This is why the first (low-temperature) absorption maximum is due to the soft DWPs, whose temperature is defined by

$$\omega_\lambda \tau_1(\Delta E)|_{\Delta E = k_B T} = 1. \tag{6.3}$$

When the temperature rises, $1/\tau_1$ grows rapidly and the contribution of the soft DWPs into the relaxation absorption decreases. At temperatures satisfying (5.26), the damping becomes so strong that there is no sense talking about the TLSs relevant to the soft DWPs.

The further growth of the relaxation absorption is caused by the other group of defects, namely rigid DWPs. If the transitions in these are assumed to be due to an activation mechanism then

$$\tau(T) \approx \tau_0 \exp[V_b/k_B T], \tag{6.4}$$

where V_b is the energy-barrier height and τ_0 is the period of the particle intrawell vibrations. The damping of these DWPs appears to be relatively weak, and the absorption maximum is shifted towards higher temperatures.

A similar explanation holds for the temperature-dependent correction to the velocity of sound. The scattering by soft TLSs (see (5.25)) dominates in the low-temperature region, while at higher temperatures the principal contribution is due to the relaxation scattering by the rigid DWPs:

$$\frac{\delta s}{s} \sim \left\langle \frac{1}{1 + \omega_\lambda^2 \tau^2(T)} \right\rangle, \tag{6.5}$$

where $\langle \ldots \rangle$ represents averaging over the potential-barrier heights.

In spite of the alleged harmony of the above scheme there remain a number of unclear issues. First, it provides no explanation of the temperature-dependent thermal conductivity or of the temperature plateau. Secondly, detailed comparison with experimental data results in rather low values of the potential-barrier height, $V_b \sim 0.05$ eV [12]. The question then arises as to whether, at such a low barrier height, the activation mechanism may be assumed to be stronger than the tunneling one. As will be shown below, tunneling in this temperature range might be more effective, even for essentially higher barriers. Furthermore, the preexponential factor in (6.4) appears in many cases (e.g. in metallic glasses [52–55]†) to be many orders of magnitude different from the characteristic vibrational timescale in solids, $\tau_0 \sim 10^{-13}s^{-1}$.

Fleurov and Trakhtenberg [21–25] carried out a detailed analysis of relaxation processes in rigid DWPs, and their interaction with the sound waves, when the barrier fluctuations are of great importance. The results of this analysis, which might explain some of the above problems, are presented in the remaining sections of this Chapter.

6.7. FLUCTUATIONAL PREPARATION OF THE BARRIER AND TWO-LEVEL SYSTEMS

The theory of TLS–phonon interaction presented in Sections 6.3–6.5 assumes the first terms of the expansion of the asymmetry parameter ϵ_a and the probability amplitude of the tunneling transition $\Delta/\hbar$ over small deformations ζ to be sufficient. Thus in this sense the theory is single-phonon. As far as the asymmetry parameter is concerned, this approximation seems to be well grounded, since no mechanism has been pointed out by which the relatively long-wave phonon vibrations can strongly alter the ϵ_a value. However, the case is different for the probability amplitude of the tunneling transitions, which is

$$\Delta/h = \nu \exp[-\tfrac{1}{2}J],$$

where the transparency parameter J of the potential barrier depends on its width and height. When a heavy particle tunnels the parameter J is large multiples of ten, and even relatively small variations of the form of the barrier, and hence of the value of J, may result in extremely large (several orders of magnitude) variations of the $\Delta/\hbar$ amplitude.

---

† A large number of amorphous structures have been synthesised that have a metallic conductivity. These are called metallic glasses. If effects due to the conduction electrons are disregarded, then many properties of these substances are quite similar to those of dielectric glasses.

This is why attention cannot be restricted to the first terms of the expansion in ζ, and the single-phonon approximation might be inapplicable. The multiquantum character of the TLS–phonon interaction is due to barrier fluctuations [56] (see also [57, 58]). Vibrations of the glass atoms cause fluctuations of the potential-barrier shape, and hence of the transparency parameter J. The probability amplitude of the tunneling transition depends exponentially on J, and so the tunneling particle finds it more favourable not to tunnel through the equilibrium barrier but to wait for a large enough fluctuation of its shape that allows easier tunneling by virtue of a decrease in J.

It can be seen from the considerations in Chapter 3 on the rate constant of solid-phase chemical reactions [58, 59] that there is a deep analogy between the calculations there and the approach to be used in constructing a multiphonon theory of kinetic processes in glasses. A double adiabatic approximation allowing for a separation of the intra- and intermolecular degrees of freedom was applied in Chapter 3, in which the influence of the spatial arrangement of the reactants on the chemical reaction rate in solids was discussed. This approximation is based on the fact that intramolecular bonds are markedly stronger than intermolecular ones.

The problem naturally arises as to whether the double adiabatic approximation applies in glasses, since there is no obvious separation of intra- and intermolecular degrees of freedom. This issue was discussed by Fleurov and Trakhtenberg [25], who came to the following conclusions. The double adiabatic approximation holds if the time for particle subbarrier transversal is smaller than the characteristic time of the barrier-shape variations. The transparency parameter J is the mechanical action in terms of $\hbar$ units, so the velocity of the particle subbarrier motion is

$$v \sim \hbar\, J/m\Delta d$$

(Δd is the barrier width), and the subbarrier transversal time is

$$\tau_b \sim m\, \Delta d^2/\hbar J.$$

This means that the double adiabatic approximation holds when the condition $\tau_b\Omega \ll 1$ (Ω is the characteristic frequency of barrier fluctuations) is valid; this may be represented as

$$\frac{m}{M}\,\frac{\Delta d}{u_0}\,\mathscr{R}^{-1/2} \ll 1. \tag{7.1}$$

$u_0^2 = \hbar/M\Omega$ is the amplitude of the zero-point vibrations of an oscillator with frequency Ω and reduced mass M; and

$$\mathscr{R} = \frac{\hbar\,(J')^2}{M\Omega}, \quad J' = \partial J/\partial \Delta d.$$

There are the following relations between the quantities in (7.1): the barrier width is most probably smaller than the interatomic spacing, $\Delta d < a$, but larger than the zero-point vibration amplitude, $\Delta d > u_0$; the values of m and M are close to each other, since in both cases atomic masses are involved. Hence the most essential parameter appears to be $\mathscr{R}$, which specifies the strength of the TLS–phonon coupling.

The value of $\mathscr{R}$ depends strongly on the frequency Ω of the barrier-shape fluctuations. There are a number of characteristic frequencies in a glass, just as in a crystal, which correspond to longitudinal and transverse acoustic phonons, optical phonons and different local vibration modes. The question naturally arises as to which of these frequencies should be substituted into the inequality (7.1), since it holds better for u soft low-frequency vibrational modes and may be violated for rigid high-frequency modes. The TLSs turn out to select themselves the soft modes.

If the frequency of the barrier fluctuations is very high, so that the subbarrier transversal time greatly exceeds their period $\tau_b\Omega \gg 1$, then the transparency parameter should be averaged over the fluctuations period [60], i.e.

$$\Delta/\hbar \sim \nu \exp[-\langle J\rangle];$$

in this case the fluctuations of the barrier and the associated effects are of minor importance. If, however, the barrier fluctuates slowly, i.e. $\tau_b\Omega \ll 1$, and the particle tunnels through a practically static barrier then it is the probability amplitude $\Delta/\hbar$ of the tunneling transition that should be averaged rather than the transparency parameter J. The fluctuations of the barrier then ensure strong TLS–phonon interaction; the softest, lowest-frequency modes for which the parameter $\mathscr{R}$ is especially large interacting with the TLSs most strongly. That is why we shall choose the lowest-frequency modes while making evaluations.

Estimates are now presented for the model system considered in Section 6.2, i.e. we assume that the oxygen atom tunnels and the fluctuations of the potential-barrier shape are due to the silicon-atom vibrations. For rigid DWPs the following parameters are assumed: the barrier height $V_b = 0.5$ eV, its width $\Delta d = 0.5$ Å and the vibration frequency $\hbar\Omega/k_B = 50$ K; then $J \approx 50$. Keeping in mind that $J' \approx$

$J/\Delta d$, we find $\mathscr{R} \approx 500$. It is because of such a large value of κ that the inequality (7.1) holds well, and hence the double adiabatic approximation may be applied in analysing the interaction of a phonon with the group of rigid TLSs.

The barrier heights in the soft DWPs may be a hundred times lower than in the rigid ones ($V_b \sim 0.005$ eV [19, 20]) while the width does not decrease so dramatically. As a result, the parameter $\mathscr{R}$ is close to unity, and the inequality (7.2) might not hold well. Therefore the two groups of TLSs interact with phonons in completely different fashions. The single-phonon theory as presented in Sections 6.3–6.5 is true for the soft TLSs. They seem to dominate at low temperatures owing to the large value of the deformation potential B. However, at temperatures of 10–30 K (see (5.26)) their contribution may be neglected because of strong damping, and attention should be turned to the group of rigid TLSs. The single-phonon approximation no longer holds for them, so we proceed to a multiphonon theory as presented in [21–25].

An individual TLS interacting with the vibrations of the medium is considered. It can be described by means of the Hamiltonian

$$H = E_1(\{q_\mu\})\, a_1^+ a_1 + E_2(\{q_\mu\})\, a_2^+ a_2 + \frac{1}{2}\Delta(\{q_\mu\})(a_2^+ a_1 + a_1^+ a_2) + H_{ph}(\{q_\mu\}). \tag{7.2}$$

Here $H_{ph}(\{q_\mu\})$ is the Hamiltonian of the phonon subsystem, q_μ are dimensionless coordinates of the phonon vibrations, a_i^+ and a_i are the quasifermion operators of creation and annihilation of the TLS states (e.g. [60]). If an expansion in terms of the deformation is carried out in (7.2) and only linear terms are retained then a Hamiltonian equivalent to the Hamiltonian (4.1)–(4.3) arises.

Since the tunneling particle is necessarily either in state 1 or state 2, the operators a_i^+ and a_i satisfy the equality

$$a_1^+ a_1 + a_2^+ a_2 = 1$$

as well as the usual commutation relations. This allows the Hamiltonian (7.2) to be transferred as

$$H = \frac{1}{2}[E_1(\{q_\mu\})) + E_2(\}q_\mu\})] - \frac{1}{2}\epsilon_a\, a_1^+ a_1 + \frac{1}{2}\epsilon_a\, a_2^+ a_2 + \sum_\mu g_\mu(b_\mu^+ + b_\mu)(a_2^+ a_2 - a_1^+ a_1) + \frac{1}{2}\Delta(\{q_\mu\})(a_2^+ a_1 + a_1^+ a_2) + H_{ph}(\{q_\mu\}), \tag{7.3}$$

where b_μ^+ and b_μ are the phonon creation and annihilation operators. Here from the start only the linear terms in the expansion of the function $\frac{1}{2}[E_1(\{q_\mu\}) - E_2(\{q_\mu\}]$ are retained.

$$g_\mu = iB_\epsilon \left(\frac{\hbar}{2\rho\omega_\mu}\right)^{1/2} \mathbf{e}_\mu \mathbf{k}_\mu, \tag{7.4}$$

where $\mathbf{e}_\mu$, $\mathbf{k}_\mu$, ω_μ are the polarization vector, wave vector and frequency of the phonon mode μ. The first term of the Hamiltonian (7.3) may result only in inessential renormalizations of the phonon spectrum, which would not be taken account of below.

Analysis of the nondiagonal processes of the TLS–phonon interaction, which are accompanied by a well-to-well particle transition, should be carried out with due account taken of the multiphonon effects of the barrier fluctuations. First, coherent tunneling well-to-well transitions in the DWPs are studied. Here the initial and final states of the phonon subsystem coincide, and so the phase is conserved in this transition. The probability amplitude of this process is

$$\Delta_i(T) = \frac{\mathrm{Tr}[\rho_{ph}(\{q_\mu\})\Delta(\{q_\mu\})]}{\mathrm{Tr}[\rho_{ph}(\{q_\mu\})]}. \tag{7.5}$$

This is a simpler expression than that appearing in the calculation of the rate constant of chemical reactions in Chapter 3. In contrast with (5.7) of Chapter 3, only one phonon density matrix $\rho_{ph}(\{q_\mu\})$ is present in (7.5), and, moreover, only its diagonal elements appear. There is also no need now to take account of energy conservation. Therefore the calculation of the $\Delta_1(T)$ amplitude by means of (7.5) is much simpler than the calculation of the tunneling reaction rate constant. Here the polaron effect (environmental rearrangement) is also neglected, i.e. the arrangement of atoms that leads to the DWP potential curve does not change in the course of the well-to-well transition. We have

$$\begin{aligned}\Delta_1(T) &= \Delta_0 \prod_\mu \int dq_\mu[\pi \coth(\tfrac{1}{2}\beta\hbar\omega_\mu/2)]^{-1/2} \\ &\times \exp\left[-q_\mu^2 \tanh(\tfrac{1}{2}\beta\hbar\omega_\mu) - \frac{1}{2}\sum_\mu A_\mu q_\mu - \frac{1}{4}\sum_{\mu\mu'} \tilde{B}_{\mu\mu'} q_\mu q_{\mu'}\right] \\ &= [\det\{\Phi_{1\mu\mu'}\}]^{-1/2}\, \hbar\nu \exp\left[-\frac{1}{2} J^*(T)\right],\end{aligned} \tag{7.6}$$

where

$$\Phi_{1\mu\mu'} = \left[\frac{1}{4}\tilde{B}_{\mu\mu'} + \tanh(\tfrac{1}{2}\beta\hbar\omega_\mu)\delta_{\mu\mu'}\right]\coth(\tfrac{1}{2}\beta\hbar\omega_\mu),$$

$$J_1(T) = \frac{1}{4}\sum_{\mu\mu'} A_\mu \left[\frac{1}{2} B_{\mu\mu'} + 2\tanh(\tfrac{1}{2}\beta\hbar\omega_\mu)\delta_{\mu\mu'}\right]^{-1} A_{\mu'},$$

$$\Delta_0 = \hbar\nu \exp\left[-\frac{1}{2} J_0\right], \quad J^*(T) = J_0 - J_1(T).$$

Thus, with the fluctuations of the potential-barrier width taken into account, the probability amplitude Δ_0 of the tunneling transition that is present in all the formulae of the single-phonon theory should now be replaced by the renormalized amplitude $\Delta_1(T)$.

Such a renormalization changes the preexponential factor only slightly, since the quantity $\det\{\phi_{1\mu\mu'}\}$ is of the order of unity (under the approximation $\tilde{B}_{\mu\mu'} = 0$ it is exactly unity). The most important variation of the probability amplitude of the tunneling transition is due to a large variation of the potential-barrier transparency parameter. To illustrate this, some estimates are now given. It is assumed as before that an oxygen atom tunnels through a barrier of height $V_b \sim 0.5$ eV and width 0.5 Å when $J_0 \approx 50$. It seems at first sight that there is no sense in considering TLSs with such parameters, since tunneling in them proceeds extremely slowly and cannot influence kinetic processes in the glass. However, this is not actually the case, and when account is taken of the fluctuations of the barrier shape it is possible to obtain quite satisfactory transition frequencies.

The fluctuations of the DWPs are assumed to be caused by an Einstein mode. Then use may be made of the results of Section 3.8:

$$\left.\begin{aligned} J^*(T) &= J_0 - \mathscr{R}/2[\mathscr{R}_1 + 2\tanh(\tfrac{1}{2}\beta\hbar\Omega)], \\ \det\{\Phi_{1\mu\mu'}\} &= 1 + \tfrac{1}{2}\mathscr{R}_1 \coth(\tfrac{1}{2}\beta\hbar\Omega), \\ \mathscr{R} &= \frac{\hbar\,(J')^2}{M\Omega}, \quad \mathscr{R}_1 = \frac{\hbar\,J''}{M\Omega}. \end{aligned}\right\} \tag{7.7}$$

If the reduced mass M of the Einstein oscillator is close to the mass of the SiO_2 molecule and the frequency $\hbar\Omega/k_B \approx 50$ K then $\mathscr{R} = 500$ and $\mathscr{R}_1 = 5$. So $J^* \approx 12$ at zero temperature, i.e. the probability amplitude of the tunneling transitions increases over many orders of magnitude owing to the fluctuations of the potential-barrier shape.

Certainly such estimates are rather approximate, since we know nearly nothing about the principal parameters of the problem (i.e. the tunneling particle mass, the mass and the frequency of the Einstein oscillator, and the parameters of the barrier) and are able to make judgements as for their values only on the basis of general physical considerations. Nevertheless, two important conclusions can be made. First, the mechanism discussed is extremely powerful. The value of the barrier transparency parameter decreases by a factor of more than 30 at low temperatures, and by even more at higher temperatures, since $J^*(T)$ tends to zero with increasing temperature. As a result, the parameter $\Delta_1(T)$, which is exponentially small at low temperatures, increases and at higher temperatures gives the main contribution to the TLS–phonon interaction (i.e. the deformation potential D in (4.5) increases). Secondly, if the barrier fluctuations are disregarded, it can appear that only TLSs with anomalously low potential barriers ($V_b < 0.1$ eV) are of importance in the kinetics of glasses. As we have just seen, this is not the case (or at least not always the case), and, in fact, the barrier height may be significantly larger which agrees better with our intuitive view of the characteristic energies in solids.

The above coherent tunneling transitions result in the formation of new superposition states, and the transitions between these occur dissipatively and are accompanied by phonon emission or absorption. Now a "purely nondissipative part" of the Hamiltonian (7.3) is distinguished. For that the term

$$\frac{1}{2}\,\Delta_1(T)(a_2^+a_1 + a_1^+a_2) \tag{7.8}$$

is added and substracted in (7.3), and the part of the Hamiltonian that is independent of the phonon coordinates is considered:

$$H_1 = -\frac{1}{2}\,\epsilon_a\,a_1^+a_1 + \frac{1}{2}\,\epsilon_a\,a_2^+a_2 + \frac{1}{2}\,\Delta_1(T)(a_2^+a_1 + a_1^+a_2). \tag{7.9}$$

The Hamiltonian (7.9) is of the same structure as (4.1), except that the renormalized amplitude $\Delta_1(T)$ that takes account of the barrier fluctuations appears here instead of the amplitude Δ_0. It can also be also diagonalized,

$$H_0' = \frac{1}{2}\,\Delta E\,a_1^+a_1 + \frac{1}{2}\,\Delta E\,a_2^+a_2$$

where $\Delta E = [\epsilon_a^2 + \Delta_1^2(T)]^{1/2}$, by means of the transformation (4.3).

The remaining part of the Hamiltonian (7.3),

$$H_{int} = \sum_\mu g_\mu(b_\mu^+ + b_\mu)(a_2^+a_2 - a_1^+a_1) + \frac{1}{2}\,[\Delta(\{q_\mu\}) - \Delta_1(T)]\,(a_2^+a_1 + a_1^+a_2), \tag{7.10}$$

is responsible for the TLS–phonon interaction. The matrix (4.3) transforms it into

$$H'_{int} = M_d(\{q_\mu\})(a_2^+a_2 - a_1^+a_1) + M_n(\{q_\mu\})(a_1^+a_2 + a_2^+a_1), \quad (7.11)$$

where

$$M_d(\{q_\mu\}) = \sum_\mu g_\mu(b_\mu^+ + b_\mu)\frac{\epsilon_a}{\Delta E} + \frac{\Delta_1(T)}{\Delta E}[\Delta(\{q_\mu\}) - \Delta_1(T)], \quad (7.12)$$

$$M_n(\{q_\mu\}) = -\sum_\mu g_\mu(b_\mu^+ + b_\mu)\frac{\Delta_1(T)}{\Delta E} + \frac{\epsilon_a}{\Delta E}[\Delta(\{q_\mu\}) - \Delta_1(T)]. \quad (7.13)$$

These formulae (7.12) and (7.13) for the operators of the TLS–phonon interaction have structures that are similar to that of (4.7), and become identical to it in the single-phonon approximation. It can easily be verified that the operators (7.12) and (7.13) give no contribution to the coherent transition (7.5) and are responsible for the transitions accompanied by phonon emission and absorption.

6.8. RELAXATION OF TLS IN THE MULTIPHONON REGIME

TLS relaxation in the single-phonon regime occurs owing to transitions between two states of the system, with the energy conservation law holding in the form

$$\hbar\omega_\mu = \Delta E, \quad (8.1)$$

where ω_μ is the frequency of the emitted or absorbed phonon. These transitions are due to the operator M_n of the nondiagonal interactions. The diagonal interaction M_d does not contribute to single-phonon relaxation because of the energy-conservation law. In this case the TLS does not change its state after the transition, and the right-hand side of (8.1) should be zero.†

A more complicated situation arises in the multiphonon regime, when a large number of phonons are simultaneously emitted or absorbed, and the energy conservation law acquires the form

$$\sum_\mu (\pm \hbar\omega_\mu) = \begin{cases} \pm \Delta E & \text{for nondiagonal processes} \\ 0 & \text{for diagonal processes} \end{cases} \quad (8.2)$$

where the summation on the left hand side is performed over all

† Hereinafter the finite damping of the levels is neglected. The influence of finite damping on kinetic processes in glasses is discussed in [25].

phonons participating in the given elementary process. The signs "+" or "−" apply on the left-hand side to phonon emission or absorption and on the right-hand side to increase or decrease of the TLS energy. It can easily be seen that the energy conservation law no longer forbids the diagonal processes. In this case the energy of all emitted phonons should merely be equal to that of all absorbed phonons.

For nondiagonal processes the relaxation mechanism consists in the TLS leaving the given state during a finite time or, on the contrary, arriving at it. Its energy then changes by ΔE, as does the phonon-bath energy. For diagonal processes the TLS remains in the same state and there is no energy transfer into the bath (or out of it). There is only emission and absorption of phonons whose total energy is zero. Then the bath state changes without a change in its total energy, i.e. only its phase varies, and hence the TLS phase varies too.

We now turn to the calculation of the relaxation time. The rate constant of the processes discussed above can be written as (cf. (5.7) in Chapter 3):

$$\left.\begin{aligned} K_{\alpha\beta} &= \frac{2\pi}{\hbar}\sum_{i,f} |\langle\{\nu_i\}|M_{\alpha\beta}|\{\nu_f\}\rangle|^2 \frac{\exp[-\beta E_i]}{\sum_i \exp[-\beta E_i]}\,\delta(E_{\alpha\beta} + E_{phi} - E_{phf}), \\ E_i &= E_\alpha + E_{phi}. \end{aligned}\right\} \tag{8.3}$$

Here $E_{\alpha\beta} = E_\alpha - E_\beta$ is the TLS energy change; E_{phi} and E_{phf} are the initial and final phonon energies;

$$M_{\alpha\beta} = M_d\sigma_z + M_n\sigma_x \tag{8.4}$$

The form of the operators M_n and M_d (see (7.12) and (7.13)) makes it clear that the calculations can be carried out in the following manner. First, $M_{\alpha\beta}$ is substituted into (8.3) for the function $\Delta(\{q_\mu\})$ and all necessary manipulations are performed; the result is then multiplied by $(\Delta_1(T)/\Delta E)^2$ for diagonal processes and by $(\epsilon_a/\Delta E)^2$ for nondiagonal processes. There is also the operator of the prime diagonal transitions $\sum_\mu g_\mu(b_\mu^+ + b_\mu)$ in (7.12) and (7.13), but it cannot affect the multiphonon transitions (besides, it has been mentioned above that the deformation potential B is small for the rigid DWPs considered here).

All the necessary calculations have been carried out in Chapter 3; use can now be made of these results in order to apply them to the present system. Substituting $\Delta(\{q_\mu\})$ in (8.3) for $M_{\alpha\beta}$ gives

$$\left.\begin{aligned} K^{(0)}(\Delta E) &= \frac{\Delta_0^2}{\hbar^2}\int_{-\infty}^{\infty} dt\{\det\Phi\}^{-1/2}\exp\left[\pm i\Delta E\left(\frac{t}{\hbar}-\frac{i\beta}{2}\right)+\mathscr{F}(T,t)\right], \\ \mathscr{F}(T,t) &= \frac{1}{4}\sum_{\mu\mu'} A_\mu (B^{-1})_{\mu\mu'} A_{\mu'}. \end{aligned}\right\} \tag{8.5}$$

Here, as before, the polaron effect (change in the medium) is neglected and $\Delta q_\mu = 0$.

An expansion similar to that used in Section 3.8 is applied in order to write the function $\mathscr{F}(T, t)$ as

$$\mathscr{F}(T,t) = J_1(T,0) + \delta J(T,t), \tag{8.6}$$

where

$$J_1(T,0) = \frac{1}{4}\sum_{\mu\mu'} A_\mu \left[\frac{1}{2}\frac{\partial^2 J}{\partial q_\mu \partial q_{\mu'}} + 2\delta_{\mu\mu'}\tanh\left(\frac{\beta\hbar\omega_\mu}{2}\right)\right]^{-1} A_{\mu'},$$

$$\begin{aligned} \delta J(T,t) = \frac{1}{2}\sum_{\mu\mu'\mu''\mu'''} & A_\mu \left[\frac{1}{2}\frac{\partial^2 J}{\partial q_\mu \partial q_{\mu'}} + 2\delta_{\mu\mu'}\tanh\left(\frac{\beta\hbar\omega_\mu}{2}\right)\right]^{-1} \\ & \times \frac{\sinh(\frac{1}{4}\beta\hbar\omega_{\mu'})}{\cosh^3(\frac{1}{4}\beta\hbar\omega_{\mu'})}\delta_{\mu'\mu''}\left[\frac{1}{2}\frac{\partial^2 J}{\partial q_{\mu''}\partial q_{\mu'''}} + 2\delta_{\mu''\mu'''}\tanh\left(\frac{\beta\hbar\omega_{\mu''}}{2}\right)\right]^{-1} \\ & \times A_{\mu'''}(1-\cos\omega_{\mu'}t). \end{aligned}$$

The expression for the function $\delta J(T, t)$ contains the factor $[\cosh(\frac{1}{4}\beta\hbar\omega_\mu)]^{-3}$, as a result of which the principal contribution at low temperatures is due to the long wave acoustic phonons, to which small deformations, and hence small values of

$$\frac{\partial^2 J}{\partial q_\mu \partial q_{\mu'}} \sim \mathscr{R}_1\left(\frac{\pi k_B T}{\hbar\omega_D}\right)^2,$$

correspond (see (7.1) of Chapter 3). Then for

$$k_B T < \hbar\omega_D/2\pi\,\mathscr{R}_1^{1/2} \tag{8.7}$$

the second derivatives $\partial^2 J/\partial q_\mu\, \partial q_{\mu'}$ in the expression for $\delta J(T, t)$ may be neglected:

$$\delta J(T,t) = -\frac{1}{4}\sum_\mu |A_\mu|^2 \frac{1-\cos\omega_\mu t}{\sinh(\frac{1}{2}\beta\hbar\omega_\mu)} \tag{8.8}$$

All the vibrational modes, both long-wave and short-wave, contribute to the function $J_1(T, 0)$ at low temperatures, the principal role being played by the zero-point vibrations. Here the quantity $\partial^2 J/\partial q_\mu \partial q_{\mu'}$ may not be neglected, and this function in the Einstein model is

$$J_1(T, 0) = \frac{\mathscr{R}}{2\mathscr{R}_1 + 4\tanh(\frac{1}{4}\beta\hbar\Omega)} . \tag{8.9}$$

Therefore (8.5) can be rewritten in the form

$$K^{(0)}(\Delta E) = \frac{\Delta_1^2(T)}{\hbar^2} \int_{-\infty}^{\infty} dt \exp\left[\pm i\Delta E\left(\frac{t}{\hbar} - \frac{i\beta}{2}\right) + \delta J(T, t)\right] . \tag{8.10}$$

Here significant multiphonon contribution has been separated that yields the renormalization of the prime probability amplitude Δ_0 of the tunneling transitions. The function det $\{\Phi\}$ is taken out of the integral at $t = 0$.

We now examine the part of the multiphonon processes described by the function $\delta J(T,t)$. Here we deal not with the renormalizations but rather with real dissipative processes of phonon emission or absorption. In the low-temperature region (8.7) use can be made of (8.8). Two situations are possible. If $\delta J(T, t)$ is small compared with unity then the exponential function (8.1) may be expanded in $\delta J(T, t)$ and the first non-vanishing terms retained. Then, applying the Debye approximation in calculating the sum (8.8), it is found that such an expansion is valid at temperatures satisfying the inequality

$$T < \frac{\hbar\omega_D}{4}\left[\frac{\pi^2 \mathscr{R}}{4}\right]^{-1/4} \equiv T_s \lesssim 10 \text{ K}. \tag{8.11}$$

Hereinafter a TLS with the same parameters as in Section 6.7 is implied.

Retaining in (8.10) only terms linear in $\delta J(T, t)$, i.e. those proportional to $\cos(\omega_\mu t)$, the integration over t is carried out. A delta function of the type

$$\delta(\Delta E \pm \hbar\omega_\mu)$$

appears, which reflects energy conservation in a single-phonon process. The inverse relaxation time

$$\frac{1}{\tau} = K_{21} + K_{12} \tag{8.12}$$

in the single-phonon regime calculated in this way coincides naturally with the result (4.9). The contribution of the single-phonon processes due to the diagonal interaction in (7.13) should also be taken into account. (The only difference is now that (4.9) contains the renormalized value $\Delta_1(T)$ instead of Δ_0. However, this is of no importance, since at low temperatures $\Delta_1(T)$ depends only weakly on T. Thus the two groups of TLSs connected with soft and rigid DWPs relax at low temperatures by means of single-phonon processes. However, the two deformation potentials B and D are small for rigid TLSs in this temperature range, whereas the deformation potential B is large for soft TLSs, which is why the quantity $\frac{1}{\tau}$ is large for these.

We now go to the higher-temperature region, $T > T_s$. The sign of the inequality (8.7) changes at these temperatures; therefore, the quantities $\partial^2 J/\partial q_\mu \, \partial q_{\mu'}$ can no longer be neglected in (8.6). The results of Section 3.8 are applied and yield, in the Einstein model,

$$K^{(0)}(\pm \Delta E) = \frac{1}{\tau^{(0)}} \exp[\pm \beta \Delta E], \tag{8.13}$$

where

$$\left.\begin{aligned} \frac{1}{\tau^{(0)}} &= \frac{\nu^2 \tau_p}{\varphi_3^{1/2}} \exp[-J_0 + \varphi_1], \\ \frac{1}{\tau_p^2} &= \frac{\partial^2}{\partial t^2} \delta J(T, t) \Big|_{t=0} = \Omega^2 \varphi_1(T) \varphi_2(T). \end{aligned}\right\} \tag{8.14}$$

The diagonal and nondiagonal contributions are then summed and the relaxation time is found:

$$\frac{1}{\tau} = \frac{1}{\tau^{(0)}} \left\{ \cosh\left(\frac{\beta \Delta E}{2}\right) - \frac{\Delta_1^2(T)}{\Delta E^2} \left[\cosh\left(\frac{\beta \Delta E}{2}\right) - 1 \right] \right\}. \tag{8.15}$$

The result (8.14) follows from (8.10) (see also (8.4) in Chapter 3) if the function $\delta J(T, t)$ in the exponent is expanded in t near the saddle point $t \approx 0$. This expansion holds if the function $\delta J(T, t)$ is large enough and determines the convergence of the integral in a narrow enough region of t. For TLSs with $\Delta E \leqslant k_B T$ this condition holds if

$$\tau_p \omega_T \ll 1, \tag{8.16}$$

where $\hbar\omega_T = k_B T$ is the characteristic frequency of the thermal phonon. Substituting τ_p (see (8.14)) into (8.16) it is found that this inequality is valid at temperatures higher than a temperature T_m defined by the equation

$$\frac{\hbar\Omega}{kT_m} - 2 \ell n \left\{ \frac{2\mathscr{R}}{(\mathscr{R}_1 + 2)^2} \left(\frac{\hbar\Omega}{k_B T_m} \right)^2 \right\}.$$

Assuming, as before, $\mathscr{R} = 500$ and $\mathscr{R}_1 = 5$, gives $k_B T_m/\hbar\Omega \approx \frac{1}{7}$.

At temperatures $T \gg T_m$ the relaxation of the rigid TLSs occurs by means of the emission of a packet that is a superposition of a large number of phonons. The duration of this packet emission and hence the packet's own duration, is of the order of τ_p, i.e. of the size of the region of t where the integral (8.10) converges.

The temperature dependence of the inverse relaxation time (8.15) is determined mainly by the exponential factor in (8.14). The exponent in this function can be written in the form

$$
\begin{aligned}
-J_0 + \varphi_1(T) &= -\frac{\mathscr{R}\tanh(\frac{1}{4}\beta\hbar\Omega)}{\mathscr{R}_1[\mathscr{R}_1 + 2\tanh(\frac{1}{4}\beta\hbar\Omega)]} \\
&\approx -\frac{\mathscr{R}\hbar\Omega}{4\mathscr{R}_1(\mathscr{R}_1 + \hbar\Omega/2k_B T)}\,\frac{1}{k_B T}\,. \qquad (8.17)
\end{aligned}
$$

The last equality in (8.17) holds at temperatures $k_B T > \frac{1}{4}\hbar\Omega$. In deriving (8.17) the relation $2\mathscr{R}_1 J_0 \approx \mathscr{R}$ has been used. This equality, which follows automatically for the parabolic barrier should generally be considered as an additional condition that allows (8.17) to become zero as $T \to \infty$. In this limit the temperature dependence of the inverse relaxation time is purely activation-like, the part of the activation energy being played not by the potential-barrier height, which is not explicitly present in (8.17), but by the quantity

$$
E_a = \mathscr{R}\hbar\Omega/4\mathscr{R}_1^2. \qquad (8.18)
$$

For our usual parameters, $E_a/k_B \approx 250$ K, whereas the barrier height $V_b/k_B \approx 5000$ K.

For $k_B T \ll \frac{1}{4}\hbar\Omega$ (8.17) tends to a limit $-\mathscr{R}/\mathscr{R}_1(\mathscr{R}_1 + 4)$ similarly to the low-temperature plateau in the chemical reactions that have been studied in the preceeding Chapters. Of special interest is an intermediate temperature region $k_B T \sim \hbar\Omega/2R_1$ in which deviations from the activation-like dependence might be observed. At $\mathscr{R}_1 \approx 5$, as chosen above, this region is very narrow, and the transition from the activation-like dependence to the low-temperature limit occurs rapidly. However, in glasses with lower $\mathscr{R}_1$ values it is possible to observe an activation-like law with a temperature-dependent activation energy. If the temperature-dependent relaxation time is plotted in Arrhenius coordinates as a function of temperature then we see a type of curve C depicted in Fig. 6.7 instead of the usual straight line A.†

† These types of curves are observed, for example in relaxation experiments in the metallic glass $Fe_{40}Ni_{40}P_{14}B_6$[62].

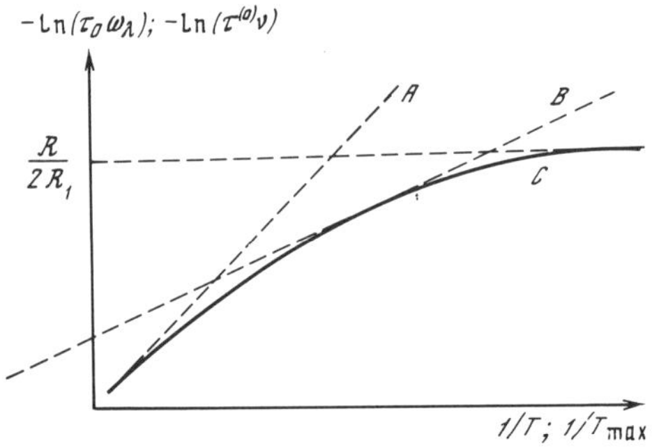

Figure 6.7 Temperature dependence of the relaxation time in the Arrhenius coordinates (solid curve).
Dashed line A shows the form of this dependence in the case when the Arrhenius law holds.

There is a broad temperature region for substances with relatively low values of the parameter $\mathcal{R}_1$ (<1) where $\mathcal{R}_1 < \hbar\Omega/2k_BT$ and the relaxation time obeys the "inverse Arrhenius law"

$$\frac{1}{\tau} \sim \exp\left[-\frac{\mathcal{R}}{2\mathcal{R}_1} + \frac{T}{T_0}\right] \tag{8.19}$$

where $T_0 = 2\hbar\Omega/k_B\mathcal{R}$, instead of the usual Arrhenius law. The former term seems to have first been introduced to denote a temperature law found in some chalcogenide glassy semiconductors (see e.g. [63]) during measurements of the temperature dependence of the luminescence intensity.

As for the interpretation of this law several opinions has been put forward. Street [63] (see also [64]) relates the inverse Arrhenius law to large-scale fluctuations of the potential, which influence the character of the electron tunneling transitions. Other authors suppose that electronic–nuclear transitions are involved here, and hence most attention should be paid to heavy-particle tunneling. Stys and Foigel [65] consider heavy-particle motion in a static DWP with a large number of levels in each well. The law (8.18) arises in their model in an intermediate region of temperatures where tunneling and activation compete on an equal footing. It can easily be seen that this approach is in some respects (static barrier, large number of levels) completely opposite to the model discussed above.

A detailed analysis of chalcogenide glasses has been performed by Phillips [17], who states that formation of short-range-order clusters is typical of these substances and the DWPs are formed in between two clusters. Combining this model with our approach above [22, 25] allows (8.18) to be applied in order to explain the inverse Arrhenius law in chalcogenide glassy semiconductors. Experiment yields $T_0 \approx 30$ K, while the inverse Arrhenius dependence is observed in the range 30 K to 200 K. This suggests that $\mathscr{R}_1$ is smaller than unity. Correspondingly, $\mathscr{R}$ should be close to unity, in contrast with the values used above. Such a diminishing of the parameters $\mathscr{R}$ and $\mathscr{R}_1$ may be achieved if it is assumed that the potential-barrier oscillations are due to the relative vibrations of the neighbouring clusters, to which a large reduced oscillator mass M corresponds. Then, in spite of the parameter $\mathscr{R}$ being close to unity, the applicability condition (7.1) for the double adiabatic approximation still holds, but now because of the large M value. Finally, the frequency of the shear interlayer vibrations $\hbar\Omega/k_B \approx 30$ K [17] is close to the observed value $T_0 \approx 30$ K, which also supports our interpretation.

6.9. TLS–SOUND INTERACTION IN THE MULTIPHONON REGIME

Two regimes of TLS-phonon interaction — resonant and relaxation — have been discussed above. This section deals with another regime, which appears when phonons interact with rigid TLSs at relatively high temperatures ($T > T_m$). To study the TLS–sound interaction in a glass it is necessary to calculate the linear response function for an external field whose role is now played by the deformations of the glass caused by the sound vibrations.

This function has the form

$$\pi^r(\omega_\lambda) - \mathrm{Av}_1 \sum_f |\langle\{\nu_1\}|M'_{\alpha\beta}|\{\nu_f\}\rangle|^2[E_{\alpha\beta} \mp (\hbar\omega_\lambda + i\delta) + E_{phi} - E_{phf}]^{\ 1}. \tag{9.1}$$

Its imaginary part determines the sound attenuation and the real part determines the renormalization of the velocity of sound. Here α and β are the particle states in the TLS, so $E_{\alpha\beta} = 0$ for diagonal processes and $E_{\alpha\beta} = \pm\Delta E$ for nondiagonal processes. The choice of sign in front of the energy difference ΔE and the energy of the sound quantum $\hbar\omega_\lambda$ depends on whether the TLS energy increases or decreases in the given elementary process and on whether the sound phonon is absorbed or emitted. So it can be seen that four different nondiagonal (Fig. 6.8a) and four diagonal (Fig. 6.8b) processes may occur.

At resonant absorption of sound the TLS energy change is ΔE, which is accompanied by the emission or absorption of a sound phonon with energy $\hbar\omega_\lambda = \Delta E$. The relaxation absorption corresponds to the

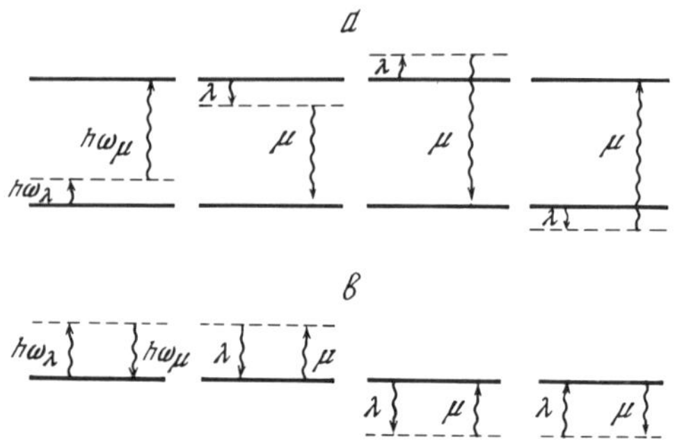

Figure 6.8 Various nondiagonal (a) and diagonal (b) processes of the interaction of the phonon $\hbar\omega_\lambda$ with a TLS Index μ stands for the multiphonon packet.

diagonal interaction when $\Delta E = 0$. This process is allowed only for nonstationary states of the TLS, i.e. as well as the sound phonon, the thermal-bath phonons also take part. This is why the relaxation interaction arises when account is taken of higher than second-order terms in the interaction operators M_n and M_d that provide for a final TLS damping in (9.1). In multiphonon processes only the second-order terms in the interaction operator are, as before, taken into account, but $\hbar\omega_\lambda \neq \Delta E$, and the energy defect is compensated at the expense of the emission or absorption of a multiphonon packet as discussed in the preceeding section.

The operator $M'_{\alpha\beta}$ in (9.1) describes the interaction of TLSs with sound waves, and in the linear approximation takes the form

$$q_\lambda \frac{\partial M_{\alpha\beta}}{\partial q_\lambda} . \tag{9.2}$$

The symbol Av_i in (9.1) stands for averaging over the initial TLS states (i.e. eight processes in Fig. 6.8) and phonons and over the ensemble of the TLSs as well. A summation over all final states is carried out.

Keeping in mind (7.12) and (7.13) for M_n and M_d, the response function (9.1) is first calculated under the assumption that

$$M'_{\alpha\beta} = q_\lambda \frac{\partial \Delta(\{q_\mu\})}{\partial q_\lambda} . \tag{9.3}$$

Using the integral representation

$$\begin{aligned}&[E_{\alpha\beta} \pm (\hbar\omega_\lambda + i\delta) + E_{phi} - E_{phf}]^{-1} \\ &= \mp \int_{-\infty}^{0} dt \exp[-it(\hbar\omega_\lambda + i\delta \mp E_{\alpha\beta} \mp (E_{phi} - E_{phf})],\end{aligned} \tag{9.4}$$

the contribution of a single TLS with energy difference ΔE to $\pi^r(\omega_\lambda)$ is examined. In addition, to be more precise, only one of the processes depicted in Fig. 6.8(a) is treated — the one with + signs in front of both ΔE and $\hbar\omega_\lambda$. Then performing all necessary integrations (e.g. (7.5)) gives

$$\left.\begin{aligned}\pi^{(1)} &= i\gamma\omega_\lambda\Delta_0^2\int_\infty^0 \frac{dt}{\hbar}\,\{\det\Phi(t+\tfrac{1}{2}i\beta\hbar)\}^{-1/2}\\ &\quad\times\exp\left[i\left(\frac{\Delta E}{\hbar}-\omega_\lambda\right)t+\mathscr{F}(T,t+\tfrac{1}{2}i\beta\hbar)\right],\\ \gamma &= \frac{1}{\omega_\lambda}\left|\left\langle q_\lambda\frac{\partial}{\partial q_\lambda}\exp\left[-\frac{1}{2}J(\{q_\mu\})\right]\right\rangle\right|^2_{q_\lambda=0} = \frac{\pi^2\hbar^2 R}{20\,\omega_D}\end{aligned}\right\} \tag{9.5}$$

The brackets $\langle\ldots\rangle$ represent averaging over the TLS orientations. The contributions of the remaining processes in Fig. 6.8(a) result from (9.5) by changing the signs in front of ΔE and $\hbar\omega_\lambda$; in the case of the diagonal processes (Fig. 6.8b) $\Delta E = 0$ should be assumed.

We now turn to the multiphonon attenuation of sound, which is determined by the imaginary part of π^r. Since the integrand in Im $\pi^{(1)}$ is even in t, the integration in (9.5) may be extended to the interval $(-\infty, +\infty)$; the substitution of the variables $t + \frac{1}{2}i\beta\hbar \to t$ may then be made. As a result (cf. (8.5))

$$\begin{aligned}\operatorname{Im}\pi^{(1)} &= K^{(0)}(\Delta E + \hbar\omega_\lambda)\,\gamma\,\omega_\lambda\\ &= \frac{\gamma\,\omega_\lambda}{2\tau^{(0)}}\exp\left[\tfrac{1}{2}\beta(\Delta E + \hbar\omega_\lambda)\right].\end{aligned} \tag{9.6}$$

Now the contributions of all elementary processes (Fig. 6.8), weighted with the function

$$\{1 + \exp[\pm\,\beta\Delta E]\}^{-1}, \tag{9.7}$$

are summed, the weighting reflecting the probability of initial-state occupation; averaging over all TLSs is then performed. As a result, the inverse length of sound attenuation (absorption coefficient) is given by the following expression:

$$\ell_m^{-1}(\omega_\lambda) = 2g(T)\left(\frac{\gamma\omega_\lambda}{\hbar^2 s\tau^{(0)}}\right)\sinh(\tfrac{1}{2}\beta\hbar\omega_\lambda), \tag{9.8}$$

where

$$g(T) = \int_{\Delta_1(T)}^{\Delta E_{max}} d\Delta E\, n(\Delta E) \left[1 + 8 \left(\frac{\Delta_1(T)}{\Delta E} \right)^2 \sinh^2(\tfrac{1}{4}\beta\Delta E) \right] \Big/ \cosh(\tfrac{1}{2}\beta\Delta E). \tag{9.9}$$

This formula holds at high temperatures $T > T_m$ and describes ultrasonic attenuation in the multiphonon regime.

The temperature dependence of the sound attenuation is mainly determined by the TLS relaxation time $\tau^{(0)}(T)$ in the limit $\Delta E = 0$. In the overwhelming majority of experimental situations, the temperatures and sound frequencies satisfy the condition $\beta\hbar\omega_\lambda \ll 1$, and

$$\sinh(\tfrac{1}{2}\beta\hbar\omega_\lambda) \approx \frac{\hbar\omega_\lambda}{2k_B T}. \tag{9.10}$$

This is why the inverse length of ultrasonic attenuation at temperatures higher than $T_a = E_a/k_B$ does not saturate but decreases as T^{-1}, i.e. the dependence $\ell_m^{-1}(T)$ goes through a maximum at $T \sim T_a$.

It should be mentioned that the approximation used above might be not very good in the vicinity of the second maximum [23–25]. The TLS attenuation rises and multiple emission of the multiphonon packets might be of importance. In taking account of only one packet, the small parameter $\tau_p\omega_T \ll 1$ is used which characterizes the probability of packet emission (see Section 6.8). But at temperatures $T \sim T_a$ the packet duration ceases to decrease with increasing temperature, and the parameter $\tau_p\omega_T$ starts to grow. This can influence some details of the behaviour of $\ell_m^{-1}(T)$ in the high-temperature region (see [25]).

We now calculate the renormalization of the velocity of sound that is due to its interaction with the TLSs. In the case of the resonant mechanism (see Section 6.5), the capture of the sound quantum is not accompanied by emission or absorption of the thermal-bath phonons. It might be said that a sound phonon is in a "bound-state" for part of the time rather than penetrating through the glass, so the velocity of sound decreases. The difference between the multiphonon and resonant regimes lies in the fact that the virtual capture of the sound phonon is accompanied by the emission or absorption of the multiphonon packet. This results in an additional strong temperature dependence of the process probability.

The correction to the velocity of sound is written in terms of the real part of the linear response function (9.1) by virtue of the relation

$$\frac{\delta s}{s} = \frac{\mathrm{Re}\,\pi^r}{2\hbar\omega_\lambda}. \tag{9.11}$$

The integral (9.5), calculated by the steepest-descent technique, is purely imaginary and of order $(\tau_p \omega_T)^{1/2}$. This is why the real part of the linear response function can be calculated only by going beyond the steepest-descent approximation and finding terms of order $\tau_p \omega_T$. Thus, calculating the integral (9.5) and averaging the result over the initial states of the TLSs gives (for details see [23, 25]), in the Einstein model,

$$\frac{\delta s}{s} = -g(T) \frac{\gamma \omega_\lambda}{\varphi_1 \varphi_2 \varphi_3^{1/2}} \left(\frac{\nu}{\Omega}\right)^2 \exp[-J_0 + \varphi_1] \sinh(\tfrac{1}{2}\beta \hbar \omega_\lambda). \qquad (9.12)$$

Equation (9.12) yields a correction to the velocity of sound due to the multiphonon mechanism of the TLS–sound interaction which similarly to the inverse length of the sound attenuation, goes through a maximum at temperature T_a. It should be noted that this temperature is, in fact, somewhat overestimated, since taking account of the finite damping of the levels [25] and the nonlinear mechanisms considered in ref. [24] might decrease it.

6.10. TWO TYPES OF TWO-LEVEL SYSTEMS

As pointed out in Section 6.2, two groups of two-level systems are possible in glasses. They are characterized by different potential-barrier parameters. The barriers in the first group are low and the zero-point vibration frequencies are relatively small. Study of their properties shows [20, 21] that the deformation potential is rather large (B ~ 0.5 eV), which corresponds to experimental data at low temperatures. Such TLSs are formed by two strongly overlapped potential curves (see Section 6.2), so that a deformation (e.g. variation of the Si-0-Si chain length) changes the form of the DWP and hence the ϵ_a value. At the same time the parameter $\mathcal{R}$ for these systems is small, since small J and J′ correspond to small barriers. The second group contains TLSs with large potential barriers, to which small deformation potential B and a large parameter $\mathcal{R}$ correspond.

The general pattern of kinetic processes in glasses can be viewed as follows. At low temperatures the leading role is played by the first group of TLSs with soft DWPs. The second group may contribute to the specific heat, for which the value of the deformation potential is of no concern. It is recalled that it was the different values of the densities of states n(0) obtained from the low-temperature specific-heat, thermal conductivity and phonon-echo measurements that have led to the idea of two types of the TLSs [28, 29].

At relatively high temperatures ($T > T_m \sim 10$ K) the ultrasonic attenuation is mainly determined by two mechanisms: the relaxation one considered in Section 6.5 and the multiphonon one. The resonant mechanism is of minor importance in this temperature range. We may also ignore the contribution of the soft TLSs owing to their large damping (see (5.28) [45, 46]). Another situation occurs for the rigid TLSs. Their prime deformation potential B is small, but the effective deformation potentials (7.12) and (7.13), which are also small at low temperatures, start growing at $T > T_m$ because the effective amplitude $\Delta_1(T)$ of the tunneling transitions grows exponentially (see [25]).

The ratio of the sound attenuation coefficients related to the relaxation and multiphonon interactions with the rigid TLSs is

$$\frac{\ell_m^{-1}}{\ell_{rel}^{-1}} \sim \frac{Ms^2\hbar\omega_D}{\Delta_1^2(T)} \frac{1+\tau^2\omega_\lambda^2}{\tau^2\,\omega_D^2} \tag{10.1}$$

Therefore, as follows from (10.1), the multiphonon mechanism is predominant at high frequencies, when not only does the condition $\tau\omega_\lambda \gg 1$ hold but also the ratio ω_D/ω is not too large.

In the low-frequency region the sound attenuation is due to the relaxation mechanism. However, the characteristic transition times $1/\tau$ are themselves connected with the multiphonon processes and obey a more complicated dependence of the type (8.14), (8.17), rather than depending on temperature according to the activation-like Arrhenius law as is usually assumed in the structural-defect model [10].

In this case the second maximum temperature of the sound attenuation is found from the equation

$$\tau(T)\,\omega_\lambda \approx 1. \tag{10.2}$$

If an activation-like dependence occurs (see (8.18)) then the maximum temperature is related to the sound frequency by

$$-\ell n(\tau_0\omega_\lambda) = E_a/k_B T_{max}, \tag{10.3}$$

in which τ_0 is the preexponential factor of (8.14) (or (8.15)). Thus, knowing the experimental dependence of T_{max} on ω_λ, we are able to calculate the τ_0 and E_a values. Then the only difference from the conventional structural-defect model (see Section 6.6) is that the role of the barrier height V_b is now played by the quantity E_a, which has another physical meaning and another energy scale. The experimentally determined magnitude ~ 500 K seems to be small for the barrier height V_b; but this is not surprising when we deal with the quantity E_a.

In the general case the dependence of $1/T_{max}$ on $\ell n\,\omega_\lambda$ is not necessarily linear, since the "activation energy" E_a itself depends on the temperature. An experiment measures only a restricted segment of the curve of $1/T_{max}$ versus $\ell n\ \omega_\lambda$,, which because of experimental errors, cannot always be reliably distinguished from a segment of a straight line (see Fig. 6.7). It is seemingly possible to use (10.3) in this case, but the τ_0 and E_a values obtained in this way are rather remotely connected with the values appearing in the theory. Such a procedure affects especially strongly the quantity τ_0, which may be changed through many orders of magnitude. The difficulties encountered in [52–55] might be due to this. There τ_0 value were found that are many orders of magnitude lower than the characteristic phonon times 10^{-13} s.

The attenuation of high-frequency sound that should be caused by multiphonon processes has not been studied experimentally to any great extent. We can point out only the work in [66, 67], which observed a quadratic frequency dependence (see (9.8), (9.10)) of the attenuation coefficient up to temperatures ~200–300 K, which seems to be a manifestation of the multiphonon mechanism of the hypersound attenuation.

6.11. THERMAL CONDUCTIVITY OF GLASSES AT RELATIVELY HIGH TEMPERATURES ($T > 10$ K)

At low temperatures ($T < T_m$) the thermal conductivity is determined by the resonant interaction of phonons with TLSs and obeys a law $\kappa \sim T^2$ (see Section 6.5). Here the principal role is played by the group of soft TLSs with large deformation potential B, the interaction of phonons with them being described within the single-phonon approach (see Section 6.5). Attempts to describe the thermal conductivity at higher temperatures appeared to be unsuccessful. A detailed discussion of this issue may be found in [68], which showed that neither the resonant nor the relaxation interaction is able to account for the experimentally observed temperature dependence of the thermal conductivity. Also, [68] did not take into account the strong damping, which according to the [46, 47], must take place at these temperatures (see (5.28) and (5.29)).† If this damping had been taken into consideration then the divergence between the theory and the experiment would have become even larger.

A way out of this difficult situation seems to be presented by taking

† Karpov and Parshin [69] draw attention to the fact that in the region around 30 K strongly asymmetric potentials may play an essential role. Anharmonic oscillators rather than TLSs correspond to these.

account of the contribution of rigid TLSs. In calculating the thermal conductivity using the equation

$$\kappa = s \sum_{\mu} \frac{(\hbar\omega_\mu)^2}{4k_B T^2 V \sinh^2(\frac{1}{2}\beta\hbar\omega_\mu)} \ell(\omega_\mu), \tag{11.1}$$

the equations obtained in Section 6.9 in the multiphonon regime should be substituted for the free path length $\ell(\omega_\mu)$. This is connected with the fact that the principal contribution to the thermal conductivity is due to high-frequency phonons $\hbar\omega_\mu \sim k_B T$, for which the relaxation interaction with TLSs is relatively weak. Therefore the thermal conductivity takes the form

$$\kappa = \frac{\tau^{(0)}(T)\, k_B^3}{\pi^2 \gamma s} T^2 \frac{G(T)}{g(T)}, \tag{11.2}$$

where G(T) is a function arising after averaging over the phonon spectrum. In the Debye approximation it has the form

$$G(T) = \int_0^{T_D/2T} dy \left(\frac{y}{\sinh y}\right)^3 .$$

More detailed formulae and a discussion may be found in [25].

The relaxation time $\tau^{(0)}(T)$ for rigid TLSs in the temperature region T_m to E_a/k_B must decrease because the amplitude of the barrier-width fluctuations grows. Owing to this, the thermal conductivity should also rapidly decrease. However this decrease is significantly weakened by the other functions in (11.2) that depend on temperature. The T^2 dependence in (11.2) corresponds to the growth of the phonon phase volume. The function G(T) depends weakly on temperature, while the function g(T) originating from the averaging over different values of the energy differences ΔE decreases rapidly with increasing temperature because the effective amplitude $\Delta_1(T)$ of the tunneling transitions grows (see Section 6.9). The fall in κ is decelerated even strongly if account is taken of the finite widths of the TLS levels in calculating $\tau^{(0)}$.

A plateau in the thermal conductivity is experimentally observed in the temperature region 10–20 K; only in a few cases is a fall seen [48, 49]. The multiphonon theory of kinetic processes in glasses presented above provides a possible qualitative explanation of this behaviour. However, detailed quantitative comparison of theory and experiment requires numerical calculations using (11.1) (11.2). For this, knowledge of both the phonon spectrum and the distribution of the parameters characterizing the TLSs is necessary. But such knowledge can be acquired only after further development of theoretical models and the acquisition of new experimental data on various kinetic characteristics of amorphous substances.

References

1. Zeller, R.C. and Pohl, R.O. (1971) *Phys. Rev.*, **B4**, 2029.
2. Ziman, J.M. (1960) *Electrons and Phonons*. The Theory of Transport Phenomena in Solids. Clarendon Press, Oxford, 1960.
3. Mason, W.P. and McSkimin, J. (1948) *J. Appl. Phys.*, **19**, 940.
4. Kittel, C. (1949) *Phys. Rev.*, **75**, 972.
5. Klemens, P.G. (1951) *Proc. R. Soc. Lond.*, **A208**, 108.
6. Dreyfus, B., Fernandez, N.C. and Maynard, R. (1968) *Phys. Lett.*, **A26**, 647.
7. Anderson, O.L. (1952) *J. Phys. Chem. Solids*, **12**, 41.
8. Hornung, E.W., Fischer, R.A., Brodale, G.E. and Gianque, W.F. (1969) *J. Chem. Phys.*, **50**, 4878.
9. Anderson, A.C., Reese, W. and Wheatley, J.C. (1963) *Rev. Sci. Instrum.*, **34**, 1386.
10. Anderson, O.L. and Bömmel H.E. (1955) *J. Am. Ceram. Soc.*, **38**, 125.
11. Krause, J.T. and Kurkjian, C.R. (1968) *J. Am. Ceram. Soc.*, **51**, 226.
12. Amorphous Solids (ed. W.A. Phillips). Springer-Verlag, Berlin, 1981.
13. Smolyakov, B.P. and Khaimovich, E.P. (1982) *Usp. Fiz. Nauk*, **136**, 317.
14. Phillips, W.A. (1972) *J. Low Temp. Phys.*, **7**, 351.
15. Anderson, P.W., Halperin, B.J. and Varma, C.M. (1972) *Phil. Mag.*, **25**, 1.
16. Hunklinger, S. In: Ultrasonic Symposium. Proceedings. IEEE, New York, p. 128.
17. Phillips, J.C. (1981) *Phys. Rev.*, **B24**, 1744.
18. Rivier, N. and Duffy, D.M. (1982) *J. de Physique*, **43**, 293.
19. Klinger, M.I. and Karpov, V.G. (1983) *Zh. Eksp. Teor. Fiz.*, **84**, 425.
20. Karpov, V.G., Klinger, M.I. and Ignatiev, F.N. (1983) *Solid State Commun.*, **44**, 333; *J. Non-Cryst. Solids*, **55**, 307.
21. Fleurov, V.N., Trakhtenberg, L.I. (1982) *Solid State Commun.*, **44**, 187.
22. Trakhtenberg, L.I. and Fleurov, V.N. (1982) *Zh. Eksp. Teore. Fiz.*, **83**, 1908.
23. Trakhtenberg, L.I. and Fleurov, V.N. (1983) *Zh. Eksp. Teor. Fiz.*, **85**, 251.
24. Fleurov, V.N. and Trakhtenberg, L.I. (1983) *Solid State Commun.*, **46**, 755.
25. Fleurov, V.N., Mikhoikin, I.D. and Trakhtenberg, L.I. (1985) *Solid State Commun.*, **55**, 531; Fleurov, V.N. and Trakhtenberg, L.I. (1986) *J. Phys.*, **C19**, 333.
26. Landau, L.D. and Lifshitz, E.M. Statistical Physics, Vol. 1, Pergamon, Oxford.
27. Cohen, M.H. and Grest, G.S. (1980) *Phys. Rev. Lett.*, **45**, 1271.

[illegible] (1977) *Phys. Rev.*, **B16**, 2879.

[illegible] 740.

[illegible] In: *Proc. Int. Conf. on Physics of MOS* [illegible] ork.

[illegible]

[illegible] *State Commun.*, **39**, 143.

[illegible] *Solid State Commun.*, **37**, 105; (1982) [illegible]

[illegible] *yst. Solids*, **38**, 371.

35. [illegible] *Non-Cryst. Solids*, **37**, 125.

36. Reichert, U., Schmidt, M. and Hunklinger, S. (1986) *Solid State Commun.*, **57**, 315.
37. Loponnen, M.T., Dynes, R.C., Narayanamurti, V. and Garno, J.P. (1982) *Phys. Rev.*, **B25**, 1161.

38. Levinson, I.V. (1983) *Pisma ZhETF*, **37**, 157.
39. Varma, C.M., Dynes, R.C. and Banavar, J.R. (1982) *J. Phys.*, **C15**, L1221.
40. Maleev, S.V. (1983) *Zh. Eksp. Teor. Fiz.*, **84**, 260.
41. Sussman, J.A. (1964) *Phys. Kond. Mat.*, **2**, 146.
42. Sussman, J.A. (1967) *J. Phys. Chem. Solids*, **28**, 1643.
43. Jäckle, J., Piché, L., Arnold, W. and Hunklinger, S. (1976) *J. Non-Cryst. Solids*, **20**, 365.
44. Joffrin, J. and Levelut, A. (1975) *J. de Physique*, **36**, 811.
45. Piché, L., Maynard, R., Hunklinger, S. and Jäckle, J. (1974) *Phys. Rev. Lett.*, **32**, 1426.
46. Gurevich, V.L. and Parshin, D.A. (1982) *Fiz. Tverd. Tela*, **24**, 1372.
47. Gurevich, V.L., Parshin, D.A. (1983) *Zh. Eksp. Teor. Fiz.*, **83**, 2301.
48. Leadbetter, A.J., Jeaps, A.P., Waterfield, C.G. and Maynard, R. (1977) *J. de Physique*, **38**, 95.
49. Xutony, P.J. and Andera, A.C. (1976) *Phys. Rev.*, **B14**, 5198.
50. Strakhna, R.E. (1961) *Phys. Rev.*, **123**, 2020.
51. Vukcevich, M.R. (1972) *J. Non-Cryst. Solids*, **11**, 25.
52. Berry, B.S. and Pritchet, W.C. (1975) *Phys. Rev. Lett.*, **34**, 1022.
53. Soshiroda, T., Koiwa, M. and Masumoto, T. (1976) *J. Non-Cryst. Solids*, **22**, 173.
54. Tyagi, S. and Lord A.E. (1979) *J. Non-Cryst. Solids*, **30**, 273.
55. Berry, B.S. (1982) *Scr. Met.*, **16**, 1407.
56. Kagan, Yu. and Klinger, M.I. (1976) *Zh. Eksp. Teor. Fiz.*, **70**, 255.
57. Kagan, Yu. and Maksimov, L.A. (1982) *Zh. Eksp. Teor. Fiz.*, **82**, 1363.
58. Klochikhin, V.L., Pshezhetskii, S.Ya. and Trakhtenberg, L.I. (1978) *Dokl. AN SSSR*, **239**, 879.
59. Trakhtenberg, L.I., Klochikhin, V.L. and Pshezhetskii, S.Ya. (1982) *Chem. Phys.*, **69**, 121.
60. Buttiker, M. and Landauer, R. (1982) *Phys. Rev. Lett.*, **49**, 1739.
61. Abrikosov, A.A. (1965) *Physics*, **2**, 5.
62. Hernando, A., García-Escorial, A., Ascasibar, E. and Vásquez, M. (1983) *J. Phys.*, **D16**, 1999.
63. Street, R.A. (1976) *Adv. Phys.*, **25**, 397.
64. Mott, N. and Davis, E. (1979) *Electron Processes in Non-Crystalline Materials*. Clarendon Press. Oxford.
65. Stys, L.E. and Foigel, M.G. (1979) *Fiz. Tekhn. Poluprovod.*, **13**, 2246.
66. Vacher, R. and Pelous, J. (1976) *Phys. Rev.*, **B14**, 823.
67. Ström, V. and Taylor, P.C. (1977) *Phys. Rev.*, **B16**, 5512.
68. Jones, D.P. and Phillips, W.A. (1983) *Phys. Rev.*, **B27**, 3891.
69. Karpov, V.G. and Parshin D.A. (1983) *Pisma ZhETF*, **38**, 536; (1985) *Zh. Eksp. Teor. Fiz.*, **88**, 2212.

CHAPTER 7
Tunneling of Heavy Particles in Crystals

Basic results of the physics of quantum crystals and quantum diffusion are presented. In crystals consisting of sufficiently light atoms particle migration may proceed at low temperatures by means of the tunneling effect. The properties of such quantum crystals differ in many respects from the properties of classical crystals. A number of peculiar features are inherent to diffusion of light particles in classical crystals, which also proceeds by means of tunneling transitions.

First, basic concepts are discussed that give an idea of the nature of quantum crystals. The absence of an unambiguous relation between the number of sites in the ideal crystal lattice and that of the constituent atoms is pointed out. A criterion for the existence of a quantum crystal is given.

A theory of quantum diffusion of noninteracting particles in an extremely narrow band of a defectless crystal is presented. The temperature dependence of the diffusion coefficient is studied. It contains two terms, corresponding to the coherent and incoherent regimes.

A more complicated pattern is also considered, which arises when account is taken of the interaction between particles and the presence of classical defects in the crystal, which cause a local band distortion. Some experiments are discussed in which quantum diffusion is observed.

The fundamentals of the physics of quantum crystal surfaces are then presented. While the surface of a classical crystal necessarily becomes atomically smooth with decreasing temperature, in quantum crystals both atomically smooth and rough surfaces might in principle exist at zero temperature. The growth coefficients of both types of surfaces behave in completely different ways at low temperatures.

A theory of crystallization waves on the atomically rough surfaces of quantum crystals is considered. The spectrum of crystallization waves is calculated. Experimental data on crystallization waves are briefly presented.

7.1. QUANTUM EFFECTS IN CRYSTALLINE SOLIDS

Chapter 6 discussed effects caused by tunneling of atoms (or groups of atoms) in amorphous solids. Their being amorphous is of principal importance, since this is where the DWPs, which are absent in crystalline solids, originate from. In fact, we considered nonequilibrium systems with extremely large relaxation times greatly exceeding the characteristic observation times. All substances in equilibrium at low temperatures crystallize in the form of a crystalline lattice. Helium is the only exception; it crystallizes, owing to quantum effects, only under a high pressure. Certainly, there are no amorphous effects in these crystals, and it is usually assumed that the particles forming an ideal crystal are localized in their sites and undergo only vibrations near their equilibrium positions with small amplitudes.

These vibrations are described in terms of phonons, and for these quantum effects are of principal importance, especially at temperatures lower than the Debye temperature T_D. We shall not dwell here upon the quantum phenomena, whose description is given in standard textbooks on solid-state physics (see e.g. [1,2]). Attention will be drawn instead to the following circumstance. In such an approach atoms are not quantum particles, since they are treated as distinguishable. They are individualized by being attributed to certain lattice sites. So the quasiclassical pattern of a crystal is only approximate, though in the overwhelming majority of cases its precision is extremely high.

Within the scope of the quasiclassical model, all the atoms in ideal defectless crystals are at low temperatures in strictly determined places — lattice sites. At finite temperatures these atoms may transfer to intersite positions by means of thermal activation, and vacancies appear at the deserted sites. So there are always a number of interstices and vacancies in the crystal, whose equilibrium density varies with temperature according to an activation-like law

$$n_V \sim \exp[-E_V/k_B T], \tag{1.1}$$

where E_V is the energy of particle transfer from site to interstice. When the temperature is lowered, the equilibrium density of vacancies and interstices rapidly decreases, and must be zero at $T = 0$ in a classical crystal.

A completely different situation occurs in quantum crystals, which are very unusual and interesting objects. In contrast with the situation in classical crystals, the behaviour of individual atoms in them must be described in terms of quantum mechanics [3–5]. In particular, it may not be asserted that a given atom is in a given lattice site; since then both the coordinate and the momentum of the atom would be simultaneously determined, the momentum being assumed to be zero (a motionless atom). The uncertainty principle states that if an atom's position is known then its momentum cannot be specified exactly. So any unambiguous connection between the number of crystal lattice sites and the number of atoms, which is characteristic of an ideal crystal, disappears. It is now not the positions occupied by the atoms that should be treated as sites, since it is not possible to specify them exactly; only the places where the atoms are most likely to be found can be stated. This means that the lattice sites are the antinodes of the atomic wave functions.

The absence of a strict connection between the number of crystal lattice sites and the number of atoms means that, even in a regular defectless lattice, the atoms may be found in arbitrary places, for example in an interstitial position. At the same time, there might be no atom in a site, i.e. there are a finite number of interstices and vacancies in the crystal, even at zero temperature.

Such unusual properties of the quantum crystals are due to the fact that an atom may leave its site because of the tunneling effect, even if this process is classically forbidden. As a result, there is a finite density of vacancies and interstices in the crystal at $T = 0$:

$$n_V \sim \exp[-1/\Lambda],$$

where $\Lambda^{-1} \sim (a/\hbar)(mV_b)^{1/2}$, a is the interatomic spacing, m is the atomic mass and V_b is the characteristic potential of the interatomic interaction, i.e. the height of the potential barrier. The parameter Λ is a measure of the quantum properties of the crystal. It is convenient to relate it to the Debye temperature. The characteristic frequency of atomic vibrations

$$\Omega \sim (\kappa/m)^{1/2}$$

is expressed in terms of the "spring rigidity"

$$\kappa \sim V_b/a^2,$$

which keeps the atoms near their equilibrium positions. At the same time, it is connected with the Debye temperature

$$T_D = \hbar\Omega/k_B.$$

As a result,

$$\Lambda \sim k_B T_D / V_b \tag{1.3}$$

The larger the quantum parameter Λ, the more pronounced are the quantum properties of the crystals. The numerical values of this parameter are largest for crystals containing the lightest and most weakly interacting atoms (e.g. $\Lambda = 0.49$ for ^3He, $\Lambda = 0.43$ for ^4He and $\Lambda = 0.28$ for H_2). For the heavier neon the quantum parameter is much smaller, $\Lambda = 0.07$, while for all other substances Λ is so small that there is no sense in considering quantum effects in them. This is why when talking about quantum crystals, we actually mean solid helium or hydrogen.

Another important situation can occur when quantum effects play an important role in the behaviour of heavy particles. This happens if some light atoms (e.g. helium or hydrogen isotopes or muons) are introduced as impurities into a crystal consisting of heavy atoms. Then the host atoms behave quite classically, whereas the impurity atoms are delocalized in the crystal and their diffusion is of an essentially quantum character.

Vacancies, interstices, foreign atoms in quantum crystals or light atoms in classical crystals can no longer be treated as ordinary defects of the crystalline structure. They are now quasiparticles, which are often called defectons. The width of the energy band for these particles is extremely small:

$$\Delta \sim \hbar\Omega \exp[-1/\Lambda],$$

(e.g. for ^3He impurities in ^4He $\Delta/k_B \sim 10^{-4}$ K, and the corresponding group velocity $v \sim a\Delta/\hbar \sim 10^{-1}$ cm s^{-1}). At finite temperatures the defectons interact with phonons, and, owing to the small bandwidth, this interaction is very peculiar and has nothing in common with, say, electron–phonon interactions in metals.

Any classical defect gives rise to a wide region around itself in which the perturbation is larger than the bandwidth Δ. As a result, even at low concentration of these defects the free motion of quantum quasiparticles (defectons) becomes impossible. A small concentration of ^3He impurities in a ^4He quantum crystal may be treated as a gas of weakly interacting particles. But the narrow bandwidth results in a large cross-section for defecton–defecton scattering, so at relatively low concentrations such a gas becomes nonideal.

Finally, there are a number of very interesting physical properties that take place on the surfaces of quantum crystals. Melting and crystallization are of a very unusual character. Under certain conditions they can proceed with virtually no dissipation. Unexpected aspects appear in such areas of classical crystallography as the problem of the equilibrium shape and the faceting of crystals. Excitation of weakly damping surface waves accompanied by the periodic melting and crystallization is possible in quantum crystals.

7.2. TRANSPORT OF PARTICLES IN EXTREMELY NARROW BANDS

The problem of defecton motion in quantum crystals or light-atom diffusion in classical crystals can be reduced to the problem of quantum-particle transport in an extremely narrow band. The band width is determined by the overlap integral of the particle wave functions in neighbouring sites:

$$\Delta_b = z\Delta \tag{2.1}$$

where z is the coordination number and Δ is the probability amplitude of the tunneling transitions between neighbouring sites. Those bands may be treated as extremely narrow whose width is small compared with the characteristic phonon energy and temperature. To describe the particle motion in such a band, it is convenient to use a density-matrix formulation such as that developed by Kagan and Maksimov [6] and Kagan and Klinger [7]. The Hamiltonian of the system is written as

$$H = H^0 + V, \tag{2.2}$$

where

$$H^0 = H_1 + H_2, \quad H_1 = H_1^0 + \langle H' \rangle, \quad V = H' - \langle H' \rangle. \tag{2.3}$$

Here H_1^0 is the Hamiltonian of the particle in a periodic potential, H_2 is the Hamiltonian of the phonon subsystem and H' is the Hamiltonian of particle–phonon interaction. The brackets $\langle \ldots \rangle$ indicate averaging over the equilibrium phonon ensemble.

Now the equation for the density matrix

$$i\hbar \frac{\partial \rho}{\partial t} = [H, \rho] \tag{2.4}$$

is averaged over the phonon variables. As a result, the equation for the particle density matrix

$$\rho_1 = \mathrm{Tr}_2 \rho \tag{2.5}$$

can be written in the form

$$i\hbar \frac{\partial \rho_1}{\partial t} = [H_1, \rho] + Tr_2\{[V, \rho]\}. \tag{2.6}$$

On the other hand, it follows from (2.4) for the total density matrix that

$$\rho(t) = e^{-iH^0 t} \rho(0) e^{iH^0 t} - \frac{i}{\hbar} \int_0^t e^{-iH_0\tau} [V, \rho(t - \tau)] e^{iH_0\tau} d\tau.$$

Then, assuming $\rho(0) = \rho_1^{(0)} \rho_2^{(0)}$ to be the initial condition ($\rho_1^{(0)}$ is the initial particle density matrix and $\rho_2^{(0)}$ is the equilibrium density matrix of the phonon subsystem), (2.6) may be rewritten as

$$\hbar \frac{\partial \rho_1}{\partial t} + i[H_1, \rho_1] = Tr_2 \int_0^t [e^{-iH^0\tau} [V, \rho(t - \tau)] e^{iH_0\tau}, V]. \tag{2.7}$$

The particle–phonon interaction is assumed to be weak, and only the second-order terms in V are retained. Therefore the expression

$$\rho(t - \tau) = e^{iH_0\tau} \rho(t) e^{-iH_0\tau} \tag{2.8}$$

found to zeroth order in V should be substituted into the right-hand side of (2.7) containing the squared interaction potential V. The resulting equation is conveniently written in the site representation:

$$\frac{\partial \rho_{1mn}}{\partial t} + i[H_1, \rho]_{mn} = -I_{mn}, \tag{2.9}$$

where

$$I_{mn} = \pi \sum_{\substack{\alpha\beta \\ s\ell}} \rho_{2\alpha}^{(0)} \delta(\epsilon_m - \epsilon_n + E_\alpha - E_\beta)$$
$$\times \{V_{ms}^{\alpha\beta} V_{s\ell}^{\beta\alpha} \rho_{1\ell n} + V_{\ell s}^{\alpha\beta} V_{sn}^{\beta\alpha} \rho_{1m\ell} - 2V_{ms}^{\alpha\beta} V_{\ell m}^{\beta\alpha} \rho_{1s\ell}\}; \tag{2.10}$$

ϵ_n is the particle energy in the site n and E_α is the energy of the phonon subsystem in a state α.

The site-nondiagonal matrix element of the operator V is proportional to the overlap integral Δ (i.e. to the probability amplitude of tunneling transitions between nearest neighbours) which is taken to be small. Therefore, to zeroth order in Δ, only site-diagonal matrix elements of the interaction operator V should be retained in (2.10). As a result,

$$I_{mn}^{(0)} = W_{mn} \rho_{1mn}, \tag{2.11}$$

where

$$W_{mn} = \pi \sum_{\alpha\beta} \rho_{2\alpha}^{(0)} \delta(\epsilon_m - \epsilon_n + E_\alpha - E_\beta) \{V_{mm}^{\alpha\beta} V_{mm}^{\beta\alpha} + V_{nn}^{\alpha\beta} V_{nn}^{\beta\alpha} - 2V_{mm}^{\alpha\beta} V_{nn}^{\beta\alpha}\}$$

characterizes the hopping probability from site m to site n. In particular, it follows from this that

$$W_{mm} = 0. \tag{2.12}$$

Therefore damping of the nondiagonal elements of the particle density matrix arises even at zeroth order in the overlap integral Δ. At the same time, the diagonal matrix elements do not decay in this approximation. Their damping arises only at second order in the overlap, and is described by a standard collision integral

$$I_{mm}^{(2)} = 2\pi \sum_{\alpha\beta s} \rho_{2\alpha}^{(0)} \delta(E_\alpha - E_\beta) |V_{ms}^{\alpha\beta}|^2 (\rho_{1mm} - \rho_{1ss}). \tag{2.13}$$

Taking into account the smallness of the overlap integral, the nearest-neighbour approximation can be applied and (2.9) can be written in the form

$$\hbar \frac{\partial \rho_{mn}}{\partial t} + i\Delta \sum_{g} (\rho_{m+g,n} - \rho_{m,n+g}) = -I_{mn}. \tag{2.14}$$

This section considers only an ideal lattice in which a small number of noninteracting particles diffuses. Then the collision integral I_{mn} depends only on the difference $\ell = m - n$, and the density matrix is conveniently represented as a function of the variables ℓ and m; Fourier transformation in variable m then yields

$$\dot{\rho}_\ell(k) + i\Delta \sum_{g} \rho_{\ell+g}(\mathbf{k}) [e^{ikg} - 1] = -W_\ell \rho_\ell(\mathbf{k}) (1 - \delta_{\ell 0}) - I^{(2)} \delta_{\ell 0}. \tag{2.15}$$

In dealing with diffusion problems, we are usually interested in studying a distribution of the particle density whose scale L is large compared with the interatomic spacing, $L \gg a$. This means that (2.15) may be analysed in the long-wave limit, i.e. at small **k** values, when the condition

$$\Delta |e^{-ikg} - 1|/W_g = \Delta \mathbf{k}g/W_g \sim a\Delta/LW_g \ll 1 \tag{2.16}$$

holds. Owing to this, (2.15) at $\ell = 0$ and $\ell = g$ becomes, after Laplace transformation in t,

$$-p\,\rho_g(\mathbf{k}) + \Delta \mathbf{k}g\, \rho_0(\mathbf{k}) = -W_g \rho_g(\mathbf{k}), \tag{2.17a}$$

$$-p\,\rho_0(\mathbf{k}) - \Delta \sum_{\mathbf{g}} \rho_{\mathbf{g}}(\mathbf{k})\,\mathbf{kg} = -\eta(\mathbf{k})\,\rho_0(\mathbf{k}), \tag{2.17b}$$

where

$$\eta(\mathbf{k}) = -2\pi i \sum_{\alpha\beta\mathbf{g}} \rho_{2\alpha}^{(0)}\,\delta(E_\alpha - E_\beta)\,|V^{\alpha\beta}_{\mathbf{m},\mathbf{m}+\mathbf{g}}|^2\,\mathbf{kg}. \tag{2.18}$$

Now the function $\rho_{\mathbf{g}}(\mathbf{k})$ is expressed in terms of ρ_0 by virtue of (2.17a) and is substituted into (2.17b). The solution of the resulting secular equation

$$p - \eta = -\sum_{\mathbf{g}} \frac{(\mathbf{kg})^2\Delta^2}{p - W_{\mathbf{g}}} \tag{2.19}$$

is sought in the form

$$p = p' + W_{\mathbf{g}}.$$

Then a quadratic equation arises for the new variable p′; its roots are

$$p_{1,2} = \tfrac{1}{2}(W_{\mathbf{g}} + \eta) \pm \left\{[\tfrac{1}{2}(W_{\mathbf{g}} - \eta)]^2 - \sum_{\mathbf{g}} (\mathbf{kg})^2\Delta^2\right\}^{1/2}.$$

As follows from (2.12) and (2.13), the quantity η is proportional to the squared overlap integral, whereas W_g does not contain this small value, i.e. $\eta \ll W_g$. Using this inequality and the inequality (2.16) gives

$$p_1 = \eta + \sum_{\mathbf{g}} (\mathbf{kg})^2\Delta^2/W_{\mathbf{g}},$$

$$p_2 = W_{\mathbf{g}}.$$

We are interested in times significantly greater than the characteristic hopping time between neighbouring sites, i.e. $t \gg \hbar\, W_g^{-1}$. This allows us to assume that all rapid processes, to which the root p_2 relates, are completed by the observation time, and we see only slow processes with characteristic times of order p_1^{-1}. Now the k dependence of p_1 is explicitly written, i.e.

$$p_1 = k^2 D.$$

The equation for $\rho_0(\mathbf{k})$ in a simple cubic lattice then takes the form of a standard diffusion equation:

$$\frac{\partial \rho_0(\mathbf{k})}{\partial t} = -k^2 D\,\rho_0(\mathbf{k}), \tag{2.20}$$

where

$$D = D_{coh} + D_{incoh},$$
$$D_{coh} = \frac{\Delta^2 a^2 z}{3W_g \hbar}, \quad D_{incoh} = \frac{\pi z a^2}{3\hbar} \sum_{\alpha\beta g} \rho_{2\alpha}^{(0)} |V_{m,m+g}^{\alpha\beta}|^2 \delta(E_\alpha - E_\beta). \tag{2.21}$$

This equation shows that the particles diffuse a distance L during the time $t \sim L^2/D$. So it can be concluded that the inequality $tW_g/\hbar \gg 1$ is identical with the inequality (2.16).

The diffusion coefficient (2.21) consists of two terms with different physical meanings. The second one, i.e. the coefficient of the incoherent diffusion, has the usual structure and its value is determined by the ratio a^2/τ, where τ is the characteristic time for site-to-site tunneling transitions. Under the approximation of nearest-neighbour hopping, the matrix element

$$V_{m,m+g}^{\alpha\beta} = \langle\{\alpha\}|\, \Delta(\{q_\mu\})|\, \{\beta\}\rangle \tag{2.22}$$

is appropriate to these transitions, and in calculating D_{incoh} the results of Chapters 3 and 6 may be used. A particular property of this process is that at low temperatures the time is determined by two-phonon processes rather than by single-phonon ones as in the TLS case. Only then is the energy conservation law fulfilled under the condition $k_B T \gg \Delta_1(T)$. This is the reason for the dependence

$$1/\tau \sim T^7, \tag{2.23}$$

which can be easily understood as follows. A phase volume $\sim T^6$ corresponds to two phonons, the squared amplitude of interaction with two long-wave phonons is proportional to T^2, one power is, however, discarded owing to energy conservation. At higher temperatures, when multiphonon processes are predominant use can be made of (7.5) from Chapter 6.

What is quite unusual in (2.21) is the first term, which describes coherent diffusion. Its physical interpretation is as follows. If the particle did not interact with phonons, it would move freely in an ideal lattice band with width $\sim \Delta$. The group velocity $v \sim a\Delta/\hbar$ corresponds to this motion. Lattice vibrations restrict the free path, since they cause oscillations of the energy levels in the individual wells, which destroy the band dynamically. Owing to the inequality $k_B T \gg \Delta$ this occurs when the levels in the neighbouring wells are removed so far that coherent tunneling of the particle between them becomes impossible. This is the process that limits the time, $1/\tau_f \sim W_g/\hbar$, of the particle free path, and hence determines the value of the coherent-diffusion coefficient:

$$D_{coh} \sim \frac{\ell_f^2}{\tau_f} \sim \frac{a^2\Delta^2}{\hbar W_g},$$

where $\ell_f \sim \tau_f v$ is the particle free path.

It should be emphasized that this estimate remains valid even at $W_g > \Delta$, when the free-path length

$$\ell_f \sim a\Delta/W_g$$

becomes smaller than the intersite spacing a. The use of the term "free-path length" is then not quite justified, because, in fact, we are dealing with hopping diffusion. The particle is mostly motionless at a site and is able to hop to another site only at those rare instances when the energy levels of the corresponding potential wells are in resonance.

The probability amplitude of coherent tunneling transitions is analysed in detail in Section 6.7. Therefore to calculate the coefficient of coherent diffusion it is only necessary to know the temperature dependence of the quantity

$$W_g = \pi \sum_{\alpha\beta} \rho_{2\alpha}^{(0)} \delta(E_\alpha - E_\beta) \,|V_{\mathbf{mm}}^{\alpha\beta} - V_{\mathbf{m+g},\,\mathbf{m+g}}^{\alpha\beta}|^2. \tag{2.24}$$

The principal contribution to (2.24) is due to double-phonon site-diagonal processes, whose amplitude is

$$V_{\mathbf{mm}} = \frac{1}{2} \sum_{\mathbf{k}_1,\,\mathbf{k}_2} \frac{B_{\mathbf{mm}}(\mathbf{k}_1, \mathbf{k}_2)}{(\omega_{\mathbf{k}_1}\, \omega_{\mathbf{k}_2})^{1/2}} b_{\mathbf{k}_1} b_{\mathbf{k}_2}. \tag{2.25}$$

Thus, using the relation

$$B_{\mathbf{mm}}(\mathbf{k}_1, \mathbf{k}_2) = B_{00}(\mathbf{k}_1, \mathbf{k}_2)\, e^{i(\mathbf{k}_1 + \mathbf{k}_2)\mathbf{m}},$$

gives

$$W_g = 4\pi \sum_{\mathbf{k}_1\,\mathbf{k}_2} \frac{|B_{00}(\mathbf{k}_1, \mathbf{k}_2)|^2}{\omega_{\mathbf{k}_1}\, \omega_{\mathbf{k}_2}} N_{\mathbf{k}_1}(N_{\mathbf{k}_2} + 1)\,[1 - \cos(\mathbf{k}_1 + \mathbf{k}_2, \mathbf{m})]. \tag{2.26}$$

Then it is easy to see that $W_g \sim T^9$. This quantity characterizes the frequency of the relative vibrations of the energy levels in neighbouring sites. Such a motion is governed by the phase difference gained by phonons in between the neighbouring sites, which results in two additional powers of temperature compared with (2.23). (This is similar to the difference between ordinary and transport lifetimes.)

Thus the diffusion coefficient may be written as

$$D = \frac{za^2\Delta_1^2(T)}{3\hbar} \{W_g^{-1} + f_{incoh}(T)\}, \tag{2.27}$$

where $f_{incoh}(T) \sim T^7$ for $T < T_m$

$$f_{incoh}(T) \approx (\hbar^2\Omega^2\varphi_1\varphi_2)^{-1/2} \quad \text{for } T > T_m.$$

A general view of the function (2.27) is given in Fig. 7.1. At low temperatures, when the quantity $\Delta_1(T)$ is practically constant, the diffusion coefficient is determined by the first term in (2.27) (coherent diffusion). It increases rapidly with the decreasing temperature [3]:

$$D_{coh} \sim T^{-9}. \tag{2.28}$$

The coefficient of incoherent diffusion $D_{incoh} \sim T^7$ is very small in this region. With increasing temperature a region in entered where the coefficient of incoherent diffusion is due to the multiphonon processes. Then the function $f_{incoh}(T)$ goes through a maximum and starts to decrease as T^{-3}, and later for $T > \frac{1}{4}T_D$, according to a slower law. In this region $W_g^{-1} \sim T^{-2}$, i.e. this function decreases everywhere more rapidly than $f_{incoh}(T)$. Therefore at high temperatures incoherent diffusion determines the whole process. However, the most important factor for the temperature dependence of the diffusion coefficient is not the crossover of regimes but the exponential growth of $\Delta_1^2(T)$ with temperature, which is due to the growth of fluctuations of potential-barrier shape [8]. As a result, the diffusion coefficient goes through a minimum, whose position is determined by competition of the exponentially growing factor $\Delta_1^2(T)$ and the power-like decreasing term in braces in (2.27). At high temperatures the diffusion becomes activation-like:

$$D \sim \exp[-T_a/T]. \tag{2.29}$$

Here, similarly to chemical reactions and to relaxation processes in glasses, fluctuating excitations of the lattice-site arrangement, causing a lowering of the barrier height, are involved, rather than overbarrier excitation of the particle.

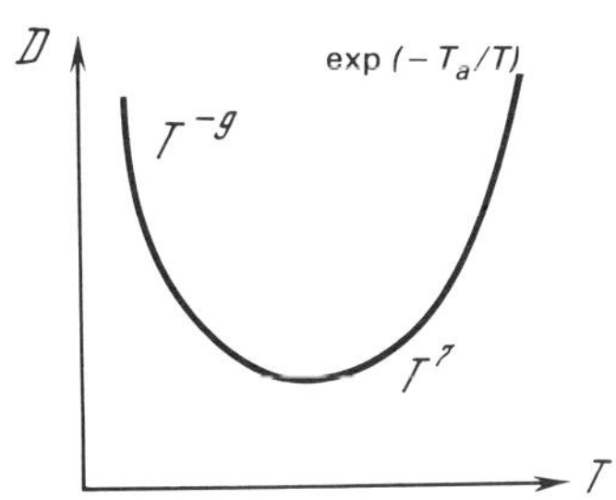

Figure 7.1 Temperature dependence of the quantum diffusion coefficient in an ideal crystal.

In principle, a polaron effect (environmental change) may also take place here, resulting in coherent bandwidth narrowing [7]. This acts against the fluctuations of the barrier, and can quantitatively influence the above model, although it does not seem to lead to qualitative changes.

7.3. QUANTUM DIFFUSION IN DEFECT CRYSTALS

The model discussed in the preceeding section is relevant to ideal defectless crystals, which, as is well known, do not exist in Nature. Even in conventional systems with broad bands (e.g. electrons in metals) a small number of defects result in an additional scattering mechanism that significantly influences the value of the diffusion coefficient at low temperatures. In extremely narrow bands this influence is much stronger and exhibits a number of peculiar features to be discussed below [9] (see also Kagan's review [20]).

For the sake of simplicity, only point defects are studied here. Each defect gives rise to a very large area of radius $r_0 \gg a$, in which static shifts of the energy levels of individual wells exceed the bandwidth $z\Delta$. The energy-level shift induced by a point defect is given by [11]:

$$|\Delta\epsilon(r)| \approx \kappa_{ip}\, \tilde{\epsilon}(a/r)^3, \tag{3.1}$$

where $\tilde{\epsilon}$ is the energy of the order of the atomic bond energy, and the dimensionless parameter

$$\kappa_{ip} \approx \left|\frac{\Delta V}{V_0}\right|_i \left|\frac{\Delta V}{V_0}\right|_p$$

is determined by the variation of the elementary cell volume ΔV occupied by the defect (i) or by the diffusing particle (p) related to the elementary cell volume V_0 in the defectless crystal. Here, for the sake of simplicity, the dependence of ΔV on the crystallographic directions is disregarded.

At low temperatures (when diffusion in an ideal crystal is purely coherent) one is dealing with the "problem of spheres" treated in percolation theory (see e.g. [12]), though in a somewhat modified formulation. The sphere radius is given by the relation

$$|\epsilon(r_0)| \approx z\Delta_1, \tag{3.2}$$

and the number of randomly situated centers around which these spheres are circumscribed, i.e. the impurity concentration, is variable. We are interested not in percolation over the spheres, but rather in percolation outside the spheres. The point is that the particle may diffuse coherently only far from the defects, where the static band distruction is weak. Here also there is a critical defect concentration at which percolation ceases completely:

$$x_c = \nu \frac{1}{\kappa_{ip}} \frac{z\Delta_1}{\tilde{\epsilon}}. \tag{3.3}$$

Here ν is a dimensionless coefficient, smaller than unity, which characterizes the percolation threshold position. The bandwidth in quantum diffusion problems is extremely small ($< 1K$), hence $x_c \ll 1$. This is why samples in which quantum diffusion is to be studied must be of extremely high quality.

Apart from the high quality of the samples, the number of diffusing particles must be small. Each particle causes shifts in the energy levels around itself, owing to which it is capable of interacting with the other particles. Correspondingly, there is a critical concentration of particles,

$$x_p^c \approx \nu \frac{1}{\kappa_{pp}} \frac{z\Delta_1}{\tilde{\epsilon}}, \tag{3.4}$$

above which they cannot be treated as a gas. The interaction between them becomes strong and they behave more like a liquid of strongly interacting particles or even like a glass.

In the overwhelming majority of experiments the conditions

$$x < x_c, \quad x_p < x_p^c$$

do not hold, and it is therefore not surprising that ideal coherent diffusion is not observed in them. Among these, in particular, are numerous experiments on hydrogen diffusion in various classical crystals.

We now consider the form of the temperature-dependent coefficient of coherent diffusion in nonideal crystals. For this it is necessary to take into account the fact that the energies ϵ_m in (2.11) are different for different lattice sites. Then $\dfrac{\Delta_1^2}{W_g}$ in the expression for the coherent diffusion coefficient (2.21) should be replaced by

$$\frac{\Delta_1^2 W_{m,m+g}}{(\epsilon_{m,m+g})^2 + (W_{m,m+g})^2}, \quad \epsilon_{m,m+g} = \epsilon_m - \epsilon_{m+g}. \tag{3.5}$$

The translational invariance, which allowed passage from the variables m and n to $\ell = m - n$, is now certainly violated. However, this is not a serious problem, especially for processes whose characteristic scale is large enough. We shall not consider here in detail the corresponding averaging, but shall present only some estimates that are sufficient to yield an understanding of the physics of these processes, both qualitatively and quantitatively.

It is first assumed that $x \ll x_c$ and $x_p \ll x_p^c$. Then the mean energy shift due to the stresses caused by the defects or diffusing particles is small. At low temperatures the free-path time of the particle is limited not by dynamical band destruction but by scattering at the excluded volumes of radius r_0 near each defect. In order to find this time, use is made of the formula

$$1/\tau = n v \sigma,$$

where n is the number of scattering centers per unit volume, v is the particle group velocity and σ is the cross-section for particle scattering at the defects. In the present case $n \sim x/a^3$, $v \sim a\Delta_1/\hbar$ and $\sigma \sim r_0^2$, so

$$W_i \sim \hbar/\tau \sim x\,(\kappa_{ip}^2 \tilde{\epsilon}^2 \Delta_1)^{1/3} \qquad (3.6)$$

should be used instead of W_g. Therefore

$$D_{coh}^{im} \approx \frac{za^2\Delta_1^{5/3}}{(\kappa_{ip}\,\tilde{\epsilon})^{2/3}}\,\frac{1}{x}. \qquad (3.7)$$

When the concentration of diffusing particles is large, i.e. $x\kappa_{ip}^{2/3} \ll x_p \kappa_{pp}^{2/3}$, the principal role is played by the particle–particle interaction, and the coherent-diffusion coefficient takes the form

$$D_{coh}^{int} \approx \frac{za^2\Delta_1^{5/3}}{(\kappa_{pp}\,\tilde{\epsilon})^{2/3}}\,\frac{1}{x_p}. \qquad (3.8)$$

Thus its value in both cases tends to a constant as $T \to 0$ (see Fig. 7.2).

Another situation is encountered in "strongly defected crystals" when $x > x_c$. (It should be recollected that $x_c \ll 1$, and such crystals might well be nearly perfect, for example from the point of view of X-ray analysis.) In this case the diffusion coefficient is described by

$$D_{coh} \approx \frac{za^2}{3}\,\frac{\Delta_1^2 W_g}{\delta\epsilon^2 + W_g^2}, \qquad (3.9)$$

which results from (2.21) by means of (3.5). Here $\delta\epsilon$ is the characteristic value of the energy difference between neighbouring sites. As shown in Section 7.2, $W_g \sim T^9$ at low temperatures. So it can readily be seen that the dependence

$$D_{coh} \sim T^9$$

is obtained instead of (2.28), i.e. the coherent-diffusion coefficient tends to zero as $T \to 0$, in contrast with the dependence T^{-9} characteristic of the coherent-diffusion coefficient in defectless crystals.

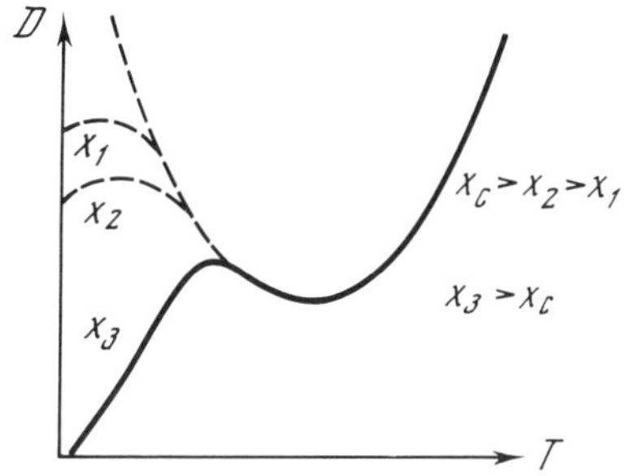

Figure 7.2 Temperature dependence of the quantum diffusion coefficient in a nonideal crystal at various defect concentrations.

This is when dealing with motion in a system with a statically strongly destroyed bond. At higher temperatures, when $W_g(T) > |\delta\epsilon|$, a cross-over to the dependence (2.28) is possible, which is characteristic of defectless crystals. The general pattern of temperature curves of the coherent-diffusion coefficient in strongly defected crystals is shown in Fig. 7.2. Addition of incoherent diffusion leaves the whole pattern qualitatively unchanged.

At present there are a number of experimental facts favouring the observation of quantum diffusion. An excellent opportunity of performing such experiments is presented by the positive-muon depolarization technique [13,14], which enables a measurement of the diffusion coefficient of the relatively light muons in various crystals to be made. Such measurements have been performed in copper [15], bismuth [16] and aluminium [17,18]. An anomalous (from the classical point of view) temperature and concentration dependence of the diffusion coefficient can be explained in terms of the model discussed above.

Another experimental area relates to ^{3}He-atom diffusion in a ^{4}He crystal, which can be studied by means of the NMR spin-echo technique [19–21]. In this case the ^{3}He atoms replace the ^{4}He atoms and diffuse by virtue of $^3\text{He} \rightleftarrows {}^4\text{He}$ exchange. It was in these experiments that the dependence $D_{coh} \sim T^{-9}$ predicted by Andreev and Lifshit [3] was most clearly revealed for the first time. The behaviour of the system was studied in detail for high concentrations of diffusing particles when the particle–particle interaction is able to decelerate the diffusion or even halt it completely.

Finally, we shall mention the experiments [22,23] on measuring orthohydrogen diffusion in parahydrogen crystals, in which anomalies were also found that connected with the quantum character of the diffusion. A more detailed discussion of these and some other experiments and their comparison with theoretical predictions can be found in Kagan's review [10].

7.4. SURFACE OF QUANTUM CRYSTALS

The quantum character of crystals results in fascinating effects on the crystal surface. Here a number of beautiful coherent dissipationless phenomena are possible, some of which are visually observable. These topics will be treated below, mainly on the basis of Andreev's and Parshin's papers [24,25].

Melting or crystallization of a classical crystal plunged into the liquid proceeds slowly at the expense of thermally activated processes (see e.g. [26,27]). At low enough temperatures the equilibrium crystals have facets, i.e. areas of plane surface perpendicular to various crystallographic axes. We consider the energy of such a surface (per unit area) versus the angle φ between the normal to the surface and the corresponding crystallographic axis. For a classical crystal the surface has the form depicted in Fig. 7.3. Each terrace represents a new monoatomic layer. At small angles φ the distance between terrace edges is large, and their interaction may be disregarded; then

$$\alpha = \alpha_0 + \frac{\beta}{a} |\varphi|, \tag{4.1}$$

where α_0 is the surface energy with zero inclination, β is the energy per unit length of the terrace edge and a is the interatomic spacing. It can be seen from (4.1) that the derivative of the surface energy with respect to the inclination angle, $\partial\alpha/\partial\varphi$, has a finite jump at $\varphi = 0$ [28]:

$$\delta \frac{\partial\alpha}{\partial\varphi} = 2 \frac{\beta}{a}. \tag{4.2}$$

These considerations can be generalized to the case when the surface inclination is determined not by one but by two angles; kinks then appear at the terrace edges (see Fig. 7.4).

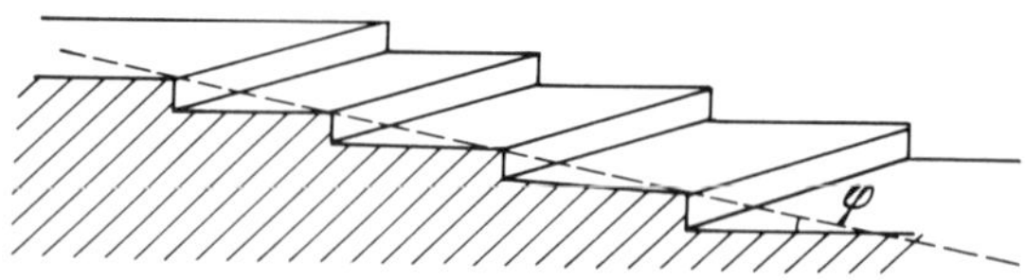

Figure 7.3 Crystal surface characterized by one inclination angle to a crystallographic plane.

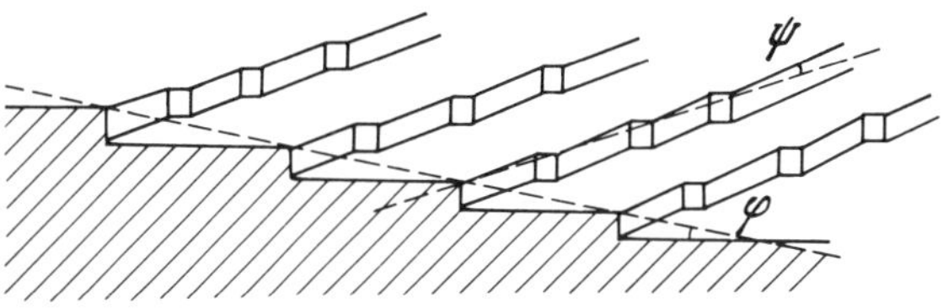

Figure 7.4 Crystal surface characterized by two inclination angles to a crystallopraphic plane.

As shown by Landau [28], crystal faceting is a consequence of the discontinuity of the function $\partial\alpha/\partial\varphi$. If this were continuous then, in the absence of an external field, the crystal would have a spherical form. For discontinuous $\partial\alpha/\partial\varphi$ the energy minimum is achieved when the crystal has plane facets whose areas are proportional to the values of the derivative jumps $\delta(\partial\alpha/\partial\varphi)$ for the corresponding crystallographic direction.

At finite temperatures the free energy

$$\tilde{\beta} = \beta - \mathrm{Ts} \tag{4.3}$$

should be considered instead of the energy β. The terrace entropy s (per unit length) is determined by the number and configuration of kinks with different signs that are generated by thermally activated processes (see Fig. 7.5). At temperatures higher than a transition temperature, which is determined by the condition

$$\tilde{\beta}(\mathrm{T}) = 0,$$

faceting becomes energetically less favourable, and an atomically rough surface with a large number of terraces and kinks appears instead of the atomically smooth surface.

Particles in a quantum crystal are delocalized, so a direct transfer of the above approach to them is inadmissible. The concepts of terrace and kink certainly remain, but they acquire an essentially quantum character. This affects the terrace energy β, which is necessarily positive

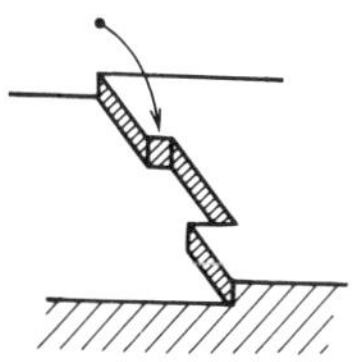

Figure 7.5 Kink motion along the terrace due to a capture of an atom from the liquid phase.

in a classical crystal. In a quantum crystal, however, the situation is different. The point is that a kink should now be treated as a delocalized quasiparticle, which may move along the terrace edge by tunneling transitions. The physical mechanism of the kink motion is capture (or abstraction) of atoms from the liquid at the corresponding "sites" at the terrace edge. So the kink motion proceeds in a one-dimensional band

$$\beta(\mathrm{p}) = \beta_0 + \frac{\epsilon}{\mathrm{a}} + \frac{\Delta}{2\mathrm{a}} \cos(\mathrm{pa}), \tag{4.4}$$

where β_0 and ϵ are the terrace and kink energies disregarding the quantum effects, Δ is the tunneling hopping integral for the kinks, i.e. the probability amplitude of capturing or abstracting an atom, and p is the quasimomentum. If the energy

$$\beta_0 + (\epsilon - \tfrac{1}{2}\Delta)/\mathrm{a}$$

corresponding to the bottom of the band is negative then a creation of pairs of kinks with opposite signs and of new terraces is energetically favourable. As a result, the surface of the quantum crystal at zero temperature becomes atomically rough. The number of terraces and kinks is no longer fixed, and is determined from the minimum-energy condition. This means that the band (4.4) is filled up to a level with a p value at which the energy $\beta(\mathrm{p})$ becomes zero and the function $\partial\alpha/\partial\varphi$ becomes smooth.

Whether the surface of a quantum crystal at zero temperature is atomically smooth or rough depends on the parameters in (4.4), which themselves depend on the crystallographic direction. This is why different surfaces of the same crystal may behave in different ways, and crystallization and melting at such surfaces may proceed differently. Crystallization of an atomically smooth surface proceeds via formation of a new crystallographic monoatomic layer. This process starts with the emergence of a germ whose energy depends on its radius as (see Fig. 7.6)

$$\mathrm{U(R)} = 2\pi\beta\mathrm{R} - \pi\mathrm{R}^2\rho_s \mathrm{a}\,\delta\mu. \tag{4.5}$$

Here the first term corresponds to the energy of the circular terrace edge, and the second one to the atoms constituting the germ; ρ_s is the mass density of the solid, $\delta\mu$ is the difference of the chemical potentials between the solid and liquid phases.

We now encounter a situation similar to that in first-order phase transitions [29,30]. The formation of a germ becomes energetically favourable only from a finite value of the radius

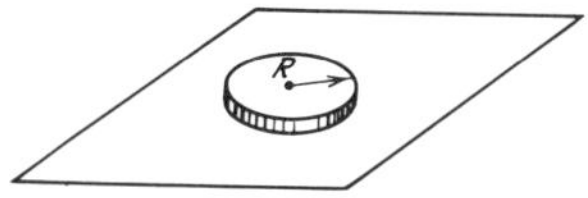

Figure 7.6 Germ of a new layer on the atomically smooth surface of the quantum crystal.

$$R > R_0 = 2\beta/\rho_s a\,\delta\mu,$$

which is determined from the condition $U(R_0) = 0$. At smaller R values the energy U(R) is positive and has the form of a curve with maximum

$$U_{max} = \pi\beta^2/\rho_s a\,\delta\mu$$

at

$$R = \tfrac{1}{2}R_0.$$

Therefore formation of a germ at zero temperature proceeds via tunneling under the barrier U_{max}.

Usually the crystallization rate is characterized by a coefficient κ, which is determined by the formula

$$\kappa = \lim_{\delta\mu \to 0} \frac{\partial v}{\partial \delta\mu},$$

where v is the crystal growth rate, i.e. the linear velocity of surface motion. As $\delta\mu \to 0$ the barrier height and width tend to infinity; hence the growth rate falls exponentially to zero. It follows from this that the growth coefficient of an atomically smooth surface of a quantum crystal becomes zero at $T = 0$. The same is also true for classical crystals whose growth proceeds by means of thermally activated processes.

A completely different situation occurs at atomically rough surfaces. These surfaces cannot exist in classical crystals at $T = 0$, but for quantum crystals permanent dissipationless creation and annihilation of kinks and terraces is characteristic. The energy barrier (U_{max}) for germ creation is now absent, since $\beta(p) \leqslant 0$. This means that crystallization and melting proceed at a finite rate for any small value of the chemical potential difference $\delta\mu$ of two phases; hence the growth coefficient κ becomes infinite in the case of an atomically rough surface of a quantum crystal as $T \to 0$.

At finite temperatures, energy dissipation takes place owing to interaction with excitations both in the crystal and the liquid. This is why the growth coefficient in fact remains finite. In the case of ^4He these excitations are phonons and rotons. For $T < 0.6$ K the principal contribution comes from phonons and $\kappa \sim T^{-4}$. At higher temperatures the interaction with rotons results in the dependence $\kappa \sim \exp[\Delta_r/k_B T]$, where Δ_r is the threshold of roton excitation. In the case of ^{3}He the energy dissipation in almost the whole temperature range (except for extremely low temperatures) is determined by the Fermi excitations, and $\kappa \sim v_F^{-1}$, where v_F is the quasiparticle velocity at the Fermi surface.

7.5. CRYSTALLIZATION WAVES

An atomically rough surface separating the crystalline and liquid phases of helium outwardly resembles a gas–liquid interface, for example it typically has a meniscus, like that of a liquid in a capillary. At low temperatures the velocity of such surface motion can be rather large, the energy dissipation being relatively weak. Therefore waves resembling capillary waves are possible on such a surface. However, this similarity is only apparent. Capillary waves are accompanied by a motion of the liquid below the interface. The waves on an atomically rough surface of a quantum crystal arise owing to periodic melting and crystallization of various parts of the surface. These waves were predicted by Andreev and Parshin [24], and later observed by Keshishev, Parshin and Babkin ([31], see also [32]).

To study the principal features of crystallization waves, a flat crystal–liquid interface is considered ($Z = 0$), whose parts may be slightly displaced in the normal direction:

$$\zeta(X_\mu, t) = \zeta(t)\, e^{ik_\mu X_\mu}. \tag{5.1}$$

Here k_μ and X_μ are the two-dimensional wave vector and radius vector in the plane $Z = 0$; $\mu = 1,2$.

The total energy of this surface is determined by the integral

$$U = \int \alpha(n)\, dS, \tag{5.2}$$

where $\alpha(n)$ is the surface energy density at the normal n to a given surface element. At small values of ζ it may be assumed that $n_z \approx 1$ and $n_\mu = \partial\zeta/\partial X_\mu$. The surface element is

$$dS = dX\, dY \left\{1 + \frac{1}{2}\left(\frac{\partial\zeta}{\partial X}\right)^2 + \frac{1}{2}\left(\frac{\partial\zeta}{\partial Y}\right)^2\right\}$$

and the energy density is

$$\alpha(n) = \alpha(0) + \frac{1}{2}\alpha_{\mu\nu}\frac{\partial\zeta}{\partial X_\mu}\frac{\partial\zeta}{\partial X_\nu}, \tag{5.3}$$

where

$$\alpha_{\mu\nu} = \frac{\partial^2 \alpha}{\partial n_\mu \partial n_\nu}.$$

So it is found that the potential energy of the crystallization waves is

$$\Delta U = \tfrac{1}{4} S\,(\alpha k^2 + \alpha_{\mu\nu} k_\mu k_\nu)\,|\zeta|^2, \tag{5.4}$$

where S is the surface area.

Interface motion with velocity $\dot{\zeta}$ is accompanied by motion of the liquid. It proceeds with a velocity that is much smaller than the velocity of sound, so liquid may be assumed to be incompressible, i.e. a velocity potential ($v = \nabla\Psi$) satisfying Laplace's equation $\Delta\Psi = 0$ may be introduced. Mass conservation at the interface is written as

$$\rho_\ell \frac{\partial \Psi}{\partial Z}\bigg|_{Z=0} = -\dot{\zeta}(\rho_s - \rho_\ell), \tag{5.5}$$

where ρ_ℓ is the liquid density. The left-hand side of (5.5) accounts for the substance flow from the liquid into the crystal, while the right-hand side accounts for the mass variation in an elementary volume near the interface when the latter moves with velocity $\dot{\zeta}$. The velocity potential Ψ decreases exponentially when moving from the surface a distance greater than the wavelength, so it follows from (5.5) that

$$\Psi = \frac{1}{k} e^{-kZ} \frac{\rho_s - \rho_\ell}{\rho_s} \dot{\zeta}$$

As a result, the total kinetic energy of the interface is

$$K = \int dX\, dY \int_0^\infty dZ\, \tfrac{1}{2}\rho_\ell |\nabla\Psi|^2 = \frac{S}{4}\, \frac{(\rho_s - \rho_\ell)^2}{k\,\rho_\ell}\, |\dot{\zeta}|^2. \tag{5.6}$$

It can be seen that the potential ΔU and kinetic K energies correspond to a harmonic oscillator with frequency

$$\omega^2(k) = \frac{\rho_\ell\, k}{(\rho_s - \rho_\ell)^2}\,(\alpha k^2 + \alpha_{\mu\nu} k_\mu k_\nu), \tag{5.7}$$

The energy $\dot{E}$ dissipated by the moving interface is $\delta\mu\,\dot{M}$, where $\dot{M}$ is the substance flow through the interface or the variation of substance mass in the crystalline phase, i.e.

$$\dot{M} = \rho_s \dot{\zeta}.$$

Then the definition (4.8) yields $\delta\mu = \dot{\zeta}\kappa$; therefore

$$\dot{E} = \int dX\, dY\, \delta\mu\, \rho_s \dot{\zeta} = \frac{S}{2}\, \frac{\rho_s}{\kappa}\, |\dot{\zeta}|^2. \tag{5.8}$$

The coefficient of crystallization-wave attenuation is therefore

$$\gamma = \frac{\rho_s \rho_\ell k}{2\kappa(\rho_s - \rho_\ell)^2}, \tag{5.9}$$

from which it follows that the waves are only weakly attenuated when the growth coefficient κ is large.

The crystallization waves are connected with periodic melting and crystallization, which cause oscillations of the phase interface. This process proceeds without any dissipation on atomically rough surfaces at $T = 0$ (since $\kappa = \infty$), and the waves are not attenuated. At finite temperatures, energy dissipation due to interaction with excitations take place. As an example, for phonons, for which $\kappa \sim T^{-4}$, the attenuation increases with temperature as $\gamma \sim T^4$.

Crystallization waves are observed in ^{4}He at temperatures below 0.5 K [31]. At higher temperatures these oscillations cease because of rapid increase in attenuation. It should be emphasized that at such temperatures the roton contribution becomes significant. Measurements of the frequency spectrum of crystallization waves show that it obeys the law $\omega \sim k^{3/2}$, which is in accord with (5.7). A meniscus is also characteristic of the surface. Here we seem to encounter a unique situation in that essentially quantum phenomena may be observed visually, which is how they have been revealed in experiments [31].

^{4}He crystals with a hexogonally close-packed crystal lattice possess facets that become atomically smooth at low temperatures [31,32]. A transition of the surface from one state to the other occurs at 0.9 K for the basal plane (0001) and at 1.17 K for the planes (1120) and (1010). This fact is also supported by the small coefficient of the sound transmition (0001) [32]. Rapid decrease of the growth coefficient with decreasing temperature is typical of these surfaces. Other facets remain atomically rough down to the lowest available temperatures, and their growth coefficient increases [32,33–35].

The surfaces of quantum crystals are peculiar two-dimensional quantum liquids, in which terrace edges and kinks are quasiparticles. Only some of their features have been discussed here, many interesting issues in this swiftly developing field having been ignored. Among these is the phase transition from an atomically smooth to an atomically rough state observed for the (0001) and (1120) ^{4}He facets. An obvious singularity of this transition springs from its being two-dimensional. Then the contribution of the fluctuations is proportional to $\ell n(L/a)$, where L is the linear size of the facet on which the phase transition occurs.

References

1. Kittel, C. (1976) *Introduction to Solid State Physics*. Wiley, New York.
2. Ziman, J.M. (1972) *Principles of the Theory of Solids*. Cambridge University Press.
3. Andreev, A.F. and Lifshitz, I.M. (1969) *Zh. Eksp. Teor. Fiz.*, **56**, 2057.
4. Andreev, A.F. (1976) *Usp. Fiz. Nauk*, **118**, 251.
5. Guyer, R.A. and Zane, L.I. (1969) *Phys. Rev.*, **188**, 445; (1970) *Phys. Rev. Lett.*, **24**, 660.
6. Kagan, Yu. and Maksimov, L.A. (1973) *Zh. Eksp. Teor. Fiz.*, **65**, 622.
7. Kagan, Yu. and Klinger, M.I. (1974) *J. Phys.*, **C7**, 2791.
8. Kagan, Yu. and Klinger, M.I. (1976) *Zh. Eksp. Teor. Fiz.*, **70**, 255.
9. Kagan, Yu. and Maksimov, L.A. (1983) *Phys. Lett.*, **A95**, 242.
10. Yu. Kagan In: *Int. Conf. on Defects in Insulating Crystals*. Riga, USSR, Zinatne Publ. House, 1981, p. 17.
11. Eshelby, D. (1956) *Solid State Physics*, Vol. 3, Academic Press, New York.
12. Shklovskii, B.I. and Efros, A.L. (1979) *Electronic Properties of Doped Semiconductors*. (In Russian.) Nauka, Moscow.
13. Gurevich, I.I. and Nikol'skii, B.A. (1976) *Experiments with positive muons*. Preprint IAE-14358, Moscow.
14. Seeger, A. (1979) In: *Hydrogen in Metals*. Springer-Verlag, Berlin.
15. Grebinnik, V.G., Gurevich, I.I., Zhukov, V.A., Manych, A.P., Meleshko, E.A., Muratova, I.A., Nikolskii, B.A., Selivanov, V.I., Suetin, V.A. (1975) *ZhETF* **68**, 1548.
16. Grebinnik, V.G., Gurevich, I.I., Zhukov, V.A., Klimov, A.I., Maiorov, V.N., Manych, A.P., Mel'nikov, E.V., Nikol'skii, B.A., Pirogov, A.V., Ponomarev, A.N., Selivanov, V.I., Suetin, V.A. (1977) *Pisma ZhETF* **25**, 322.
17. Borghini, M., Niinikoski, T.O., Soulie, J.C., Hartmann, O., Karlsson E., Norlin, L.O., Pernestal, K., Kehr, K.W., Richter, D., Walker, E. (1978) *Phys. Rev. Lett.* **40**, 1723.
18. Fiory, A.T., Lynn, K.G., Parkin, D.M., Kossler, W.J., Lankford, W.F. and Stronach, C.E. (1978) *Phys. Rev. Lett.*, **40**, 968.
19. Mikheev, V.A., Esel'son, B.N., Grigor'ev, V.N., Mikhin, M.P. (1978) *Fiz. Nizk. Temp.*, **3**, 385; (1978) *Zh. Eksp. Teor. Fiz.*, **74**, 2311.
20. Allen, A.R. and Richards, M.G. (1978) *Phys. Lett.*, **A65**, 36.
21. Grigor'ev, V.N., Esel'son, B.N. and Mikheev, V.A. (1973) *Pisma ZhETF*, **18**, 289.
22. Schweizer, R., Washburn, S. and Meyer, H. (1979) *J. Low Temp. Phys.*, **37**, 289.
23. Schweizer, R., Washburn, S. and Meyer, H. (1978) *Phys. Rev. Lett.*, **40**, 1035.
24. Andreev, A.F. and Parshin, A.Ya. (1978) *Zh. Eksp. Teor. Fiz.*, **75**, 1510.
25. Andreev, A.F. (1982) In: *Quantum Theory of Solids* (ed. I.M. Lifshitz), p. 11. Mir, Moscow.
26. Chernov, A.A. (1961) *Usp. Fiz. Nauk*, **73**, 277.
27. Jackson, K.A. (1975) In: *Crystal Growth and Characterization* (ed. R. Veda and J.B. Multin), p. 21. North-Holland, Amsterdam.
28. Landau, L.D. (1969) In: *Collected Papers*, Vol. 2, p. 117. (In Russian.) Nauka, Moscow.
29. Lifshitz, I.M. and Kagan, Yu. (1972) *Zh. Eksp. Teor. Fiz.*, **62**, 385.
30. Iordanskii, S.V. and Finkel'shtein, A.M. (1972) *Zh. Eksp. Teor. Fiz.*, **62**, 403.
31. Keshishev, K.O., Parshin, A.Ya. and Babkin, A. (1981) *Pisma ZhETF*, **31**, 63; (1979) *Zh. Eksp. Teor. Fiz.*, **80**, 716.
32. Parshin, A.Ya. (1982) *Physica*, **B109/110**, 189.
33. Castaing, B., Baliber, S. and Laroche, C. (1980) *J. de Physique*, **41**, 897.
34. Landau, J., Lipson, S.G., Määttänen, L.M., Balfaur, L.S. and Edward, D.O. (1980) *Phys. Rev. Lett.*, **45**, 31.
35. Avron, J.E., Balfour, L.S., Kuper, C.G., Landau, J., Lipson, S.G. and Schulman, L.S. (1980) *Phys. Rev. Lett.*, **45**, 814.

CHAPTER 8
Dispersive Transport

A theory for the transport of excess electrons in dielectric amorphous solids is presented in this chapter. First, distinctions between two types of transport are discussed: Gaussian transport, which takes place in ordered media, and dispersive transport, which takes place in disordered media.

The continuous-time random-walk approximation (CTRWA) for electron diffusion in disordered media is described. An essential role is played in this model by a space averaging that allows the application of the random-walk theory developed initially for describing particle diffusion over a regular lattice.

It is explained how, in the CTRWA, Markovian diffusion over a disordered medium becomes non-Markovian diffusion in an ordered medium. The CTRWA is shown to apply when the characteristic length over which the electron hops is greater than the mean distance between the traps the electrons move. This happens if the diffusion is due to the thermal activation of electrons into the conduction band (trap-controlled diffusion).

For hopping diffusion, when electron transport proceeds by means of tunneling transitions between nearest neighbours, and the CTRWA does not hold, a percolation model of dispersive transport is developed. First, purely nondiagonal disorder is considered, when the electron binding energies are equal for all the traps and only the trap-to-trap spacing varies. The traps are subdivided into a diffusion cluster, over which proper electron diffusion takes place and isolated centers, which only exchange electrons with the traps in the diffusion cluster. As a result, it is possible to describe electron transport by means of a modified diffusion equation with sources.

Finally, the percolation model is generalized so as to account for diagonal disorder, when the electron binding energies in the traps also vary.

8.1. GAUSSIAN AND DISPERSIVE TRANSPORT

Electron transport that takes place in solids over localization centers, either via tunneling or activation-like transitions, play an important role in many phenomena in physics, chemistry and biology. This chapter is confined to some issues of electron dispersive transport with which we deal in physics. Electron transport in physico-chemical and biological systems will be discussed in the following chapters.

We shall first consider the term "dispersive transport", which is usually counterposed to the term "Gaussian transport". In order to understand the meaning of these terms, the following situation is considered. Let an electron packet move in a solid dielectric or semiconductor by means of electron hopping from one localization center (trap) to another. Structural point defects may serve as such traps. They can be also created artificially by introducing specific impurities. We shall not for the time being specify how the electron passes from one trap to another, whether by means of activation or via tunneling, for the distinctions between the two kinds of transport have nothing to do with this. The traps are assumed to be arranged in an ordered lattice. The electron motion is then described by the standard diffusion equation, and a packet that was initially localized in a narrow region spreads in the course of time over a broader region according to the well-known law

$$n(r) \sim \exp[-r^2/Dt],$$

where D is the diffusion coefficient, r is the space coordinate and t is time. Such a packet has a Gaussian shape, and its dispersion increases proportionally to $t^{1/2}$. If an external electric field is applied to the sample then the packet drifts, retaining its shape. This kind of transport is therefore known as Gaussian transport.

A more complicated situation occurs in disordered systems. Then the characteristic hopping time between pairs of neighbouring traps may vary within a very broad range. As a result, the usual diffusion equation appears to be inapplicable for the description of the electron motion (this will be discussed in detail later). So the electron packet retains its Gaussian shape and is characterized by an anomalous dispersion, which is why the term "dispersive transport" has arisen as applied to this kind of systems. A similar situation arises in the study of electron transport in some physico-chemical processes. The so-called polychronic kinetics has been developed to account for these processes (i.e. kinetics of processes with a broad spectrum of characteristic times). It will be presented in the next chapter.

In this chapter we shall investigate how electron diffusion in disordered system proceeds. Now the question of how an electron transfers from one trap to another, either via tunneling or by means of activation, plays the principal role. The physical character of the diffusion and the mathematics used for its description depends on the answer to this question. In this connection two models are now elaborated: the CTRWA model [1–3] and the percolation model [4–9]. The first, intended originally for describing the hopping (i.e. tunneling) dispersive transport, is not well formulated for this case. However, it may be applied satisfactorily in studying trap-limited (i.e. activation-like) electron transport [4,10–12]. Hence the necessity arises of developing an alternative model capable of describing hopping dispersive transport. Thus we turn to the percolation model.

These two approaches differ from each other both conceptually and mathematically. To elaborate them, completely different mathematical approaches have been used. An averaging procedure is applied in CTRWA that allows the problem of electron diffusion in a disordered medium to be reduced to a problem of diffusion in an ordered medium. When this is done, however, the usual probability of trap-to-trap hopping is replaced by a distribution of hopping times. As a result, a technique developed in the theory of the random walks over regular lattices [13] now becomes applicable. At the same time, in order to escape a dangerous spatial averaging, percolation theory is used as a basis of the percolation model.

The development of theoretical studies in this field was encouraged by work on improvements in modern copying devices. Experiments were carried out, measuring nonstationary electron diffusion in a number of amorphous substances in external electric fields. Different substances, i.e. amorphous As_2S_3 [3,14], As_2Se_3 [15,16] and ??? [17,18], were studied. A large number of references to experimental papers on dispersive transport can be found in the reviews [19,20].

8.2. CONTINUOUS-TIME RANDOM-WALK APPROXIMATION (CTRWA)

This approximation was proposed by Scher and Lax [1,2] and Scher and Montroll [3] in order to describe the time dependence of current flow in so-called time-of-flight experiments. These are usually arranged as follows: a number of excess electrons is injected by means of an electric discharge or light flash in one of the planes of a flat parallel-sided plate made of an amorphous dielectric (Fig. 8.1). Driven by an external

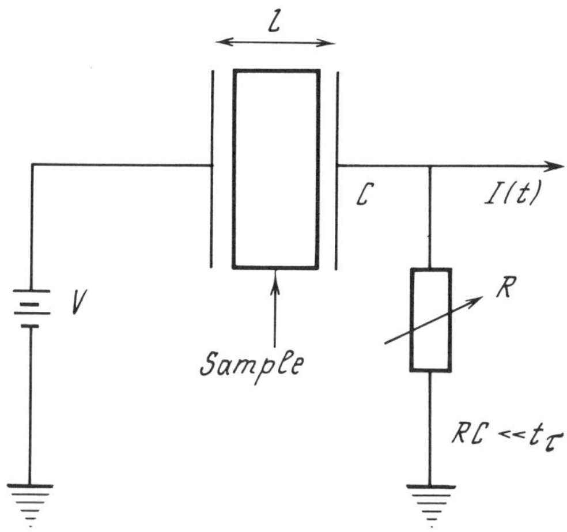

Figure 8.1 A scheme of the time-of-flight experiment.

transversal electric field, they diffuse to the opposite plane. Ideally, all the electrons move with equal velocities, and the time dependence of the current flow in the external circuit can be plotted as a stepwise function as shown in Fig. 8.2. Here it is implied that the excess electron packet with practically zero transverse size goes through a distance L during the time of flight t_τ from one plane to the other. During this period $0 < t < t_\tau$, a finite and constant current is observed in the external circuit, and for $t > t_\tau$ the current in the circuit is zero.

The following mechanisms can be pointed out that result in deviations from the ideal behaviour: (a) a finite delay time RC in the circuit (Fig. 8.1); (b) electron losses due to capture by deep scavengers; (c) deviations of the local drift mobility from its mean value

$$\mu = L^2 t_\tau / V \tag{2.1}$$

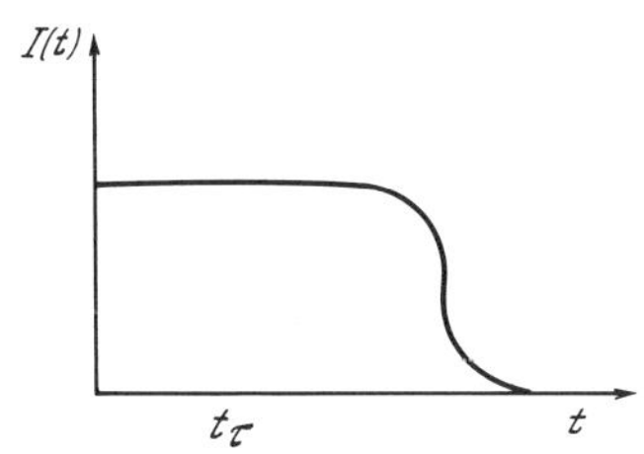

Figure 8.2 The ideal dependence of the current on time in the time-of-flight experiment in the case of ordered medium.

because of inhomogeneity of the sample; (d) local electric fields generated by the electrons captured by the deep scavengers; (e) spreading of the electron packet up to a transverse size comparable to L; and (f) Coulomb interaction of electrons. The effects of nearly all of these mechanisms can be practically eliminated by properly fitting the experimental conditions. (a) and (b) are insignificant if

$$RC \ll t_\tau \ll \tau_D, \tag{2.2}$$

where τ_D is the characteristic time of electron capture by the deep scavengers. Mechanisms (c) and (d) change the current, but cannot distort the step considerably. By properly fitting the intensity and duration of the light flash, the transverse size of the sample, and the electric field strength, it is possible to make the effects of (e) and (f) negligible.

In spite of these facts, the experimentally observed pattern is very far from ideal (see Fig. 8.3). Immediately after the light flash, a short peak of current flow is seen, then a shoulder and a very long tail. The existence of this tail indicates that we are dealing with a statistical process that yields an extremely broad spread of the times of flight of individual electrons. A "universality" property should also be mentioned, which has been observed in [17]. It states that the shape of the transient normalized current flow $I(t)/I(t_\tau)$ varies weakly when the electric field E or the transverse size L of the sample are varied.

The statistical spread of the times of flight is clearly connected in some way with the disorder that is inherent in amorphous materials. This induced the authors of [1–3] to propose a continuous-time random-walk approximation (CTRWA), to be discussed below, for description of the experimental data. The conventional theory of particle diffusion in ordered media deals with the motion of "Gaussian packets" of electrons is based on the mathematical theory of Markovian processes. Its principal equation is of the form [13]

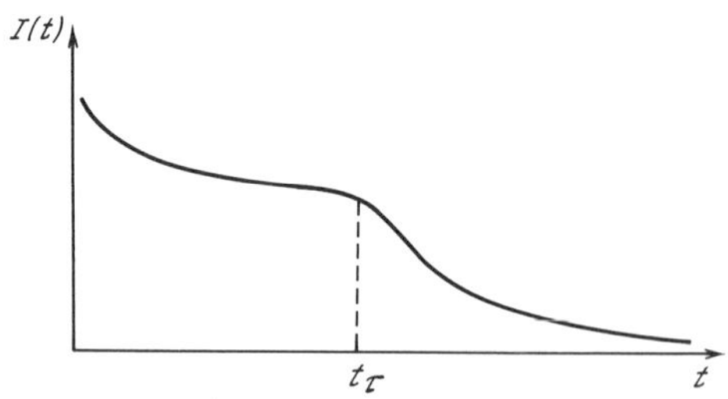

Figure 8.3 Characteristic dependence of the current on time in the time-of-flight experiment in the case of disordered medium.

$$\frac{dP(s,t)}{dt} = \frac{1}{\tau_0} \sum_{s'} [p(s - s') P(s',t) - p(s' - s) P(s,t)], \qquad (2.3)$$

where the function P(s,t) characterizes the probability of finding a particle at the site s at time t. The first term on the right hand side represents the incoming particles hopping from sites s′ to site s, whereas the second term represents the outgoing particles from the site s to sites s′. In the general case, it is also possible to consider non-Markovian diffusion processes, which are described by the generalized equation

$$\frac{dP(s,t)}{dt} = \int_0^t \varphi(t - \tau) \sum_{s'} [p(s - s') P(s',\tau) - p(s' - s) P(s,\tau)]\, d\tau \quad (2.4)$$

Here the function $\varphi(t)$ specifies the "memory" of the system. Its presence means that the evolution of the system for $t' > t$ depends not only on its state at the observation moment $t' = t$, but also on its prehistory, i.e. on its evolution for $t' < t$. This issue will be discussed in more detail in the next section. The Laplace transform of the function $\varphi(t)$ is connected with the Laplace transform of the distribution of the hopping probabilities by means of the simple relation

$$\varphi^*(p) = \frac{p\Psi^*(p)}{1 - \Psi^*(p)}. \qquad (2.5)$$

For Markovian processes there is only one characteristic time τ_0, and

$$\Psi(t) = \frac{1}{\tau_0} \exp[-t/\tau_0]. \qquad (2.6)$$

Equations (2.5) and (2.6) therefore yield

$$\varphi(t) = \frac{1}{\tau_0} \delta(t),$$

and hence (2.4) coincides in this case with (2.3).

Particle diffusion over traps in a disordered medium cannot be characterized by a single hopping time τ_0, since the parameter τ_0 in (2.3) now depends on the site numbers between which the particle hops. The authors of [1–3] formulated a model in which the electron diffusion in a disordered medium is nevertheless described by (2.4). The amorphous substance is viewed as a medium in which traps are randomly distributed. Electrons move in this medium by means of trap-to-trap hopping. This hopping may be either due to tunneling under the barrier or the electrons may be thermally activated from some sites (traps) into the conduction band, move in it for a while, and then be captured by other traps (trap-controlled transport). In any case the characteristic hopping time varies markedly for different pairs of traps between which the electron hops. These variations are especially pronounced for the tunneling transitions, when the characteristic time depends exponentially on the trap-to-trap distance. In this case, even technically very homogeneous substances should be treated as strongly disordered systems, for the spread of hopping times may be over many orders of magnitude.

In order to make use of the mathematics elaborated in the random walk theory, especially for a description of particle motion over a lattice [13], the whole space filled by the amorphous substance is subdivided into identical cells arranged in a regular lattice. The cell size is assumed to be sufficiently large to contain a large number of randomly situated traps. The fluctuations of the trap density in the cells are therefore small enough that these cells can be treated as being equivalent from this point of view as well.

In the course of its motion, the electron hops subsequently from one trap to another and transfers finally from one cell into another. Now, in dealing with motion over the cell lattice, the time passing from the electron entrance into a cell up to its exit should be treated as a hopping time. Disorder of the trap distribution within the cell is reflected by the form of the hopping-time distribution function $\Psi(t)$, which is the probability of the event that an electron leaves the cell after a time t passes after its entrance (calculations of the function $\Psi(t)$ will be presented below). The above assumptions allow the problem of electron motion over a disordered medium to be reduced to that of motion over a regular lattice, the disorder now being characterized by the hopping-time distribution function $\Psi(t)$.

The problem is analysed by means of the random-walk theory, which leads, in particular, to (2.4) for non-Markovian diffusion. It is assumed that a simple cubic lattice of cells is constructed in such a way that one of its axes with the coordinate s_1 is directed along the electric-field vector $\mathbf{E}$. The electric current caused by the electron packet moving in the amorphous substance can be found if the mean coordinate s_1 of the packet is known:

$$I(t) \sim \frac{d\langle s_1\rangle}{dt}, \tag{2.7}$$

where

$$\langle s_1\rangle = \sum_s s_1 P(s,t). \tag{2.8}$$

For a Gaussian packet (for which the hopping probability is characterized by a single time τ_0) random-walk theory results in

$$\langle s_1(t)\rangle = \bar{s}t/\tau_0, \tag{2.9}$$

where $\bar{s}$ is the mean particle displacement along the s_1 axis per hop. Similarly, the dispersion of the packet is

$$\sigma(t) = \{\langle s_1^2(t)\rangle - \langle s_1(t)\rangle^2\}^{1/2}, \tag{2.10}$$

which in the Gaussian case is

$$\sigma(t)/\langle s_1(t)\rangle \sim (t/\tau_0)^{-1/2}. \tag{2.11}$$

Therefore, according to (2.9) and (2.11), the electron packet moves uniformly and its width grows owing to diffusion. Here the electric current does not depend on time until the packet reaches the opposite face of the condenser (see Fig. 8.1). Then the current falls rapidly to zero. So a pattern results that is close to the ideal one (Fig. 8.2). All these statements hold not only for the exponential function $\Psi(t)$, (2.6), but also for any other function for which the first two moments, $\bar{t}$ and $\overline{t^2}$, exist, where

$$\overline{t^n} = \int_0^\infty dt\, t^n \Psi(t),$$

and under the condition that only times satisfying the conditions $t > \bar{t}$, $(\overline{t^2})^{1/2}$ are considered.

A completely different situation arises if the function $\Psi(t)$ decreases at large times slowly and the second moment $\overline{t^2}$ is infinite. For example, consider the function

$$\Psi(t) = \left[A\left(\frac{t}{\tau_1}\right)^{1+\alpha}\Gamma(1-\alpha)\right]^{-1}, \quad 0 < \alpha < 1, \tag{2.12}$$

in which τ_1 is a characteristic time scale, A is a constant and $\Gamma(1-\alpha)$ is the gamma function. According to Schlesinger's calculations [21] which used the Tauberian theorem [22] for the asymptotic behaviour of the Laplace transform of the function (2.12), the mean coordinate is determined by the equation

$$\langle s_1(t)\rangle \approx \bar{s}\frac{A}{\Gamma(1+\alpha)}\left(\frac{t}{\tau_0}\right)^{\alpha}, \tag{2.13}$$

while the dispersion is

$$\frac{\sigma(t)}{\langle s_1(t)\rangle} \approx \left\{\frac{2\Gamma^2(1+\alpha)}{\Gamma(1+2\alpha)} - 1\right\}^{1/2} \tag{2.14}$$

Therefore the diffusing electrons now have a very unusual behaviour. The mean coordinate is proportional to a fractional power of time, and the ratio of the dispersion and the mean coordinate does not depend on time at all. Hence the motion of the packet as a whole and its spreading are indistinguishable. Differentiating (2.13) with respect to t, it is found that the current depends on t according to the law

$$I(t) \sim t^{-(1-\alpha)} \tag{2.15}$$

There is no plateau, in contrast with the Gaussian case — instead the current decreases with increasing time, tending asymptotically to zero at $t \to \infty$ even if no carrier absorption by the opposite face of the plate takes place. Moreover, the results (2.13) and (2.14) imply that the maximum of the electron density distribution is practically not displaced, the packet only widens asymmetrically in the direction of the electric field; therefore the mean coordinate $\langle s_1(t) \rangle$ and the dispersion depend on time in the same way.

To describe the experiment completely, account should be taken of the finite size of the sample and of the fact that the carriers that have reached the right face of the condenser no longer contribute to the current flow. The function $P_b(s_1, t)$ is therefore considered that determines the probability for a carrier that left the left face of the condenser ($s_1 = 0$) at $t = 0$ to arrive at a cell $s_1 \neq 0$ at $t \neq 0$. This event is clearly possible only if the carrier has never crossed the right boundary ($s_1 = N_1$) during the time from 0 to t.

The probability of arriving at s_1 from 0 along a path crossing the right face can be represented as the product of the probability $F(N_1, \tau)$ of reaching the boundary for the first time at an intermediate moment $0 < \tau < t$ and the probability of arriving at the point s_1 from the right boundary along any path during the remaining time $t - \tau$. Using the quasiperiodic boundary conditions for the function $P(s_1 - N_1, t - \tau)$ gives

$$P_b(s_1, t) = P(s_1, t) - \int_0^t d\tau \, P(s_1, t-\tau) \, F(N_1, \tau). \tag{2.16}$$

Equation (2.16) is then multiplied by $s_1 - N_1$ and summed over s to give

$$\langle s_1(t) \rangle_b = \langle s_1(t) \rangle - \int_0^t d\tau \, \langle s_1(t-\tau) \rangle \, F(N_1, \tau). \tag{2.17}$$

At the initial stage, when the particle has had no time to make enough steps necessary to reach the right boundary, the second term on the right-hand side of (2.17) may be neglected, and the results obtained above without taking account of the electron absorption by the right face of the condenser may be used. This approximation is valid if the inequality $\langle s_1(t)\rangle \ll L$ holds, i.e. during the time span

$$t < t_\tau = \left(\frac{\Gamma(\alpha + 1)}{A}\right)^{1/\alpha} \left(\frac{L}{\bar{s}}\right)^{1/\alpha} \tau_1. \tag{2.18}$$

In the opposite case, for $t \gg t_\tau$ the electron absorption at the right boundary is of importance. The second term on the right of (2.17) grows, and compensates nearly completely for the first term, so that their difference tends to zero. To study the electric-current behaviour in this case, it is convenient to take the Laplace transform of (2.18) and to use the asymptotic behaviour of the function

$$F^*(N,p) \approx 1 - ap^\alpha \tag{2.19}$$

at small p (as found in [21]).

In contrast with Gaussian transport, when the current, after the electron packet reaches the right boundary, starts decreasing exponentially with time, in the case of dispersive transport the electrons arrival at the boundary and their absorption occur much more slowly, since the electron mobility decreases in the course of time. Equations (2.17)–(2.19) result in the following time-dependent function for the current:

$$I(t) \sim \begin{cases} t^{-(1-\alpha)} & (t < t_\tau), \\ t^{-(1+\alpha)} & (t > t_\tau). \end{cases} \tag{2.20}$$

From this arises one of the most significant conclusions of the theory, which can be directly verified experimentally: the sum of the indices of the two asymptotics is -2, regardless of the type of substance, the latter being specified by the α value.

Now the question to be discussed is why kinetic processes in disordered media should be described by using the distribution $\Psi(t)$ with the asymptotics (2.12). It was initially assumed in [1–3] that electron transport proceeds via tunneling. However, it later became clear that the CTRWA does not hold in this case, but may be applied to trap-limited transport. Nevertheless, $\Psi(t)$ will be calculated here for the hopping tunneling transport, which will help us to better understand the differences between the two cases. These results will also be used below. The function $\Psi(t)$ for trap-limited electron transport will be calculated in the next sextion.

The quantity Q(t) is considered that determines the probability of finding a particle in a given cell at $t > 0$ if it has been there at $t = 0$. $Q(t)$ satisfies the equation

$$\frac{dQ}{dt} = -Q \sum_j W(r_j), \tag{2.21}$$

where

$$W(r_j) = \nu \exp[-2r_j/a] \tag{2.22}$$

is the rate constant of the tunneling transitions between traps separated by distance r_j. Then the solution of (2.21) is averaged over all possible configurations of traps [1–3]:

$$Q(t) = \exp\{-\int d^3r\, p(r)\, (1 - \exp[-W(r)t])\}, \tag{2.23}$$

where $p(r)$ is the probability density of finding a trap at a distance r.

In order to find the function $\Psi(t) = dQ(t)/dt$, the integrand in (2.23) is replaced by a stepwise function

$$1 - \exp[-W(r)t] \approx \theta(\tfrac{1}{2} a\, \ell n\, \nu t - r),$$

and the trap distribution is assumed to be random, i.e. $p(r) = c_t$, where c_t is the relative trap concentration. As a result,

$$\Psi(t) = 4\pi\, \nu c_t\, \ell n^2 \nu t\, (\nu t)^{-[1 + (4\pi c_t/3)\, \ell n^2 \nu t]} \tag{2.24}$$

where

$$\nu t \gg 1.$$

Therefore the distribution $\Psi(t)$ is close to the function (2.12), the parameter $\alpha = (4\pi c_t/3)\, \ell n^2 \nu t$ being dependent on time. However, this dependence is logarithmically small, and it may be neglected for a large enough time interval. Sometimes it is important to take account of the time dependence of α. For example, the function with asymptotics (2.12) does not even have a first moment, whereas this does exist for the function (2.24). We shall repeatedly meet the function (2.24), both in questions of dispersive transport and in the next chapter, where electron scavenging will be studied.

8.3. WHEN DOES THE CTRWA HOLD?

The most important approximation used in the CTRWA model [1–3] is a peculiar averaging procedure, which allows the problem of electron motion in a disordered medium to be reduced to a problem of electron motion in an effective ordered medium in which the site-to-site hopping probability is specified by a distribution of times rather than by a single time and the function $\Psi(t)$ is essentially nonexponential. As a result, the mathematics of the random-walk theory is now applicable, which is equivalent to solution of a kinetic equation of the type (2.4) describing a non-Markovian process with a memory function $\varphi(t)$.

Before continuing our speculations, the definition of a Markovian process is recalled. The system may occur in different states α. Its state is known at time t. The evolution of the state is Markovian if the prediction of the state at a subsequent time $t'' > t$ does not require any knowledge of its state at an earlier time $t' < t$. In other words, the conditional probability

$$P = P(\alpha, t'' \mid \beta, t; \gamma, t') \tag{3.1}$$

of finding the system in state α at time t'', if it is known to have been in state β at time $t < t''$ and in state γ at $t' < t$ does not depend on γ for a Markovian process. If this is not the case then, the process is non-Markovian, and the system retains, in some way, a memory of its previous evolution. It is this property that is described by the function $\varphi(t - \tau)$.

If coherent effects (e.g. Anderson localization [23]) are ignored then the electron random walks are completely Markovian. Let an electron be at site i. The probability of hopping to another site j does not depend on the site it has come from earlier. This assertion holds for both ordered and disordered systems, which is why the question arises as to why and under what conditions does (2.4) for a non-Markovian process in an ordered system describe a Markovian process in a disordered system?

In brief, Pollak's answer [4] to this question can be formulated as follows: the equation for the non-Markovian process in an ordered system comes not from the non-Markovian character of the process, but rather from the "coarsening" averaging procedure, when a certain amount of information is lost. This assertion can easily be understood by means of a simple illustration. A Markovian process of electron motion along a one-dimensional chain of sites is considered (Fig. 8.4). Now for the sake of simplicity the electron is assumed to hop after a

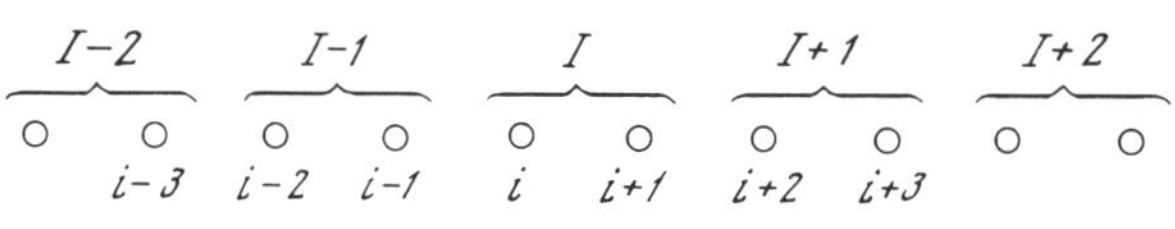

Figure 8.4 One-dimensional chain of sites in which groups of two sites are distinguished.

certain time interval Δt to a neighbouring site, left or right, with probability $\frac{1}{2}$. In this case the only nonzero conditional probabilities are

$$\left.\begin{aligned} P(i,t\,|\,i-1,\, t-\Delta t) &= 1/2, \\ P(i,t\,|\,i+1,\, t-\Delta t) &= 1/2. \end{aligned}\right\} \tag{3.2}$$

It is also clear that information about where the electron has been at the time $t - 2\Delta t$ is not needed.

We now turn to a coarser description, when groups containing pairs of neighbouring sites (see Fig. 8.4) are considered. The motion over these groups is non-Markovian. In fact, it can easily be verified that the conditional probabilities

$$\left.\begin{aligned} P(I,t\,|\,I-1,\, t-\Delta t;\; I-1,\, t-2\Delta t) &= \tfrac{1}{4}, \\ P(I,t\,|\,I-1,\, t-\Delta t;\; I,\, t-2\Delta t) &= \tfrac{1}{2}, \\ P(I,t\,|\,I-1,\, t-\Delta t;\; I-2,\, t-2\Delta t) &= 0 \end{aligned}\right\} \tag{3.3}$$

can be found only if it is known from which site the electron came at time $t - 2\Delta t$. The meaning of this result is clear enough. When treating the electron motion in detail we always know which site it is visiting now, and hence the state of the system is completely determined. On having "coarsened" the description, the site group number visited by the electron does not determine the state of the system completely, since it remains unknown which site of the site group is actually visited. Therefore part of the information connected with the internal structure of the group is lost. To compensate for this loss, it is necessary to know the state of the system at time $t - 2\Delta t$.

Generally, this may be formulated more formally. The fact that the site occupation functions f_i can be found, if their values at earlier times are known, follows from the possibility of describing their evolution by means of Boltzmann equations, i.e. by means of a system of first-order differential equations. In the coarsened description the occupation functions f_I for the site groups are linear combinations of the functions f_i. However, there are fewer functions f_I than functions f_i, and the more sites are included in a group the fewer functions there are. The theory of linear differential equations states that one may lower the number of variables by correspondingly raising the order of the differential equations. Therefore the evolution of the functions f_I is described by a system of higher-order differential equations, and knowledge of $f_I(t)$ at a certain instant does not give a complete set of initial conditions. It is also necessary to know either the derivatives $f_I^{(n)}(t)$ at the same time or the values $f_I(t')$ at other times. It is clear that the total number of equations in the set of initial conditions must coincide with the total number of sites. In the limit of groups containing macroscopically large numbers of sites there is a passage from initial conditions given at discrete times to a continuous time range $0 < \tau < t$. This is the property that was meant when talking about retaining memory in the CTRWA.

This approximation can be used for macroscopically uniform materials. When dividing the material into cells, their size h should be chosen to be large enough so that the fluctuations of the site distribution are small and all the cells can be treated as equivalent. This requirement settles a scale h above which the substance looks uniform. The trap (site) number in a cell of size h is very large, and the electron motion over cells is non-Markovian. Here on the one hand, we have complicated the problem by replacing a Markovian process by a non-Markovian one, while, on the other hand, we have simplified it significantly by considering an effectively uniform medium.

The question that should be answered now is as follows: under what conditions and to what extent does the information loss caused by this averaging procedure affect the validity of the physical results obtained under the CTRWA? To gain a better understanding of this question, transport processes are considered in the case when an electron transfers from trap to trap via an excitation into the conduction band (trap-limited transport). Such a situation often occurs in impure semiconductors, in which the binding energy in a trap is comparable to the temperature.

As shown by Pollak [4], electron transport in such a system may be quite adequately described by the CTRWA. The kinetic equations for the trap occupation numbers f_i and the conduction band occupation number f_c are

$$\frac{\partial}{\partial t} f_i = f_c w_{ci} - f_i w_{ic}, \tag{3.4a}$$

$$\frac{\partial}{\partial t} f_c = \sum_i (f_i w_{ic} - f_c w_{ci}). \tag{3.4b}$$

Here w_{ci} and w_{ic} are the rate constants of the band-to-trap and backward transitions. w_{ci} corresponds to a transition with decreasing energy, and is not connected with the activation; w_{ic} corresponds to a transition with increasing energy, and remains small at low enough temperatures, $w_{ci} \gg w_{ic}$. Therefore it may be assumed that

$$\frac{\partial f_c}{\partial t} \approx 0$$

and

$$f_c = \sum_i f_i \frac{w_{ic}}{\sum_j w_{cj}}. \tag{3.5}$$

Substituting (3.5) into (3.4) gives

$$\frac{\partial f_i}{\partial t} \approx \sum_{j \neq i} f_j \frac{w_{jc}\, w_{ci}}{\sum_j w_{ck}} - f_i w_{ic}. \tag{3.6}$$

Now the rate constant of the trap-to-trap transition is

$$w_{ji} \equiv w_{jc}\, w_{ci} \Big/ \sum_k w_{ck}. \tag{3.7}$$

Then using the equality

$$\sum_j w_{ij} = w_{ic} \sum_j \frac{w_{cj}}{\sum_k w_{ck}} = w_{ic} \tag{3.8}$$

gives

$$\frac{\partial f_i(t)}{\partial t} = \sum_j (f_j(t)\, w_{ji} - f_i(t)\, w_{ij}). \tag{3.9}$$

w_{ji} in (3.7) can be written in the form $w_j p_{ij}$, where $w_j = w_0 \exp[-E_j/k_B T]$ is the activation rate constant and p_{ij} is the probability for the carrier to reach the trap j if it has left the trap i. It can easily be verified that

$0 \leqslant p_{ij} \leqslant 1$ and $\sum_j p_{ij} = 1$.

Now, according to the CTRWA, the medium is divided into cubic cells. Then all the traps that are in one cell comprise a group. The size of the cell should not be smaller than h. The probability of transition from cell I to cell J is

$$p_{IJ} = \sum_{j \in J} p_{ij} \qquad (i \in I). \tag{3.10}$$

This definition makes sense only if p_{ij} depends weakly on i for all $i \in I$. Then the quantity p_{IJ} may be used in (2.4). This is the case if the probability of transition to trap i does not depend on its location within cell I and intragroup transitions are sufficiently rare. These conditions are satisfied if the mean length of the trap-to-trap transition is larger than the cube size, i.e. $\ell > h$. If, however, $\ell \ll h$ then the carrier may arrive at cell I from another cell K, the transfer taking place between traps lying near the K–J interface. In the initial period after the transfer the probability p_{IJ} is large if $J = K$, and small if $J \neq K$. If the carrier remains within cell I for a long time then it can diffuse deep into the cell, thus causing a decrease in p_{IK} and an increase in p_{IJ} ($J \neq K$). Therefore for $\ell \ll h$ the quantities p_{IJ} cannot be treated as constants, and their time dependence must be taken into account. Such a situation cannot be described by a single function $\Psi(t)$ as in the CTRWA.

If the electron hops from trap to trap via the conduction band then the average hopping length ℓ is of the order of the electron free path in the conduction band with respect to its capture by traps. This quantity can be larger than the mean trap-to-trap distance, and may even be so large that the condition $\ell \gg h$ is satisfied. A distribution function $\Psi(t)$ can then be introduced, and the results obtained in Section 8.2 under the CTRWA become valid.

The form of the function $\Psi(t)$ for trap-limited electron transport [4,11,12] is now found. It is assumed that the electron can initially be found in any trap of a given cell with equal probability. This means that the dependence of the probability w_{ci} on the trap depth ϵ_i is neglected. In order to find the function $\Psi(t)$, the quantity

$$w_i \exp[-w_i t] \tag{3.11}$$

must be averaged over all values of w_i:

$$\Psi(t) = \int dw\, p(w)\, w \exp[-wt]. \tag{3.12}$$

Here p(w) is the probability density of finding a value w. It is connected, through its definition, with the density of impurity states, $f(\epsilon) = N_t^{-1} p(\epsilon)$, where N_t is the total concentration of traps. If p is independent of ϵ then

$$p(w) \quad w^{-1},$$

so that

$$\Psi(t) \sim t^{-1}.$$

If, however, there is a dependence of $p(\epsilon)$ on ϵ then

$$\Psi(t) \sim t^{-(1+\alpha)}. \tag{3.13}$$

The parameter α can be found using the relation

$$p(w) = w\, p[-\ell n(w/w_0)] \tag{3.14}$$

for the probability densities. Equation (3.13) arises if the function (3.14) in a certain range of ϵ values is approximated by the formula

$$p\left[-\ell n\left(\frac{w}{w_0}\right)\right] \sim K\left(\frac{w}{w_0}\right)^{\alpha}, \tag{3.15}$$

where K is a constant, so that

$$\alpha = k_B T \left. \frac{d\,\ell n\, p(\epsilon)}{d\epsilon} \right|_{\epsilon = -k_B T\,\ell n(tw_0)} . \tag{3.16}$$

It is also possible to drop the assumption above that w_{cj} does not depend on ϵ_j. Then

$$\Psi(t) = \int d\epsilon\, p(\epsilon)\, w_c(\epsilon)\, w(\epsilon) \exp[-w(\epsilon)t] \tag{3.17}$$

(where $w_c(\epsilon) = w_{cj}$), instead of (3.12). Assuming that $w_c(\epsilon)$ depends on ϵ more weakly than exponentially, an equation for α is obtained that is analogous to (3.16):

$$\alpha = \left. \frac{d\,\ell n[p(\epsilon)\, w_c(\epsilon)]}{d\epsilon} \right|_{\epsilon = -k_B T\,\ell n(tw_0)} . \tag{3.18}$$

All of the above results may be treated as proof of the applicability of the CTRWA for the description of dispersive transport in systems in which the electron transport proceeds via the conduction band and hence the condition $\ell \gg h$ is satisfied. Meanwhile, the characteristic hopping length in the systems, where the electron hops from trap to trap directly by means of tunneling without any participation of the conduction band, is of the order of the mean trap-to-trap distance

$$\ell \sim (3/4\pi N_t)^{1/3}. \tag{3.19}$$

Thus the condition

$$h \gg (3/4\pi N_t)^{1/3} \tag{3.20}$$

on h that specifies the region that may be treated as macroscopically uniform holds. Therefore $h \gg \ell$ and the CTRWA averaging procedure does not apply in this case.

This result seems to be very instructive. It is known that the theory of disordered systems often uses various averaging procedures. Care should be taken in applying these, and justifications sought in each particular case. The above analysis proposed by Pollak [4] is just such a justification for the case of dispersive transport. The result can be formulated as follows: if the condition $\ell > h$ holds then the system is in a sense "self-averaging", i.e. the averaging in the mathematical calculations is dictated by the nature of the physical system. On the other hand, for $\ell \ll h$ the situation is quite the opposite, and application of the formal averaging is not physically reasonable.

Klafter and Silbey [24], discussed the validity of CTRWA and the possibility of introducing the function $\langle P_i(t) \rangle$ averaged over trap configurations, which characterizes the probability of finding an electron at the site i; they then proved the applicability of (2.4) to it.

The principal question here is whether the averaged probability function $\langle P_i(t)\rangle$ can be used. If this question is positively answered beforehand then the assertion that this function satisfies (2.4) follows. Pollak's statement is that the averaged function $\langle P_i(t)\rangle$ may *not* always be used. It is worth pointing out a similar issue in Anderson localization [23]. This effect cannot be obtained by considering an averaged one-particle Green function. Either the averaging must be renounced [23] or more complicated averaged objects, such as two-particle Green functions must be introduced (see e.g. the review [25]).

8.4. PERCOLATION MODEL OF DISPERSIVE TRANSPORT. BOND DISORDER.

If an electron hops from trap to trap via quantum-mechanical tunneling then Pollak's arguments (see Section 8.3), which allowed the application of the CTRWA to trap-limited transport, do not apply. In fact, the characteristic hopping length in that case is of the order of the mean distance between the traps

$$\ell \sim \left(\frac{4\pi}{3} N_t\right)^{-1/3}, \tag{4.1}$$

and hence

$$h \gg \ell. \tag{4.2}$$

The inequality (4.2) follows from the obvious requirement that a space domain may be treated as uniform only if it contains a large number of impurities. The necessity therefore arises for a model that can replace the CTRWA when the condition (4.2) holds. Pollak [4] proposed to base such a model upon percolation ideas. Later, this approach was also used by Zvyagin [6,7] for a description of the initial stage of the process. Finally, a percolation model of dispersive transport was worked out in detail by Chekunaev *et al.* [8,9]. (These results are the basis of this section.) Similar approaches were used by Kogan and Shklovskii [26,27] when analysing the low-frequency noise in hopping conduction systems.

Amorphous matter is considered as a medium with randomly distributed localization centers (traps). Excess electrons, injected in some way (see the description of the experiment in Section 8.2), occupy a small fraction of these centers and may tunnel from one center to another. In this section we shall neglect the energy distribution of traps;

i.e. the electron energy is assumed not to change after each hop, and the only variable specifying the process is the distance r between the neighbouring localization centers. Such disorder is often called bond disorder. In fact, the binding energies are always distributed within a range $\delta\epsilon$, and the above assumption is indeed valid if the temperature is high enough for the inequality

$$k_B T \gg \delta\epsilon \tag{4.3}$$

to hold. A theory of dispersive transport at low temperatures (opposite inequality to (4.3)) is presented in Section 8.5, where this criterion is formulated more precisely.

Because of (4.3), the electron energies in traps i and j may be assumed to be equal. Then the rate constant of the electron tunneling transition between centers i and j decreases with increasing r as

$$W_{ij} = W_{ji} = \nu \exp[-2r/a], \tag{4.4}$$

where ν is a characteristic frequency and a is the parameter of the electron wave-function localization. Any dependence of ν and a on i and j is neglected.

The characteristic time $\tau = W^{-1}$ of the electron tunneling transition depends exponentially on the random parameter r. The velocity of the electron motion along various paths therefore varies very strongly. Here the term "path" implies a sequence of traps visited by an electron that tunnels from one trap to another. If a path is chosen along which the traps lie close to each other and the characteristic hopping times are small then the velocity of electron motion is rather high. At the same time, another path might have been chosen with the same geometry and length but with fewer traps. Then the characteristic distances and moreover the hopping times are much larger and the electron moves very slowly along this path. Its velocity may be less by orders of magnitude.

The situation is quite different in a medium with an ordered distribution of traps. Then the time Δt for particle motion along a path L is determined by

$$\Delta t \sim \tau L/r_0 \tag{4.5}$$

and does not depend on its geometrical shape or its spatial location. Because of this, the motion in ordered systems may be characterized by a universal parameter, i.e. the mobility

$$\mu_0 \sim \frac{e\nu r_0^2}{k_B T} \exp\left[-\frac{2r_0}{a}\right], \tag{4.6}$$

and owing to the equivalence of all the paths, the particle flow is distributed uniformly over them. Inhomogeneity is possibly connected with the sample geometry, initial conditions and external fields. Such an inhomogeneity, however, is on a macroscopic scale, and at any rate its characteristic size is larger than h. In a disordered medium there are "rapid" and "slow" paths (under the condition $h \gg \ell$) caused by fluctuations of the spatial distribution of traps. As a result, a microscopically inhomogeneous current flow arises whose scale is less than h. Electrons predominantly move along the rapid paths, while the slow paths do not practically contribute to the current flow, forming stagnation domains, which serve as sources and sinks. It is clear that such a pattern cannot be adequately described within the CTRWA, and an alternative approach must be applied that does not make use (at least in the essential initial stages of the analysis) of any averaging. Percolation theory is very suitable for this. It allows separation of the rapid and slow paths and leads to a description of electron diffusion in such a system.

To distinguish a set of rapid paths along which the charge is largely transferred speculations are made that are similar to those of [28–30] (see also [31]), where the conductivities of strongly disordered media have been calculated. A certain value r_m is set in the problem of spheres (see e.g. [31]). The traps separated by a distance $r_{ij} \leqslant r_m$ are assumed to be bound. Otherwise, they are not bound. Trajectories passing via bound traps are assumed to be rapid. All the other trajectories are slow. So the principal question now is how to choose the optimal value of r_m. It is clear that it must be larger than the critical percolation radius,

$$r_m \geqslant r_c. \tag{4.7}$$

Otherwise, all the rapid paths are finite and the macroscopic current flow along them is zero. It is still zero if $r_m = r_c$ is chosen, since the density of the infinite cluster in this case is zero. A finite current flow may arise if the rapid paths are distinguished using a value $r_m > r_c$. Then, on one hand, the density of the infinite cluster grows with r_m, resulting in a growth of the current flowing through it. On the other hand, carriers have to hop still larger distances and starting from $r_m \geqslant r_d$ any further increase of r_m no longer has a significant influence on the magnitude of the current flowing through the infinite cluster. The infinite cluster with $r_m = r_d$ we call the diffusion cluster (DC). (In the theory of hopping conductivity of strongly disordered systems it is often called the critical network.) It follows from its definition that the current of electrons flowing over it coincides practically with the total current flow through the sample.

In order to understand by how much r_d must exceed r_c, the following considerations are made. At a given value of r_m the current flowing through the infinite cluster can be evaluated by means of the integral

$$I \sim \int_{r_c}^{r_m} f(\xi) \exp[-2\xi/a]\, d\xi. \tag{4.8}$$

Here the function $f(\xi)$ determines the "density of paths" in the infinite cluster $r_c \leqslant \xi$. It is clear that near the percolation threshold $f(\xi)$ depends on $(\xi - r_c)/r_c$ according to a power law with a certain critical index. Therefore the convergence of the integral is determined by the exponential factor characterizing the trap-to-trap hopping rate. So it can be seen that the DC is the infinite cluster satisfying the bond criterion

$$r_{ij} \leqslant r_d \approx r_c + a.$$

Thus it may be assumed that hopping electron diffusion proceeds nearly exclusively over the DC. The accuracy of this assertion is higher the stronger the dependence of the characteristic hopping time on the distance at $r_{ij} \sim r_d$. Fortunately, the condition $a \ll r_c$ usually holds well in dispersive-transport experiments, and hence the application of percolation techniques seems to be quite reliable.

Electron hopping diffusion thus proceeds nearly exclusively over the DC. It is a network with characteristic cell size of order L. Those localization centers that lie inside the cells only exchange electrons with the DCs, and no current flows over them. Such centers will be called isolated centers (IC). The proposed mechanism of electron transport is outlined schematically in Fig. 8.5. The DC is shown by the wavy solid line. Letters A and B designate two arbitrary ICs. If at $t = 0$ an electron is in the IC A then at $t > 0$ it may take the shortest path to a center A′

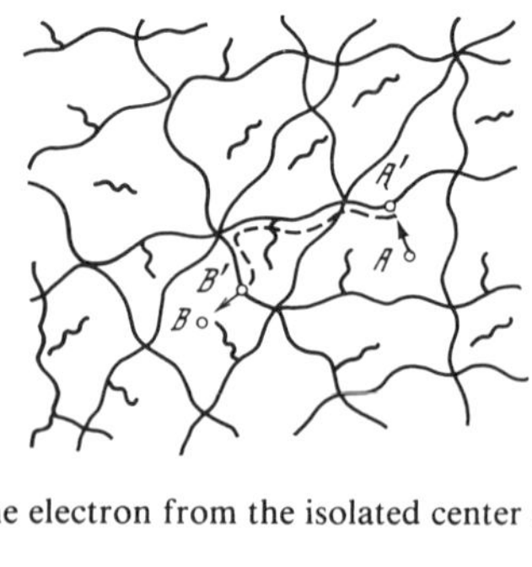

Figure 8.5 Motion of the electron from the isolated center A to the isolated center B via the diffusion claster.

belonging to the DC and start moving in the direction shown by the dashed wavy line. On having reached B', the electron may either go on moving in the DC or leave it for the IC B. In principle, it might have tunneled directly from A to B or passed from A to B along a path that does not belong to the DC. However, such processes are extremely slow, and they will be neglected.

In spite of the fact that the current flow is largely determined by the motion of carriers over the DC, ICs that permanently exchange electrons with the DC should not be completely ignored. Owing to this relatively slow process they play roles as sources and sinks for the electrons moving over the DC.

A mathematical description of this process is needed. For this sake a definition of the distance R between an IC and the DC is introduced. Among all the paths leading from a given IC A to a trap belonging to the DC a trajectory is chosen that requires the shortest time to pass it (see Fig. 8.6). Then, among all the hops to be performed by the electron when moving along this minimal trajectory, the longest is chosen. The length R of this hop is called the distance between the IC A and the DC. Therefore, the exchange rate between the IC and the DC is determined practically completely by the chosen optimal hop, i.e.

$$W_{ex}(R) \sim \nu \exp[-2R/a]. \tag{4.9}$$

This seems to be valid owing to the exponentially strong dependence of the tunneling rate constant of the trap-to-trap distance at $r_d \gg a$.

We now turn to electron exchange between the DC and ICs. First the number of ICs, $N(R \leqslant R_0)$, that lie not further from the DC than R_0 is evaluated. All these traps belong to an infinite cluster with $r_m = R_0$,

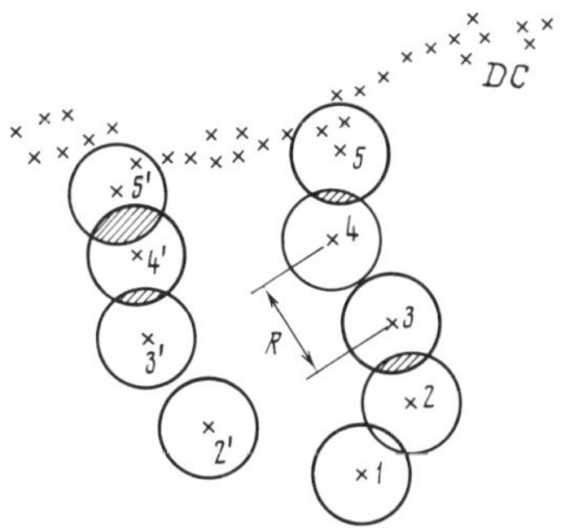

Figure 8.6 Illustration of the definition of the distance, R, from the isolated center 1 to the diffusion claster.

and the traps that are further from the DC than R_0 do not belong to this cluster. Therefore

$$N(R \leqslant R_0) \sim \left(\frac{R_0 - r_c}{r_c}\right)^{\beta}, \tag{4.10}$$

where $\beta < 1$ is the critical index. This estimate is correct if $R_0 - r_c \gg a$ and $(R_0 - r_c)/r_c \ll 1$. Then the number of ICs within the range $(R_0, R_0 + dR)$ is given by the expression

$$\frac{dN(R \leqslant R_0)}{dR}\, dR \sim \frac{\beta}{r_c}\left[\frac{R_0 - r_c}{r_c}\right]^{\beta}. \tag{4.11}$$

However, for larger distances R the Poisson distribution may be used. In this case the number of traps further from the DC than R_0 is determined by the formula

$$N(R \geqslant R_0) \sim \exp\left[-\frac{4\pi}{3} N_t R_0^3\right], \tag{4.12}$$

which holds at $R_0 \gg r_c$.

If the Poisson formula is used then it implies that the first hop from the IC is the critical hop determining at $R \gg r_c$ the distance from the IC to the DC. This situation is presented by an isolated trap surrounded by a volume of size R free of other traps. If a cluster of several traps is surrounded by a free volume then it is not the first hop that will play the role of the critical one. The probability of such a cluster at $R \gg r_c$ is small, and the corresponding contribution is not taken account of in (4.12).

The kinetics of electron exchange between the DC and ICs at distance R is described by the equation

$$\frac{dN_i(t)}{dt} = W_{ex}(R_i)\,[\eta(t) - N_i(t)]. \tag{4.13}$$

Here $N_i(t)$ is the probability of finding an electron in the IC at time t; $\eta(t)$ is the probability of finding an electron in a center belonging to the DC. Strictly speaking, it is necessary to consider the probabilities of finding an electron in two centers i and j between which the critical hop is performed. Moving towards the DC, the center j is reached that belongs to an infinite cluster with $r_m < R$. Intermixing of electrons in this infinite cluster is carried out during a time shorter than the characteristic time of the critical hop. Therefore, it may be assumed that the probability N_j of finding the electron at the center j is practically equal to η. The same holds for intermixing of electrons within the finite cluster to which the center i might belong. If the number of centers in this cluster is M then the quantity

$$W_{ex}(R) = \frac{\nu}{M} \exp\left[-\frac{2R}{a}\right] \tag{4.14}$$

must be taken as the exchange rate constant (this formula is derived in [27]).

We now turn to the derivation of diffusion equations for electrons moving over the DC in the presence of an external electric field. Let n(r) and N(r) be the electron densities in the DC and IC respectively, averaged over a scale larger than the correlation length L. The continuity equation is

$$\frac{\partial}{\partial t}[n(\mathbf{r},t) + N(\mathbf{r},t)] = -\frac{1}{e}\,\mathrm{div}\,\mathbf{j}, \tag{4.15}$$

where e is the electron charge. Since only the DC electrons carry the current, it is written as

$$\mathbf{j} = -e\,D_0\nabla n(\mathbf{r},t) + e\mathbf{E}\mu n(\mathbf{r},t), \tag{4.16}$$

where

$$D_0 \sim \nu\, r_d^2 \exp\left[-\frac{2r_d}{a}\right], \quad \mu = \frac{e\,D_0}{k_B T}$$

are the diffusion coefficient and the mobility of the DC electrons. Substituting (4.16) into (4.15) gives

$$\frac{\partial}{\partial t}[n(\mathbf{r},t) + N(\mathbf{r},t)] = D_0\Delta n(\mathbf{r},t) - \mu\mathbf{E}\nabla n(\mathbf{r},t). \tag{4.17}$$

In order to exclude the quantity N(r,t) from (4.17), (4.13) for electron exchange between the DC and IC is used. This equation is solved with respect to $N_i(t)$, and the result is averaged over a volume Ω with characteristic size larger than L but smaller than the sample dimension. The result is

$$N(r,t) = \int_0^\infty \Psi(t-\tau)\, n(\mathbf{r},\tau)\,d\tau + \frac{1}{\Omega}\sum_i N_i(\mathbf{r},0)\exp\{-W_{ex}(R_i)t\}, \tag{4.18}$$

where†

† Calculation of $\Psi(t)$ by means of the distribution (4.12) yields the some result as under

$$\Psi(t) = \frac{1}{\Omega N_{DC}} \sum W_{ex}(R_i) \exp\{-W_{ex}(R_i)t\},$$

and N_{DC} is the density of centers in DC.

Equation (4.18) is now substituted into (4.17) and the equation of electron diffusion over the DC is found:

$$\frac{\partial n(\mathbf{r},t)}{\partial t} + \frac{\partial}{\partial t} \int_0^t \Psi(t-\tau)\, n(\mathbf{r},\tau)\, d\tau - D_0 \Delta n(\mathbf{r},t) + \mu \mathbf{E} \nabla n(\mathbf{r},t)$$

$$= -\frac{d}{dt} Q(t), \tag{4.19}$$

where

$$Q(t) = \frac{1}{\Omega N_{DC}} \sum_i N_i(r,0) \exp[-W_{ex}(R_i)t].$$

The second term on the left-hand side of (4.19) describes electron exchange between the DC and ICs. The term on the right-hand side of the equation is of the same origin and depends significantly on the initial electron distribution over the ICs. Other terms are standard for diffusion equations. Equation (4.19) holds for time

$$t \gg t_0 \sim \nu^{-1} \exp[2r_d/a]$$

large enough for intermixing of the DC electrons to occur. How the electric current behaves at smaller times will be discussed separately at the end of this section.

The standard time-of-flight experiments, as described in Section 8.2, are now studied, and the (4.19) is solved for this case. It is assumed that initially, at $t = 0$, an excess electron density is created at a distance ℓ_0 from the left face of the condenser. Then

$$n(x,t=0) = n_0\, \delta(x - \ell_0)$$

for the DC and

$$N(x,t=0) = N_0\, \delta(x - \ell_0)$$

for the ICs. If it is assumed that the electrons are initially randomly distributed over the traps, then (4.19) becomes

---

the CTRWA (see (2.34)). The results of calculation by means of the distribution (4.10)–(4.11), valid in a restricted time interval, are presented in [8]. It should, however, be emphasized that the function $\Psi(t)$ plays different roles in the two models and has different physical meanings.

$$\frac{\partial n(x,t)}{\partial t} + \frac{\partial}{\partial t}\int_0^t \Psi(t-\tau)\, n(x,\tau)\,d\tau - D_0\frac{\partial^2}{\partial x^2} n(x,t) + E\mu\frac{\partial}{\partial x} n(x,t)$$

$$= n_0 \Psi(x)\, \delta(x-\ell_0), \tag{4.20}$$

with the boundary conditions $n(0,t) = n(\ell,t) = 0$. In deriving the right-hand side of (4.20), the relation

$$N_0/n_0 = N_t/N_{DC}$$

has been used.

This equation is conveniently analysed by taking the Laplace transform of n(x,t) with respect to t. Then integrating the solution of the resulting linear differential equation of second order in x within the limits 0 to ℓ the Laplace transform of the time-dependent current flow is found

$$I^*(p) = \frac{eE\mu n_0}{\ell p}\left\{1 - \exp[\alpha_1(\ell-\ell_0)]\,\frac{\sinh \alpha_2\ell_0}{\sinh \alpha_2\ell} - \exp[-\alpha_1\ell]\,\frac{\sinh \alpha_2(\ell-\ell_0)}{\sinh \alpha_2\ell}\right\}, \tag{4.21}$$

where

$$\alpha_1 = \frac{E\mu}{2D_0}, \quad \alpha_2 = \left[\alpha_1^2 + \frac{\epsilon(p)}{D_0}\right]^{1/2}, \quad \epsilon(p) = p + p\Psi^*(p).$$

Unfortunately, it is not possible to calculate the inverse Laplace transform of the function (4.21) and find the time-dependent electric current in the general case. However, this can be done in some limiting cases corresponding to special relations between the parameters in (4.21). The strong-field limit is considered, $\alpha_1\ell_0 \gg 1$. Since $\ell \gg \ell_0$, the inequality $\alpha_2\ell \gg 1$ follows automatically. $\tau_d \sim \ell_0^2/D_0$ is the time needed by a particle to diffuse the distance ℓ_0 separating it from the left condenser face; $\tau_f \sim \ell_0/\mu E$ is the time necessary to pass the same distance driven by the electric field E. Then the strong-field condition takes the form $\tau_f \ll \tau_d$. This means that the particle driven by the electric field towards the right condenser face is not able to diffuse to the left face. On the other hand, $\alpha_2\ell \gg 1$, and the particle motion towards the right condenser face is governed by the electric field rather than by diffusion.

Two characteristic intervals may be distinguished, in which the electric current depends differently on time. At small times $t < (\ell/\ell_0)\,\tau_f$, when the electron packet has not reached the right face, the inequality $(\alpha_2 - \alpha_1)\,\ell \gg 1$ holds and from the inverse Laplace transform of (4.21),

$$I(t) = eEn_0\mu/\ell. \tag{4.22}$$

The current is thus proportional to the electric field, and does not depend on time. It corresponds to an electron packet driven by the electric field and affected relatively little by the diffusion. In this sense the situation does not differ very much from the motion of a Gaussian packet.

By the time of flight $t \sim (\ell/\ell_0)\,\tau_f$, the electron packet reaches the right condenser face and no longer contributes to the current. For $t \gg (\ell/\ell_0)\,\tau_f$ the current is caused only by the electrons that were initially at ICs far removed from the DC and have only now begun to reach the DC. It is clear that the magnitude of the current should be proportional to their rate of flow. In fact, as follows from (4.21), the current

$$I(t) = en_0\Psi(t) \tag{4.23}$$

depends on time and is independent of the electric field.

The region of applicability of the diffusion equation (4.19) is restricted to the times $t > t_0$. At smaller times the current is due to electron transitions between closely situated centers ($r_{ij} < r_d$). Its magnitude can be found by means of the following simple consideration [32]. Two pairs of traps connected by the vector **r** are considered, and each one is assumed to possess an electron that is initially located in such a way that it hops against the field in one pair and along the field in the other. Then

$$\left.\begin{aligned} \frac{dN_1}{dt} &= -W_{ex}(r)\,N_1, \\ \frac{dN_2}{dt} &= -W_{ex}(r)\,N_2 \exp\left[-\frac{e\,Er}{k_BT}\right], \end{aligned}\right\} \tag{4.24}$$

where N_i is the probability that the electron transition in the ith pair has not occurred. At small enough values of r ($e\mathbf{E}\mathbf{r}/k_BT \ll 1$) the sum of the currents in these two pairs is found, and the result is then averaged over the direction and magnitude of the vector **r**.

Since it is the pairs with small r that mainly contribute to the current, their density N_p may be regarded as being independent of r. The total current flow is then

$$I \approx \frac{4}{3}\,\frac{e^2EN_p}{k_BT}\left(\frac{a}{2}\right)^5\frac{\ell n^4\nu t}{t}. \tag{4.25}$$

Zvyagin [6,7] studied the initial time-dependent current flow in more detail, having applied percolation theory in order to be able to account for the electrons that at $t < t_0$ had been able to make several "rapid" hops ($r < r_d$).

8.5. PERCOLATION MODEL OF DISPERSIVE TRANSPORT. SITE DISORDER.

The percolation model of dispersive transport presented in Section 8.4 accounts only for the bond disorder caused by a spread of distances between the electron localization centers. The site disorder, i.e. a spread of the electron binding energies in traps, is not taken into account. As a result, a pattern has arisen that comprises features of both Gaussian and dispersive transport. At times smaller than the time of flight the current remains constant, as in the Gaussian case. However, at times larger than the time-of-flight the behaviour of the current differs from the exponential decay that it would have in the Gaussian transport. Meanwhile, the dispersive character of the electron transport is clearly seen in many cases at all times in experiments carried out in good dielectrics in which the electrons hop from trap to trap exclusively owing to tunneling.

This section presents a percolation model of dispersive transport for site-disordered systems in which the spread of the electron binding energies is of importance. It is clear that such a spread always exists, and can effect the electron diffusion if the temperature is small compared with its characteristic scale. Only at high enough temperatures can the situation analysed in Section 8.4 occur. This problem was treated by means of computer simulations in [33–35] (see also the review [20]). However, it will be shown below that the principal results can in fact be obtained analytically [9].

Electron diffusion over traps with binding energies spread in a certain range differs strongly from diffusion in the purely bond-disordered systems. In this case, not only must the spatial location of the electron be followed, but also its energy. So the diffusion in this sense becomes four-dimensional. The binding-energy spread is characterized by a density of states $f(\epsilon)$ in a finite range $(-\epsilon_b, +\epsilon_b)$. A symmetric distribution with single maximum at $\epsilon = 0$ is considered for the sake of simplicity (see Fig. 8.7). At first glance, the impression might arise that the principal contribution is due to the electrons moving over the traps with energies close to the maximum of the density of states or to the maximum of the electron energy distribution (see e.g. [36]). However, this is not the case.

First a region of low temperatures and large enough times is

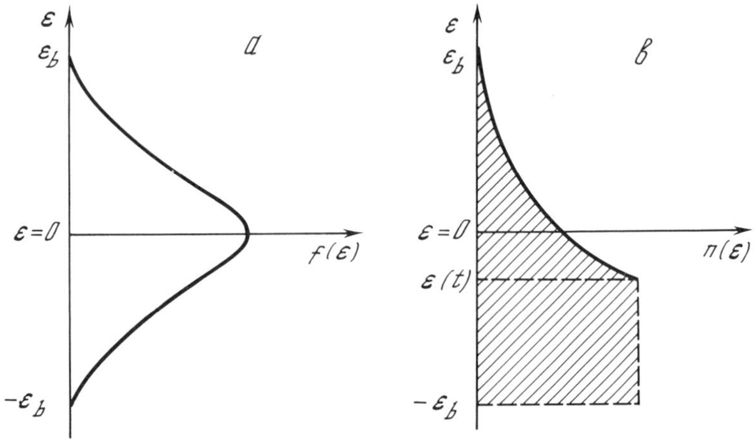

Figure 8.7 Distribution of the site binding energies (a) and dependence of the probability to find an electron in a site on its energy (b) at the instant t. Electrons above the level $\epsilon(t)$ are thermalized.

considered when the electrons may be treated as locally thermalized, i.e. the probability of finding an electron in a given trap depends on its energy according to a Boltzmann-like distribution (strict criteria are presented below). In the region lying below the maximum of $f(\epsilon)$, the density of traps decreases with decreasing energy ϵ, and the distance between them increases, resulting in a decrease of the electron mobility over such traps. However, the occupation of these traps increases, so the competition of these two counteracting factors determines the energy level that gives the contribution to the current. This speculation makes sense only at low enough temperatures, when electron hops with large energy gains are hardly likely.

This issue is now analysed in greater detail. For this an energy interval around ϵ of width Δ is considered. The electrons that hop over the traps with energies lying within the range $(\epsilon - \frac{1}{2}\Delta, \epsilon + \frac{1}{2}\Delta)$ give rise to a current flow determined by

$$j(\epsilon) = e\,E\mu\, f(\epsilon)\, n_0\, \Delta(\epsilon) \exp[-\beta(\epsilon + \epsilon_b)], \tag{5.1}$$

where

$$\mu = e\,R^2(\epsilon)\,\beta\nu \exp\left[-\frac{2R(\epsilon)}{a} - \beta\Delta\right] \tag{5.2}$$

is the mobility of the electrons hopping over the distance $R(\epsilon)$ with energy change Δ, $n_b \exp[-(\epsilon + \epsilon_b)]$ is the electron density in the traps with energy ϵ, n_b is the electron density at $\epsilon = -\epsilon_b$, and $\beta = 1/k_BT$. A relation between $R(\epsilon)$ and $\Delta(\epsilon)$ can be found by means of a consideration similar to that used in the derivation of Mott's law [36] (see also [31]). The average spacing between the traps with energies in the range $(\epsilon - \frac{1}{2}\Delta, \epsilon + \frac{1}{2}\Delta)$ is determined by

$$R^{-3}(\epsilon) \sim f(\epsilon)\,\Delta(\epsilon). \tag{5.3}$$

Substituting (5.3) into (5.2) and minimizing the exponent with respect to $R(\epsilon)$, the following relations arise

$$R(\epsilon) \sim a\left(\frac{\beta}{f(\epsilon)\,a^3}\right)^{1/4}, \quad \beta\Delta(\epsilon) \sim \left(\frac{\beta}{f(\epsilon)\,a^3}\right)^{1/4}. \tag{5.4}$$

These trap-to-trap spacings and energy changes specify the hops that are mainly performed by the electrons with the energy ϵ. Hops with other R and Δ values are hardly likely and do not effect the mobility. However, the energy ϵ to which the principal current contribution is related is not yet determined. So (5.4) and (5.2) are substituted into (5.1) and the current flow transported by the electrons moving over the traps with energies in the range $(\epsilon - \frac{1}{2}\Delta, \epsilon + \frac{1}{2}\Delta)$ is found:

$$j(\epsilon) \approx e^2 n_b E\left(\frac{\beta^3}{a}\right)^{1/4} \nu \exp\left[\frac{1}{4}\,\ell n\, f(\epsilon) - g\left(\frac{\beta}{f(\epsilon)\,a^3}\right)^{1/4} - \beta(\epsilon + \epsilon_b)\right], \tag{5.5}$$

where g is a dimensionless factor of order unity.

On minimizing the exponent in (5.5), the following equation is obtained:

$$\frac{f'}{4f}\,[1 + 2\beta\Delta] = \beta, \tag{5.6}$$

whose solution determines the optimal percolation level (OPL) ϵ_{OPL}, which is the main contributor to the current flow. At low enough temperatures, when $k_BT \ll f/f'$, the term 1 in the brackets in (5.6) may be neglected. If the density of states is also approximated in the energy range of interest by the function

$$f(\epsilon) = \frac{N_t}{2k_BT_0}\exp\left[-\frac{|\epsilon|}{k_BT_0}\right] \tag{5.7}$$

then the solution of (5.6) is

$$\epsilon_{OPL} = -3k_BT_0\,\ell n\left(\frac{T\,R_{av}}{T_0 a}\right), \tag{5.8}$$

where $R_{av} = (\frac{4}{3}\pi\, N_t)^{-1/3}$ is the mean trap-to-trap spacing. Then substituting (5.8) into (5.5) gives the expression for the OPL contribution to the current flow:

$$j(\epsilon_{OPL}) \approx \frac{e^2 n_b E\nu}{a\, k_B T_0} \exp\{-\beta[\epsilon_b + \epsilon_{OPL}(T)] - 4\beta k_B T_0\}. \tag{5.9}$$

This result describes the temperature dependence of the current flow in a system in which a small constant local electron density is maintained. It is a generalization of Mott's law [36] for a nondegerate electron gas in a system with variable density of states. The temperature dependence now obeys an activation-like law with an activation energy that is dependent on T.

The solution (5.8) holds for temperatures at which ϵ_{OPL} lies far from both the maximum of $f(\epsilon)$ and the lower boundary of the density of states: $-\epsilon_b < \epsilon_{OPL} < 0$. As the temperature increases, ϵ_{OPL} tends to zero, and for

$$\frac{T}{T_0} > \frac{a}{2R_{av}} \tag{5.10}$$

the current flows mainly over the traps lying in the vicinity of the $f(\epsilon)$ maximum. Then (5.7) is not valid, but in the energy region where $f'(\epsilon) = 0$ the density of states may be assumed to be constant. So (5.5) yields

$$j \sim \exp[-\beta\epsilon_b], \tag{5.11}$$

i.e. the magnitude of the current is determined by the activation to the level $\epsilon \approx 0$. This situation will hold at all temperatures satisfying the inequality (5.10), since the electrons moving over the traps with large positive energies above the $f(\epsilon)$ maximum contribute negligibly to the current flow owing to the small density of these traps.

A stationary situation has been discussed so far, when a constant local density of electrons is maintained and these have been thermalized. However, the situation is different in the time-of-flight experiment, especially at the initial stage. Injected electrons occupy the traps uniformly, regardless of their energy. Hence initially the main contribution to the current flow comes from electrons moving over the traps with $\epsilon \sim 0$, i.e. near the maximum of the density of states.

At high temperatures, this situation does not change in the course of time, for then $\epsilon_{OPL} \sim 0$. At lower temperatures ($T/T_0 < a/2R_{av}$), however, the level at which the current flow drops with time until it reaches its stationary value $\epsilon \sim \epsilon_{OPL}$. By a time t at which this process is not yet over, there is an energy $\epsilon(t) > \epsilon_{OPL}$ corresponding to the level at which the current flows. Above this level the energies lie where the electrons are thermalized, below it the trap occupation depends weakly on the energy (see Fig. 8.7b).

The dependence $\epsilon(t)$ is now found from the following consideration. The probability that an electron occupying a trap with energy ϵ will pass to another trap with a lower energy is evaluated using the formula

$$W(\epsilon) = \nu \exp[-(\tfrac{4}{3}\pi N(\epsilon) a^3)^{-1/3}], \qquad (5.12)$$

where $N(\epsilon) = N_t \int_{-\epsilon_b}^{\epsilon} f(\epsilon)\, d\epsilon$ is the number of traps with energies below ϵ. It is clear that the electrons in traps with the given energy thermalize by the time

$$t \approx W^{-1}(\epsilon). \qquad (5.13)$$

Solving (5.13) with respect to $\epsilon(t)$ gives the desired dependence, which for the density of states (5.7) has the form

$$\epsilon(t) = -3k_BT_0 \ell n \left[\frac{a}{2R_{av}} \ell n(\nu t)\right]. \qquad (5.14)$$

Finally, on comparing (5.14) with (5.8) it can be seen that the level $\epsilon(t)$ reaches ϵ_{OPL} by time

$$t_\theta \approx \nu^{-1} \exp[-\beta k_B T_0], \qquad (5.15)$$

which may be taken as a characteristic local thermalization time.

The time dependence of the current flow in the initial period $t < t_\theta$ can be calculated in the same way as was done for purely bond-disordered systems. In this case the current flow is determined by the electron transitions between closely situated pairs of traps, and is described by (4.31). Now, however, the mean trap-to-trap spacing in the OPL $r_{d\,OPL} \sim a\,\beta\, k_BT_0$ should take the place of $r_d \sim R_{av}$. Correspondingly, the duration of the initial period is characterized by a larger time, t_θ rather than t_0.

At large times $t > t_\theta$ the electron thermalization level $\epsilon(t)$ becomes lower than ϵ_{OPL}, where the current mainly flows. Further thermalization is now due to the exchange of electrons between the traps lying below ϵ_{OPL} by means of their intermediate excitation to the OPL. To describe this process, traps with $\epsilon \in (\epsilon_{OPL} - \frac{1}{2}\Delta,\ \epsilon_{OPL} + \frac{1}{2}\Delta)$ are considered. The treatment of Section 8.4 allows a DC to be made from these traps. As for the ICs, the situation is now more complicated than in purely bond-disordered systems, since there are two types of ICs: some are spatially separated from the DC, while the others, lying outside the OPL, are energetically separated.

The diffusion equation in such a system is derived in the same way as in Section 8.4. The only difference is that two equations should be written instead of (4.13) describing the kinetics of the electron exchange between the ICs and the DC. One corresponds to the electron exchange with the ICs with energies within the OPL, $|\epsilon_i - \epsilon_{OPL}| < \frac{1}{2}\Delta$:

$$\frac{\partial N_i(t)}{\partial t} = \nu \exp\left[-\frac{2R_i}{a}\right]\{N_i(t) - \eta(t)\exp[\beta\Delta]\}, \qquad (5.16)$$

where R_i is the distance from the ith IC to the DC, as defined in Section 8.4. The second equation describes the exchange with the ICs lying outside the OPL, $|\epsilon_i - \epsilon_{OPL}| > \frac{1}{2}\Delta$:

$$\frac{\partial N_i(t)}{\partial t} = -\overline{W}\{N_i(t) - \eta(t)\exp[\beta|\epsilon_i - \epsilon_{OPL}|]\}, \qquad (5.17)$$

$$\overline{W} = \nu \exp\left[-\frac{2R_{DC}}{a} - \beta(\epsilon_{OPL} - \epsilon_i)\right],$$

where $R_{DC} \sim (\frac{4}{3}\pi N_{DC})^{-1/3}$ is the mean distance between the centers in the DC. It can easily be verified that this equation describes the onset of thermodynamic equilibrium between the IC and OPL centers.

Here (5.17) is written proceeding from the following consideration. Only low temperatures are of interest, when the Boltzmann exponent decreases with increasing ϵ more rapidly than the density of states increases. The contribution of the ICs with $\epsilon_i \gg \epsilon_{OPL}$ is therefore neglected. The density of traps with $\epsilon_i \ll \epsilon_{OPL}$ is then much smaller than the density of the DC traps, hence the mean distance between the ICs with $\epsilon_i \ll \epsilon_{OPL}$ and the DC traps is equal to the mean distance between the DC traps. Finally, (5.17) implies that a direct hop from the IC to a DC trap is important, rather than a hop via an intermediate IC. It was shown in [9] that processes with intermediate ICs might only change preexponential factors to a limited extent.

Thus a diffusion equation with sources of the type (4.19) again arises. The difference is only in the functions Q(t) and $\Psi(t)$, which now contain two terms. The first term is due to the OPL traps. Its presence results from the bond disorder discussed in detail in Section 8.4. In this case

$$-\frac{dQ_b}{dt} = -\Psi_b(t),$$

which allows for the transformation of (4.19) into (4.20).

The second term results from site disorder, i.e. the spread of the electron binding energies in the ICs, and traps with $\epsilon_i \ll \epsilon_{OPL}$ contribute to it. Then

$$-\overline{W}\,Q_s(t) = \Psi_s(t), \tag{5.18}$$

and the function $Q_s(t)$ is given by

$$Q_s(t) = \int d\epsilon\, f(\epsilon) \exp\{-\overline{W}t \exp[\beta(\epsilon - \epsilon_{OPL})]\}. \tag{5.19}$$

Without dwelling on the calculation of this integral, presented in [9], the final result for the density of states (5.7) is

$$Q_s(t) = \tfrac{1}{2}\pi^{1/2} e^{\epsilon_{OPL}/k_BT_0} \left(\frac{T}{T_0}\right)^{T/T_0 - 1/2} (\overline{W}t)^{-T/T_0}. \tag{5.20}$$

A similar expression can be obtained for other functions $f(\epsilon)$ satisfying the condition $\beta f'/f \gg 1$ for the values $\epsilon \sim \epsilon_0$ mainly contributing to the integral (5.19). Then a characteristic temperature

$$k_BT_0 \sim f(\epsilon)/f'(\epsilon)|_{\epsilon \sim \epsilon_0}$$

will appear in (5.20). It depends logarithmically weakly on time, since the quantity ϵ_0 is time-dependent (cf. Section 8.3).

Now, by comparing the expressions for Ψ_b and Ψ_s corresponding to the bond- and site-disorder contributions, it can easily be verified that at low temperatures, $T < T_0$, the bond-disorder contribution may be neglected. At higher temperatures, $T > T_0$, the opposite holds, and the bond disorder is the principal contributor, i.e. the situation considered in Section 8.4 occurs.

Returning now to the analysis of the low-temperature region, (4.19) is solved and an expression for the current flow in the standard time-of-flight experiment is found. It differs from (4.22) only in the form of the functions Q and Ψ.

For times less than the time of flight the current flow takes the form

$$I^*(p) = \frac{eE\mu n_0}{\ell}\,\frac{2 - pQ_s^*(p)}{p(1 + \Psi_s^*(p))}. \tag{5.21}$$

The Laplace transform of the function $\Psi_s(t)$ at small p is found by means of the Tauberian theorem (see e.g. [22])

$$\Psi_s^*(p) = \Gamma\left(1 - \frac{T}{T_0}\right)\left(\frac{T}{T_0}\right)^{T/T_0 - 1/2}\left(\frac{p}{\overline{W}}\right)^{T/T_0 - 1}. \tag{5.22}$$

$\Gamma(x)$ is the gamma function. It can be seen from (5.22) that for $p < \overline{W}$ (this inequality holds if $t > t_\theta$) the quantity $\Psi_s^*(p)$ increases and becomes greater than unity, while the quantity $pQ_s^*(p)$ decreases and becomes less than unity. Therefore, the inverse Laplace transform, also performed by means of the Tauberian theorem, yields for (5.21)

$$I(t) = \frac{eE\mu}{\pi\ell}\left(\frac{T}{T_0}\right)^{1/2 - T/T_0}(\overline{W}t)^{T/T_0 - 1}\sin\left(\frac{\pi T}{T_0}\right). \tag{5.23}$$

At times larger than the time of flight

$$I^*(p) = en_0(2 - pQ_s^*(p)).$$

In this case the small quantity $pQ_s^*(p)$ cannot be neglected in comparison with 2, since after the inverse Laplace transform has been performed the term independent of p gives $\delta(t)$, so the behaviour of the current flow at large times is determined just by the small correction:

$$I^*(p) = en_0\,\overline{W}\left(\frac{T}{T_0}\right)^{T/T_0 + 1/2}(\overline{W}t)^{-1 - T/T_0}. \tag{5.24}$$

Therefore, having studied a model which is quite different from the CTRWA model and having applied completely different mathematics, we have nevertheless come to the conclusion that the current flow depends on time according to the law (cf. (2.20))

$$I(t) \sim \begin{cases} t^{-(1-\alpha)} & (t < t_\tau), \\ t^{-(1+\alpha)} & (t > t_\tau), \end{cases} \tag{5.25}$$

where $\alpha = T/T_0$. The same value of α is obtained if the density of states (5.7) is substituted into (3.16). In fact, this coincidence is not very surprising. In spite of some important differences in the formulations of the problem in the percolation and CTRWA models, there are some analogies. In both cases the dispersion of the electron-binding energies has to be taken into account. In the CTRWA model this is necessary to prove its applicability. In the percolation model without energy dispersion the current flow behaves as in the Gaussian case for times less than the time of flight.

In both models an energy region is distinguished in which electron transport takes place. In the CTRWA model it is the conduction band, into which electrons are excited from the traps. In the percolation model these are the traps with the OPL energies forming the DC. At the same time there are differences in these models. First, the temperature dependence of the electron mobility in the conduction band and the OPL are completely different. Electrons in the conduction band are quasifree, and their mobiliy in this case is determined by phonon or impurity scattering. Electron motion in the OPL is due to tunneling trap-to-trap hopping, the mobility in this case is as calculated above. What is most striking is the difference in the purely bond-disordered systems, where any dispersion in the trap-limited transport is absent while the percolation model yields Gaussian transport at $t < t_\tau$ and dispersive transport at $t > t_\tau$.

The high-temperature limits of the two models are also quite different. In the CTRWA model the electrons are excited into the conduction band, where the transport becomes purely Gaussian. In the percolation model, with increasing temperature the OPL is displaced closer to the maximum of the density of states and broadens. In this limit a system arises, as described in Section 8.4, in which the bond disorder is of principal importance. Thus the transport retains in some respects its non-Gaussian character. Along with this, as pointed out in [9], for

$$T_0 > T > T_0 a/2R_{av}$$

and

$$t_\tau < t < \nu^{-1} \exp\left[\left(\frac{T}{T_0}\right)^{1/2} \left(\frac{2R_{av}}{a}\right)^{3/2}\right]$$

the bond-disorder case is partially realized. It should then be assumed that

$$\alpha = \left(\frac{a}{2R_{av}}\right)^3 \ell n^2 \nu t$$

in (5.28).

In conclusion, the regions of applicability of the two models of electron transport in disordered media considered in this chapter have been compared. The CTRWA seems to work well in semiconductors with relatively narow forbidden energy gaps. In this case the electrons move over the traps whose binding energies may be comparable to or even smaller than room temperature. In dielectrics the energy of electron binding to defects is, as a rule, of atomic scale and any thermal excitation to the conduction band (if any) is practically impossible. In this case the percolation model should certainly be used. For the CTRWA to be valid the following conditions should hold. On one hand, the mean electron binding energy $\bar{\epsilon}$ and its spread $\delta\epsilon$ must be larger than the temperature,

$$\bar{\epsilon}, \quad \delta\epsilon > k_B T,$$

otherwise all the electrons will be excited to the conduction band (if $\epsilon < k_B T$) or the dispersion of the characteristic activation times will be too small (if $\delta\epsilon < k_B T$). On the other hand, activation into the conduction band must be more probable than direct trap-to-trap tunneling, i.e.

$$k_BT > \bar{\epsilon}a/2R_{av}.$$

Otherwise hopping will take place and the percolation model must be applied.

Therefore it can be seen that there is a restricted temperature range in which the CTRWA holds. In general such a range need not exist. For example, in doped semiconductors in which the binding-energy spread is due to overlap of the electron wave functions, i.e.

$$\delta\epsilon \sim \bar{\epsilon}\exp[-2R_{av}/a],$$

the above conditions cannot be simultaneously valid.

In amorphous systems, however, the binding-energy spread may be large, and the CTRWA model has a wide region of applicability. Systems with large values of $\delta\epsilon$ are of special interest (e.g. amorphous SiH [37]). They may have a mobility edge ϵ_m (see Section 1.6), and the role of the conduction band is played by the delocalized states lying above it, $\epsilon > \epsilon_m$. The role of the traps is played by the Anderson-localized states lying at lower energies $\epsilon < \epsilon_m$. At low enough temperatures, when $\epsilon_{OPL} < \epsilon_m$, the electron transport is governed by the percolation mechanism. With increasing temperature, ϵ_{OPL} is displaced to the higher-energy region, and for $\epsilon_{OPL} \gtrsim \epsilon_m$ there is crossover to the CTRWA. Comparing these two energies, the crossover temperature [9] is found to be

$$T_m \sim T_0 \exp[\epsilon_m/3k_BT_0]$$

At this temperature hopping conductivity is converted into band conductivity, and the function $\ell n\ \mu$ has a cusp.

References

1. Scher, H. and Lax, M. (1973) *Phys. Rev.*, **B7**, 4491.
2. Scher, H. and Lax, M. (1973) *Phys. Rev.*, **B7**, 4502.
3. Scher, H. and Montroll, E.W. (1975) *Phys. Rev.*, **B12**, 2455.
4. Pollak, M. (1977) *Phil. Mag.*, **36**, 1157.
5. Marshall, J.M. (1978) *Phil. Mag.*, **38**, 335.
6. Zvyagin, I.P. (1979) *Phys. Stat. Solidi (b)*, **95**, 227.
7. Zvyagin, I.P. (1980) *Phys. Stat. Solidi (b)*, **98**, K159.
8. Chekunaev, N.I., Berlin, Yu.A. and Fleurov, V.N. (1982) *J. Phys.*, **C15**, 1219.
9. Chekunaev, N.I. and Fleurov, V.N. (1984) *J. Phys.*, **C17**, 2917.
10. Shmidlin, F.W. (1977) *Phys. Rev.*, **B16**, 2362.
11. Noolandi, J.J. (1977) *Phys. Rev.*, **B16**, 4466.
12. Arkhipov, V.I. and Rudenko, A.I. (1977) *Phys. Lett.*, **A61**, 55.
13. Montroll, E.W. and Weiss, G. (1965) *J. Math. Phys.*, **6**, 167.
14. Shutov, S.D., Iovu, M.A. and Iovu, M.S. (1979) *Fiz. Tekhn. Poluprovod.*, **13**, 956.
15. Pfister, G. and Scher H. (1977) *Phys. Rev.*, **B15**, 2062.
16. Kolomiets, B.T., Lebedev, E.A. and Kazakova, L.P. (1978) **12**, 1771.

17. Pai, D.M., Damodar, M. and Scharfe, M.E. (1972) *J. Non-Cryst. Solids*, **8/10**, 752.
18. Pfister, G. (1974) *Phys. Rev. Lett.*, **33**, 1474.
19. Pfister, G. and Scher, H. (1978) *Adv. Phys.*, **27**, 747.
20. Marshall, J.M. (1983) *Rep. Prog. Phys.*, **46**, 1235.
21. Shlesinger, M.F. (1974) *J. Stat. Phys.*, **10**, 421.
22. Feller, W. (1966) *An Introduction to Probability Theory and its Applications*, Vol. 2. Wiley, New York.
23. Anderson, P.W. (1958) *Phys. Rev.*, **109**, 1492.
24. Klafter, J. and Silbey, R. (1988) *Phys. Rev. Lett.*, **44**, 55.
25. Sadovskii, M.V. (1981) *Usp. Fiz. Nauk*, **133**, 223.
26. Shklovskii, B.I. (1980) *Solid State Commun.*, **33**, 273.
27. Kogan, Sh.M. and Shklovskii, B.I. (1982) *Fiz. Tekhn. Poluprovod.*, **15**, 1049.
28. Ambegaokar, V., Halperin, B.I. and Langer, J.S. (1971), *Phys. Rev.*, **B4**, 2612.
29. Shklovskii, B.I. and Efros, A.L. (1971) *Zh. Eksp. Teor. Fiz.*, **60**, 867.
30. Pollak, M. (1972) *J. Non-Cryst. Solids*, **11**, 1.
31. Shklovskii, B.I. and Efros, A.L. (1979) *Electronic Properties of Doped Semiconductors*. (In Russian.) Nauka, Moscow.
32. Pollak, M. and Gebale, T.H. (1961) *Phys. Rev.*, **122**, 1742.
33. Schönherr, G., Bässler, H. and Silver, M. (1981) *Phil. Mag.*, **B44**, 47.
34. Schönherr, G., Bässler, H. and Silver, M. (1981) *Phil. Mag.*, **B44**, 369.
35. Schönherr, G., Eiermann, R., Bässler, H. and Silver, M. (1980) *Chem. Phys.*, **52**, 287.
36. Mott, N. and Davis, E. (1979) *Electron Processes in Non-Crystalline Materials*. Clarendon Press, Oxford.
37. Tiedje, T., Cebulka, J.M., Morel, D.L. and Abeles, B. (1981) *Phys. Rev. Lett.*, **46**, 1425.

CHAPTER 9
Tunneling Electron Scavenging

This Chapter is devoted to the study of basic ideas concerning electron scavenging by means of tunneling transitions. After a brief account in the first section of the history of the topic and formulation of the main problems encountered, the next two sections consider different approximations that are used in analysing single-step reactions of tunneling electron scavenging.

The fourth section examines the effect of an electric field in varying the potential-barrier shape and altering the electron-scavenging kinetics.

A discussion of diffusion-controlled reactions follows. The fifth section considers this question specifically for liquids, in which ordinary diffusion of both donors and acceptors is possible, in line with direct tunneling of the electrons between them.

In the sixth section the scavenging of particles moving over a regular lattice with randomly distributed scavengers is studied, when the kinetics at large times is determined by relatively rare fluctuations of the spatial distribution of scavengers.

Two last sections consider the more general issue of electron-hopping diffusion in a system of randomly situated centers, some of which are scavengers.

9.1. INTRODUCTION

This chapter deals with a class of phenomena relevant to tunneling transport of electrons and related in many respects to the subject of the preceeding chapter, i.e. to electron dispersive transport. The topic is concerned with electron scavenging by traps that are deep enough to exclude the electron from further participation in the process. Strictly speaking, these processes also occur in dispersive transport. But there they are treated as subsidiary and experimentalists make strenuous efforts to minimize them. Here, in contrast, they are the main focus of our study, and we shall seek conditions under which electron scavenging is of principal importance.

The study of these processes is of great interest in many areas of chemistry and biochemistry. Reduction–oxidation reactions are accompanied by electron transfer between the reactants. This transfer proceeds, as a rule, when the molecules are in close contact and the distance between them is equal to the sum of the gas-kinetic or crystallographic radii. However, in solids at low temperatures, when the reactants find difficulty in getting in to contact owing to their low mobility, processes may become of principal importance in which electrons are transferred over large distances, up to several tens of Ångströms.

One of the earliest examples was the observation of electron transfer in relatively complicated biological system, i.e. from cytochrome C to the oxidized form of chlorophyll in *Chromatium D* bacteria [1] over a broad temperature range (4.2–300 K). The rate constant of this process does not depend on temperature at $T < 130$ K, and obeys an activation-like law at higher temperatures. The seemingly self-evident explanation, that the low-temperature region is the result of tunneling electron transfer, and the high-temperature region the result of activation, did not appear to be quite adequate. In fact, there is electron tunneling in the whole temperature range, since the tunneling temperature T_t [2] for such a light particle exceeds several thousand degrees. In order to understand the temperature dependence of the electron transfer rate constant, it is necessary to introduce ideas of the electronic–nuclear transitions [3–5]. This will be discussed in more detail in the next chapter.

It is more convenient to employ simpler chemical systems as examples when making a detailed and careful study of the electron-transfer reactions [6–12]. Up to the present time, many experimental papers have appeared in this field. A detailed discussion of results has been given recently in the book of Zamaraev and Khairutdinov [13].

All the experiments on electron scavenging are in fact very similar. A sample, which as a rule is an amorphous or polycrystalline dielectric, is exposed at low temperatures to a light or some higher-energy radiation. The excited electrons then start to move through the sample until they become thermalized and localized in traps, which are always present in large numbers in disordered substances. As a result, there arises a concentration of molecules (or atoms) that have lost electrons, and structural defects that have captured electrons. Their presence can be detected, for example, by an ESR signal if the loss or capture of an electron results in the formation of paramagnetic centers. The trapped electrons e_{tr}^- may later tunnel and return to the centers from which they have been excited. The tunneling-transition rate constant is

$$W(r) = \nu \exp[-2r/a]. \tag{1.1}$$

By measuring the kinetic curves of the process of electron recapture it is possible to determine important parameters such as the frequency ν, the electron localization radius a and the mean distance of electron tunneling transfer $\bar{r}$.

The first process for which an experimental proof was found that the electron really does tunnel over an appreciable distance was the transfer of a trapped electron to the anion O^- [10]. The trapped electrons were created in a 10 M solution of NaOH in H_2O at 77 K in the course of 4.5 MeV electron irradiation. The time dependence of the concentration of the trapped electrons in the course of the reaction

$$O^- + e_{tr}^- \longrightarrow O^{2-}$$

was registered by an ESR signal at different temperatures. The mean distance between e_{tr}^- and O^- as estimated from the width of the ESR lines was greater than 14Å. This fact, in line with the observed independence of the process rate of temperature (the temperature plateau), favours the tunneling character of the electron transfer.

Kinetic curves of the e_{tr}^- concentrations obeyed the law (Fig. 9.1)

$$\frac{n(t)}{n(t_0)} = 1 - M \lg\left(\frac{t}{t_0}\right), \tag{1.2}$$

where $n(t_0)$ is the e_{tr}^- concentration at a time t_0 taken as a reference point and M is a numerical coefficient. The logarithmic law (1.2) was explained in [10] (see also [14]) by virtue of the fact that the distances in the different pairs of reacting particles were spread. Thus the process is characterized by a broad range of constants rather than by a single rate constant, and hence there is a broad range of characteristic times. Because of this, the kinetics of this type of process is now known as a polychronic, (sometimes the term "polychromatic" is also used).

The paper [10] became a prototype of a large number of papers published during recent years. The monograph [13] presents the results obtained in the experiments in which a solid host is deliberately doped by molecules of various compounds, from which electrons are then excited. They serve as acceptors in the process of recapture. In [15,16] molecules capturing the excited electrons, i.e. scavengers, are also deliberately inserted.

Usually, the electron-recapture reaction at low temperatures is treated as a single-step process, which allows for successful interpreta-

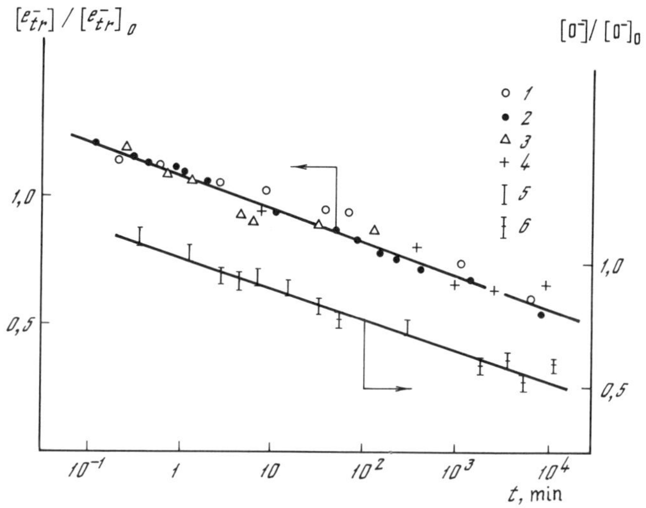

Figure 9.1 Kinetic curves of the e_{tr}^- and 0^- concentration decrease in the glassy water solution of 10M NaOH e_{tr}^-: 1 — β-radiolysis, dosage 3.10^{19} eV/cm^3, T = 77K; 2 — β-radiolysis, dosage $3.6 \cdot 10^{20}$ eV/cm^3, T = 77K; 3 — β-radiolysis, dosage $1.2 \cdot 10^{20}$ eV/cm^3, T = 120K; 4 — γ-radiolysis, dosage $1.2 \cdot 10^{20}$ eV/cm^3, T = 120K; 0^-: 5 — β-radiolysis, dosage $1.2 \cdot 10^{20}$ eV/cm^3, T = 77K; 6 — γ-radiolysis, dosage $1.2 \cdot 10^{20}$ eV/cm^3, T = 77K.

tion of numerous experimental data on electron scavenging in glasses [9–21]. In single-step electron scavenging the process is assumed to proceed by virtue of a single tunneling transition from an acceptor to the donor (i.e. to the molecule from which it has been excited). The possibility of a multistep process, where the electron makes several tunnelings and visits other traps, is thus excluded.

The problem of multistep tunneling electron scavenging now attracts a great deal of attention. The amount of experimental data favouring electron tunneling diffusion on its way from the acceptor to the donor is still not large. Among these an ultraviolet shift in 1-propanol irradiated by electrons at 77 K [22] should be mentioned. This shift might be explained by the fact that as the electron moves towards the charged donor, the traps visited by it become deeper.

Much more attention has been paid to the problem by theorists. This interest has been particularly stimulated by the obvious relationship of electron diffusion between acceptors and donors and the dispersive transport discussed in Chapter 8. Thus Hamill and Funabashi [24–26] have made use of the CTRWA model. A similar approach was developed in [27–29]. However, the consideration presented in Chapter 8 shows that the CTRWA may be applied only to trap-limited electron transport, which is due to thermal excitation. This approach cannot describe tunneling diffusion, and so the necessity of an alternative approach arises.

This theoretical issues was discussed in the literature fairly long ago. Among the papers on it was that of Balagurov and Vaks [30], which considered particle motion over a regular lattice with randomly distributed scavengers. The principal result of this paper is that the long-term kinetics of electron scavenging is determined by the low-probability fluctuations of the spatial distribution of scavengers. However, there are no regular lattices in real experiments and not only is the scavenger distribution disordered, but the sites visited by the electrons are also randomly situated in space. Therefore fluctuations of the spatial distribution of sites are also of importance. Such a situation was analysed by Chekunaev *et al.* [31], using ideas similar to those forming the basis of the percolation model of dispersive transport.

The large variety of theoretical approaches to this topic reflects, in fact, the diversity of experimental possibilities in that, in principle the experimental conditions can be chosen, in such a way that the conditions of applicability of a particular theoretical model can be realized. Unfortunately, this problem is far from being completely solved. Experimentalists cannot even always determine which type of the process they are observing. We hope that the different approaches presented in this chapter, together with their conditions of applicability, will promote a clarification of the issue.

9.2. APPROXIMATION OF ISOLATED PAIRS

Owing to the excitation of an electron from a molecule A and its subsequent capture by another molecule B, positively (A^+) and negatively (B^-) charged centers form (acceptors and donors respectively). This process and the resulting distribution of the centers A^+ and B^- is described by using two characteristic lengths. The free path length ℓ_1 characterizes the distance travelled by the electron from the center A^+ to the trap B^- that captures it. It is clear that this length is determined by the thermalization time and the velocity of the electron. The average distance between two centers of the same type, A^+ and B^-, is connected with the irradiation dosage. It is assumed in this section that the concentrations N and n of both types of centers are equal; then

$$\ell_2 \sim (\tfrac{4}{3}\pi N)^{-1/3}. \tag{2.1}$$

The kinetics of single-step electron scavenging depends principally on the relation between two lengths. If

$$\ell_2 \gg \ell_1 \tag{2.2}$$

then the distributions of the centers A^+ and B^- are heavily correlated. Owing to the inequality (2.2), each electron "remembers" from which molecule A^+ it has been excited. It is removed by a small distance of the order of ℓ_1, while all the other molecules A^+ are much further away than ℓ_1. Thus it is clear that during recapture the electron returns to its parent molecule with nearly unit probability, i.e. recapture within each pair A^+ and B^- takes place independently.

The time dependence of the concentration of the centers A^+ and B^- on time can be found by means of a simple consideration. Let N(r,t) be the concentration of pairs A^+B^- separated by distance r at time t. It is easy to see that the time dependence of this quantity is described by the following differential equation:

$$\frac{dN(r,t)}{dt} = -W(r)\,N(r,t), \tag{2.3}$$

whose solution is

$$N(r,t) = N(r,0)\exp[-W(r)t]. \tag{2.4}$$

The quantity N(r,t) at an initial time t = 0, taken as the reference point, is written in the form

$$N(r,0) = N_0\,\eta(r),$$

where N_0 is the initial total concentration of pairs and $\eta(r)$ is the initial distribution of the distances between reagents, normalized to unity

$$\int_0^\infty \eta(r)\,dr = 1. \tag{2.5}$$

A decrease in the e_{tr}^- concentration starts at t = 0 when the sample irradiation is over. To find the total concentration of pairs at t = 0 (2.4) is averaged over the distance r:

$$N(t) = N_0 \int_0^\infty dr\,\eta(r)\exp\left\{-\nu t\exp\left(-\frac{2r}{a}\right)\right\}. \tag{2.6}$$

Here (1.1) is used for the tunneling transition rate constant W(r).

The form of the initial distribution $\eta(r)$ and hence N(t) depend on how the pairs have been generated. If the spread of the free paths of excited electrons around its mean value ℓ is not large then the uniform distribution within the range $\ell_1 - \frac{1}{2}\Delta\ell/2$ to $\ell_1 + \frac{1}{2}\Delta\ell$ may be assumed to be the simplest one, i.e.

$$\eta(r) = \begin{cases} 1/\Delta\ell & (|r - \ell| \leqslant \frac{1}{2}\Delta\ell), \\ 0 & (|r - \ell| > \frac{1}{2}\Delta\ell). \end{cases} \tag{2.7}$$

Then substituting (2.7) into (2.6) gives

$$N(t) = \frac{N_0}{\Delta\ell} \int_{\ell_1 - \Delta\ell/2}^{\ell_1 + \Delta\ell/2} dr \exp\{-\nu t \exp[-2r/a]\}. \tag{2.8}$$

Usually both the electron free path and its dispersion markedly exceed the electron localization radius, i.e.

$$\ell_1 \gg a, \quad \Delta\ell \gg a. \tag{2.9}$$

Hence there exists a very wide interval of t,

$$\nu^{-1} \exp\left[\frac{2(\ell_1 - \frac{1}{2}\Delta\ell)}{a}\right] \ll t \ll \nu^{-1} \exp\left[\frac{2(\ell_1 + \frac{1}{2}\Delta\ell)}{a}\right],$$

in which the integrand in (2.8) may be replaced by the Heaviside step function $\theta[r - \frac{1}{2}a \ln \nu t]$. Then the integral (2.8) is easily performed to give

$$N(t) = \frac{N_0}{\ell} (\ell_1 + \tfrac{1}{2}\Delta\ell - \tfrac{1}{2}a \ln \nu t). \tag{2.10}$$

The physical meaning of the result (2.10) is quite obvious. By time t electron recapture in all the pairs with reactant-to-reactant distances $r > \frac{1}{2}a \ln \nu t$ is over, and only the pairs with large distances $r > \frac{1}{2}a \ln \nu t$ remain. If the distance distribution in the pairs is characterized by (2.7) then the number of remaining pairs is given by (2.10). Such a strict separation of pairs corresponds to the replacement of the integrand in (2.8) by the Heaviside function. In fact, however, there is a transition region at $r \sim \frac{1}{2}a \ln \nu t$ in which only a fraction of pairs has reacted. However, owing to the inequality (2.9) and the exponential dependence of the tunneling rate constant W(r) on the distance r, the corresponding corrections to (2.11) are negligible.

The above approach was used in [10,14] to interpret experimental results on electron recapture in a frozen aqueous solution of NaOH. Later (2.10) was applied to intepret the results of numerous experiments (see e.g. [13]), leading to the determination of the parameters a and ν.

Although the approximation of a uniform initial distribution (2.7) might be the basis for a qualitative and, in many cases, quantitative description of the process, the distribution never has precisely such a form in real conditions. It is pointed out in [23] how this distribution can be found from the experimental kinetic curves on pair recapture.

The following substitution of variables in (2.6) is made:

$$x = \nu \exp(-2r/a).$$

Equation (2.6) is then differentiated with respect to time:

$$\frac{1}{N_0} \frac{dN(t)}{dt} = \frac{a}{2} \int_0^\infty dx\, \eta \left(\frac{a}{2} \ell n \frac{\nu}{x} \right) \exp(-xt)\, dx. \tag{2.11}$$

The left-hand side of (2.11) is an experimentally observed quantity. If it is known over a wide enough time interval then it is possible to find the dependence $\eta(\frac{1}{2}a\, \ell n(\nu/x))$ and hence $\eta(r)$ by means of the inverse Laplace transformation.

9.3. APPROXIMATION OF PAIR COLLISIONS

This section considers a situation that is in a sense opposite to that discussed in the preceeding one. The electron free path is assumed to be large, i.e.

$$\ell_2 \ll \ell_1. \tag{3.1}$$

Then the electron no longer "remembers" from which center A^+ it was excited, and may tunnel to any other center in the course of recapture. To describe the kinetics of electron scavenging in this case it is convenient to employ a technique similar to the virial expansion used in the kinetic theory of gases (see e.g. [32]). This approach was applied to this problem in [33–36].

The concept of the conditional concentration is introduced. Let $n(\mathbf{r}_1, t; \{\mathbf{r}_i\}, \{\mathbf{R}_j\})$ $(i \neq 1)$ be the conditional concentration of donors at time t at point $\mathbf{r}$ if it is known that the points $\{\mathbf{r}_i\}_{i \neq 1}$ are occupied by other donors and the points $\{\mathbf{R}_j\}$ are occupied by acceptors. It is natural to suppose that the vectors $\{\mathbf{r}_i\}$ and $\{\mathbf{R}_j\}$ do not coincide. Similarly, the definition of the conditional concentration of acceptors can be introduced. For the totality of donor and acceptor conditional concentrations a chain of equations may be written. For example, the conditional concentration of donors $n(\mathbf{r}_1, t; \mathbf{R}_1)$ is connected with the conditional concentration of acceptors $N(\mathbf{R}_2, t; \mathbf{r}_1, R_1)$ by the equation

$$\frac{\partial}{\partial t} n(\mathbf{r}_1, t; \mathbf{R}_1) = -W(|\mathbf{r}_1 - \mathbf{R}_1|)\, n(\mathbf{r}_1, t; \mathbf{R}_1) - \int d^3R_2\, W(|\mathbf{R}_2 - \mathbf{r}|)\, N(\mathbf{R}_2, t; \mathbf{r}_1, \mathbf{R}_1)\, n(\mathbf{r}_1, t; \mathbf{R}_1). \tag{3.2}$$

This reflects the fact that the electron density at the donor with coordinate $\mathbf{r}_1$ may change by virtue of the electron escaping either to the acceptor at the point $\mathbf{R}_1$ (the first term in (3.2)) or to any other acceptor at the point $\mathbf{R}_2 \neq \mathbf{R}_1$. Since it is now known that there is a donor at the point $\mathbf{r}_1$ and an acceptor at the point $\mathbf{R}_1$, the probability of finding an acceptor at the point $\mathbf{R}_2$ is characterized by the conditional concentration $N(\mathbf{R}_2, t; \mathbf{r}_1, \mathbf{R}_1)$. An equation can be written that connects this with the conditional concentration of donors. The structure and meaning of this equation is analogous to (3.2). It is also clear that such equations remain valid when all donor quantities are exchanged with acceptor ones, and vice versa. In (3.2) the approximation has been made that electron scavenging is a single-step process, since it does not contain terms describing diffusion of electrons over the centers B or holes over the centers A.

To solve the chain of equations, which may be continued *ad infinitum* some additional approximations are necessary, enabling it to be derived. Usually the approximation called the "self-consistent field" approximation is used. It is assumed that

$$N(\mathbf{R}_2, t; \mathbf{r}_1, \mathbf{R}_1) = N(\mathbf{R}_2, t; \mathbf{r}_1), \tag{3.3}$$

$$n(\mathbf{r}_2, t; \mathbf{r}_1, \mathbf{R}_1) = n(\mathbf{r}_2, t; \mathbf{R}_1). \tag{3.4}$$

Then (3.2) is reduced to the effectively one-dimensional equation [37]

$$\frac{\partial}{\partial t} n(r,t) = -\left[W(r) + 4\pi \int_0^\infty d\rho \rho^2 W(\rho) N(\rho,t)\right] n(r,t). \tag{3.5}$$

Similarly, for the conditional concentration of donors the following equation can be written:

$$\frac{\partial}{\partial t} N(r,t) = -\left[W(r) + 4\pi \int_0^\infty d\rho \rho^2 W(\rho) n(\rho,t)\right] N(r,t). \tag{3.6}$$

Ordinary concentrations are connected with the conditional concentrations in (3.5) and (3.6) by means of the limiting relations

$$n(t) \doteq \lim_{r\to\infty} n(r,t), \quad N(t) = \lim_{r\to\infty} N(r,t). \tag{3.7}$$

Since in the course of recapture equal numbers of donors and acceptors "disappear",

$$\frac{\partial n(t)}{\partial t} = \frac{\partial N(t)}{\partial t}, \tag{3.8}$$

The condition (3.8) together with (3.5) and (3.6) give

$$N(r,t)\,n(t) \;=\; n(r,t)\,N(t). \tag{3.9}$$

Now the differential equation

$$\frac{\partial}{\partial t}\,\theta(r,t) = -\,W(r)\,\theta(r,t) \tag{3.10}$$

is written for a new dimensionless variable

$$\theta(r,t) \;=\; \frac{n(r,t)}{n(t)} = \frac{N(r,t)}{N(t)}. \tag{3.11}$$

Solving (3.10), the dependence of the donor concentration on time is found

$$n(t)/n(0) = \frac{1 - N_0/n_0}{1 - (N_0/n_0)\exp\left\{(1 - N_0/n_0)\int\limits_0^\infty [1 - \exp\{-W(r)t\}]\,N(r,0)4\pi r^2\,dr\right\}}, \tag{3.12}$$

where n_0 and N_0 are the initial concentrations of donors and acceptors respectively.

If it is assumed that $N_0 \gg n_0$ and that the acceptors have been initially distributed uniformly and independently of the donors, i.e. $N(r,0) \;=\; N_0$, then the integral in (3.12) can be calculated in the same way as a similar integral in the preceeding chapter (see Section 8.2). This gives

$$n(t) = n_0 \exp[-\tfrac{1}{6}\pi a^3 N_0\, \ell n^3\, \nu t]. \tag{3.13}$$

The same result can be obtained by means of simple quantitative considerations. A donor a distance R from the nearest acceptor is considered. The probability of finding such a donor is given by the Poisson distribution

$$f(R) \sim 4\pi R^2 \exp\left[-\frac{4\pi}{3}\,R^3 N_0\right]. \tag{3.14}$$

It is clear that by time t electrons from all the donors with $R < \tfrac{1}{2}a\,\ell n\,\nu t$ will be transferred to acceptors. Therefore the number of remaining electrons is determined by (3.14), into which $R = \tfrac{1}{2}a\,\ell n\,\nu t$ should be substituted, which gives (3.13).

The chain of equations of the type (3.2), electron diffusion being disregarded, contains the complete information on electron recapture. To solve this chain, the approximations (3.3) and (3.4) are used which result in the closed system (3.5) and (3.6). The only point that remains unclear is when these equations and the resulting solutions are applicable.

Equation (3.5) no longer contains the requirement $R_1 \neq R_2$. This means that fictitious contributions appear in the second term on the right-hand side of this equation, which take account of processes already included in the first term. Thus (3.5) leads to a situation where two acceptors occur at the same point. The associated error is small if the acceptor concentration is small. There is a similar inaccuracy in (3.6); therefore the concentration of donors must be also be small. Both these requirements are usually met under the real experimental conditions.

The problem of applicability of the approximations (3.3) and (3.4) is more complicated. For example, (3.3) implies that the presence of an acceptor at the point $\mathbf{R}_1$ does not affect the conditional concentration $N(R_2, t; r_1)$. Although a strict mathematical analysis of the issue is lacking, we shall try to point out situations where the approximation (3.3) seems reasonable. If the concentration of donors is much greater than that of acceptors, $N_0 \ll n_0$, then the acceptors will compete for donors, and hence the processes for different acceptors will be independent. In the opposite case, $N_0 \gg n_0$, each donor is influenced by a large number of acceptors, and the appearance of one more acceptor does not lead to any significant change. Similar speculations apply to (3.4).

If the concentrations of donors and acceptors are equal, $N_0 = n_0$, then mutual correlations of the acceptor and donor distributions arise in the course of electron-recapture kinetics. This is why analysis of electron-recapture kinetics by means of (3.5) and (3.6) may be applicable only at the very beginning of the process, when these correlations are still insignificant.

An example of a situation where the approximations (3.3) and (3.4) do not apply is the case of isolated pairs discussed in Section 9.2. In fact, the donors and acceptors are assumed there to be distributed in pairs, with relatively large distances between these pairs, i.e. their distributions are strongly correlated. Owing to this the second term on the right-hand side of (3.2) may be discarded. This results in an equation similar to (2.3).

9.4. INFLUENCE OF AN EXTERNAL ELECTRIC FIELD ON THE KINETICS OF ELECTRON SCAVENGING

An external electric field acting on a particle in a potential well changes the shape of the barrier and thus changes the probability of tunneling. This mechanism reveals itself frequently in various physical effects. Ionization of atoms in strong electric fields [38], Landau–Zener break-down and the Franz–Keldysh effect (see e.g. [39]) are examples. It is thus not surprising that an electric field can also influence tunneling electron scavenging by increasing its rate. This effect was first observed by Bullot and Albrecht [40] in a study of the recapture of electrons excited at low temperatures by flash photolysis of glassy hosts. An explanation of these data was proposed in [41,42] (see also [13]).

Here this problem is considered for the pairwise distribution. The principal effect of the electric field on the kinetics of tunneling electron scavenging is assumed to be connected with a variation of the tunneling transition rate constant. Its dependence on distance is mainly determined by the exponential factor, in which the exponent takes the value

$$-\frac{2r}{a_E} = -\frac{2}{\hbar}\int_0^r [2m\,U(\rho)]^{1/2}\,d\rho \tag{4.1}$$

and

$$U(\rho) = U_0(\rho) - eE\rho\cos\theta.$$

Here $U_0(\rho)$ is the potential barrier in the absence of the external electric field E, e is the electron charge, θ is the angle between the direction of electron transfer and the direction of the electric field, and r is the barrier width. If the field is weak enough to be treated as a perturbation then

$$a_E^{-1} = a^{-1}\left(1 - \frac{eEr\cos\theta}{4U_0}\right), \tag{4.2}$$

where U_0 is a characteristic energy determining the scale of the potential barrier $U_0(\rho)$:

$$U_0 = \left(\frac{1}{r}\int_0^r [U(\rho)]^{1/2}\,d\rho\right)^2. \tag{4.3}$$

Therefore the tunneling rate constant in the presence of the external electric field has the form

$$W(r,\theta) = \nu \exp\left[-\frac{2r}{a} - \frac{r^2}{2aU_0}\, eE\cos\theta\right]. \tag{4.4}$$

In deriving this formula, the electric field is assumed to be small; hence it holds only for tunnelings over distances that are not too large: $r < 4U_0/eE$.

The further calculation of electron recapture is quite similar to that carried out in Section 9.2. The only difference is that the conditional concentration $n(r,\theta,t)$ now depends on the angle between the direction of electron tunneling and the direction of the electric field. If the initial pair distribution is assumed to be isotropic, i.e.

$$n(r,\theta,t) = \frac{1}{2} n_0\, \eta(r) \sin\theta, \tag{4.5}$$

then the kinetics of electron scavenging is described by the formula

$$n(t) = \frac{1}{2} n_0 \int_0^\infty dr \int_0^\pi d\theta \sin\theta \exp[-W(r,\theta)t]\, \eta(r). \tag{4.6}$$

It makes sense to consider not the quantity n(t) but rather its time derivative, especially as experiments studying the influence of the electric field on the kinetics of electron scavenging measure the luminescence intensity I(t), which is the total number of electrons captured per unit time and hence is proportional to dn(t)/dt:

$$I(t) = \tfrac{1}{2} A \int_0^\infty dr \int_0^\pi d\theta \sin\theta\, W(r,\theta) \exp[-W(r,\theta)t]\, \eta(r). \tag{4.7}$$

Here A is a proportionality coefficient.

The convergence of the integral (4.8) is determined by the exponential factors; therefore the relatively smooth function $\eta(r)$ may be taken out of the integral at the point $r = \frac{1}{2} a \ln \nu t$ owing to the smallness of the electric field, this procedure is only valid for times that are not very large:

$$t < \nu^{-1} \exp\left(\frac{8U_0}{eEa}\right). \tag{4.8}$$

At larger times the main contribution to the integral (4.7) is given by the electron tunneling over such great distances that the condition of a small electric field no longer holds.

Retaining in the integration the leading terms in the small parameter $4U_0/eE$ leads to the expression

$$I_0(t) = A \frac{a}{2t} \eta(\tfrac{1}{2} a \ell n\, \nu t), \tag{4.9}$$

which is close to the result obtained at $E = 0$. A nontrivial result is obtained if the electric field is switched on at a later time $\tau > 0$ rather than at the initial time $t = 0$. Then the dependence of the luminescence intensity on time is found at $t > \tau$ by calculating the integral

$$I_1(t) = \tfrac{1}{2} A \int_0^\infty dr \int_0^\pi d\theta \sin\theta\, W(r,\theta) \exp[-W(r)\tau] \exp[-W(r,\theta)(t-\tau)]\, \eta(r). \tag{4.10}$$

The integration in this case results in

$$I_1(t) = A \frac{\eta(\tfrac{1}{2} a \ell n\, \nu t)}{2\, \ell n\, b} \frac{1}{x\tau} \ell n \frac{1 + bx}{1 + x/b}, \tag{4.11}$$

where

$$x = \frac{t - \tau}{\tau}, \quad b = \exp\left[\frac{a\, \ell n^2\, \nu t}{4U_0} eE\right].$$

The influence of the electric field on the kinetics of electron scavenging is conveniently revealed by comparing the luminescence intensities at $E = 0$ (see (4.9)) and at $E \neq 0$ (see (4.11)) at $t = \tau$, i.e. at the time of switching on the electric field. Then the luminescence intensity increases stepwise, the relative height of the step being

$$\frac{I_1(\tau)}{I_0(\tau)} = \frac{b - 1/b}{2\, \ell n\, b}. \tag{4.12}$$

Here the parameter b in (4.12) depends, generally speaking, on time. However, this dependence is weak and it may be disregarded in time range in which b is treated as a constant. To evaluate this time range, the quantity b is written in the form

$$b = (\nu t)^\gamma,$$

where $\gamma = (aeE/4U_0)\, \ell n\, \nu t$. Owing to (4.8) $\gamma \ll 1$; therefore b may be treated as constant in the interval $(t, t + \Delta t)$ if $\Delta t < t/\gamma$.

Equation (4.12) can be applied to find the parameter b experimentally. Thus, for example, according to the [40], where the kinetics of the luminescence of photoexcited matrices of 3-methylpentane doped with tetramethyl-*p*-phenylendiamine was examined at 77 K, the luminescence intensity became twice as small on switching on an electric field $E = 4 \times 10^5\ Vm^{-1}$. From this it is easy to find that $b = 8.7$ [41].

If b is known then it is possible to find the distribution of pair distances $\eta(r)$. As can be seen from (4.11), the relation

$$\frac{\eta(R_t)}{\eta(R_\tau)} = \frac{I_1(t)}{I_1(\tau)} \frac{(b + 1/b)x}{\ell n \left(\frac{1 + bx}{1 + x/b}\right)} \tag{4.13}$$

holds for $t > \tau$. Here $R_t = a \ell n \nu t$. For $t < \tau$ the corresponding relation can be obtained from (4.9):

$$\frac{\eta(R_t)}{\eta(R_\tau)} = \frac{t}{\tau} \frac{I_0(t)}{I_0(\tau)}, \tag{4.14}$$

which gives the expression for the same quantity as (4.13), but in the absence of an external electric field.

Here we finish our presentation of the basic theoretical results relevant to single-step kinetics of electron scavenging. The results presented in Sections 9.2–4 lead not only to qualitative understanding and quantitative description of many experiments in various physicochemical systems, but also allow the parameters characterizing the spatial distribution of the reactants or individual tunneling acts to be found. The most interesting quantities are ν and a, which determine the characteristic frequency of the electron motion inside its potential well and the localization of the electron wave function. These parameters are found by direct comparison of theoretical and experimental kinetic curves. They have been measured for a large number of molecules deliberately inserted as acceptors. The corresponding data are presented in [13,15].

It follows from these data that the parameter a, which gives an estimate of the localization radius of the electron wave function, is actually close to the Bohr radius, 0.5Å, and differs from it by a factor of not more than two or three.

The situation concerning the preexponential factor ν is more complicated. It is usually implied that this has the meaning of the "number of attempts to penetrate the barrier", and its order of magnitude is close to the characteristic electronic frequency 10^{15}–10^{16} s^{-1}. However, the experiment shows that ν varies within a very broad range from $\nu = 10^{10.9}$ s^{-1} for BrO_3^- in a frozen aqueous solution of NaOH to $\nu = 10^{23.3}$ s^{-1} for CrO_4^{2-} in the same solution.

To interpret such a strong deviation of the preexponential factor ν from the characteristic electronic frequencies, it is necessary to give up the naive model of tunneling transitions and pass to more complicated models based on the theory of radiationless transitions [43–45]. The process is then considered as an electronic–nuclear transition, the possible participation of excited states also being taken into account. As we shall see below, a possible multistep character of electron scavenging might also influence the observed value of ν.

9.5. DIFFUSION-CONTROLLED ELECTRON SCAVENGING. MUTUAL DIFFUSION OF DONORS AND ACCEPTORS

In analysing the kinetics of electron scavenging, we have restricted ourselves to the single-step approximation, supposing that the electron tunnels from the center B^- directly to the center A^+. Although in many cases this approximation appears to be sufficient for an adequate description of experimental data, there are situations where a theory is necessary that does not contain such a strong restriction. A multistep character of the process of electron scavenging might be due either to diffusion of the centers A^+ and B^- or to tunneling hopping diffusion of the electrons over the centers B^-. In the first case, which will be analysed in this section, the mobility of the relatively heavy entities (acceptors or donors) must be high enough. Such conditions can be realized in liquids or in solids at not very low temperature.

Before dealing with the general issue of diffusion-controlled electron scavenging in liquids, it makes sense to consider a simpler problem. It is supposed that the electron is captured only via direct contact of the centers A^+ and B^- and that electron tunneling transfer plays no role. The kinetics of such a process is described by the spherically symmetric diffusion equation

$$\frac{\partial P}{\partial t} = D\left[\frac{1}{r^2}\frac{\partial}{\partial r}\left(r^2\frac{\partial P}{\partial r}\right) + \frac{\alpha}{r^2}\frac{\partial P}{\partial r}\right], \qquad (5.1)$$

where $\alpha = e^2/\epsilon k_B T$ is the Onsager radius [46], e is the electron charge, ϵ is the dielectric permeability of the substance, $P(r,t)$ is the conditional concentration of donors carrying the electrons, and D is the coefficient of mutual diffusion of donors and acceptors. The origin of the second term on the right-hand side of (5.1) is as follows. In the presence of the electric potential $\varphi(\mathbf{r})$, a drift term $e\mu\nabla\varphi\cdot\nabla P$ always appears in the diffusion equation (μ is the particle mobility, which is connected with the diffusion coefficient by the Einstein relation $\mu = D/k_B T$). If it is recalled that in this case the electric field is due to the mutual attraction of the charged donors and acceptors, i.e. $\varphi(r) = -e^2/\epsilon r$, then it is obvious that this second drift term arises in the (5.1).

This problem (without Coulomb interaction) was analysed by Smoluchowsky in connection with the problem of particle coagulation

[47]. He assumed that there is a coagulation center of radius R to which floating particles stick. The initial concentration of these particles was P_0 and their distribution was uniform. The coagulation center was also assumed not to influence the distribution of the particles far removed from it. As a result, the Smoluchovsky boundary conditions arise:

$$\left.\begin{array}{lll} P(r,t) = 0 & (r < R, & t \geqslant 0), \\ P(r,t) = P_0 & (r > R, & t = 0), \\ P(r,t) \to P_0 & (r \to \infty, & t \geqslant 0). \end{array}\right\} \tag{5.2}$$

The solution of this problem

$$P(r,t) = P_0 \left[1 - \frac{R}{r} + \frac{2}{\pi^{1/2}} \frac{R}{r} \int_0^{(r-R)/(4Dt)^{1/2}} e^{-\xi^2}\, d\xi \right] \tag{5.3}$$

reduces in the stationary limit (as $t \to \infty$) to

$$P_\infty(r) = P_0(1 - R/r). \tag{5.4}$$

The current flow of particles through a sphere of radius R is then

$$I = 4\pi DR^2 \left. \frac{\partial P}{\partial r} \right|_{r=R} = 4\pi DRP_0,$$

which is connected with the rate constant of the process in the stationary limit

$$k = \frac{I}{P_0} = 4\pi DR. \tag{5.5}$$

The radius R appearing in (5.5) for the reaction rate constant is often called the reaction radius. In the above case it coincides with the acceptor radius. However, this is far from always being so. For example, the Coulomb interaction of the electron with the acceptor can be taken into account. Then it is found that the stationary solution of (5.1),

$$P_\infty(r) = P_0 \left\{ 1 + \frac{R}{\alpha} (1 - e^{\alpha/r}) \right\},$$

yields a reaction radius

$$R_{ef} = \alpha \frac{e^{\alpha/R}}{e^{\alpha/R} - 1}. \tag{5.6}$$

In the limit $\alpha \gg R$ the effective reaction radius is completely determined by the Coulomb interaction, $R_{eff} = \alpha$ [46,48,49].

An even more complicated situation arises if account is taken of the fact that the electron may tunnel from donor to acceptor. Then the equation

$$\frac{\partial P}{\partial r} = D\left[\frac{1}{r^2}\frac{\partial}{\partial r}\left(r^2\frac{\partial P}{\partial r}\right) + \frac{\alpha}{r^2}\frac{\partial P}{\partial r}\right] - W(r)P \qquad (5.7)$$

should be applied. This was analysed in [50–53] without accounting for the Coulomb interaction and in [54–56] in the general case. If this equation is compared with (3.6) for single-step tunneling recapture of electrons, then the following can be seen. On the one hand, the diffusion and drift of the donors and acceptors and the electron tunneling between them is now taken into account. On the other hand, it does not allow for the possibility of electrons escaping to "other" acceptors (second term on the right-hand side of (3.6)). This process was considered in [57]. Here we restrict ourselves to a simpler version, and discuss in detail the kinetics of electron scavenging described by (5.7).

If the Coulomb interaction is disregarded then the stationary problem for (5.7) has the solution

$$P_\infty(r) = \frac{aP_0}{r}\left[K_0(x) - \frac{K_0(x_0)\,I_0(x)}{I_0(x_0)}\right], \qquad (5.8)$$

where

$$x = a\left(\frac{\nu}{D}\right)^{1/2}\exp\left(-\frac{r}{a}\right), \quad x_0 = a\left(\frac{\nu}{D}\right)^{1/2}\exp\left(-\frac{R}{a}\right),$$

$K_0(x)$ and $I_0(x)$ are the McDonald and Bessel functions of imaginary arguments. This expression allows calculation of the effective reaction radius

$$R_{eff} = a\left\{\ell n\left[\frac{a\gamma}{2}\left(\frac{\nu}{D}\right)^{1/2}\right] + \frac{K_0(x_0)}{I_0(x_0)}\right\}, \qquad (5.9)$$

where $\gamma = e^C$ (C is Euler's constant).

If the rate of tunneling transitions is small, $x_0 \ll 1$, then the expression (5.9) coincides with the Smoluchowsky result, i.e. $R_{eff} = R$. In the opposite limiting case, $x_0 \gg 1$,

$$R_{eff} = a\,\ell n\left(\tfrac{1}{2}\gamma\, x_0\right) + R. \qquad (5.10)$$

The reaction rate is now determined by electron tunneling transfer rather than by diffusional approach of the reactants. As a result, the effective reaction radius may grow considerably, since the reaction will proceed without direct contact of the reactants. Such a situation was qualitatively considered in [10,14].

In studying the kinetics of electron scavenging, not only is the rate constant obtained in the quasistationary limit as $t \to \infty$, but the whole kinetic curve of the electron concentration dependence on time is also of interest. This problem for the case $\alpha > R$ was considered analytically in [54] (it was solved numerically in [56]). It also makes sense to give up the assumption of a uniform initial electron distribution around acceptors and to study a more general case, describing this distribution by the function $\eta(r_0)$ (see Section 9.2). We assume that initially the electron is at a point $\mathbf{r}_0$ (the acceptor is at the reference point), i.e.

$$P(\mathbf{r},t)|_{t=0} = \delta(\mathbf{r}-\mathbf{r}_0). \tag{5.11}$$

In this case it is convenient to use the method of "prescribed diffusion" [48,49] and seek a solution in the form

$$P(\mathbf{r},t) = (4\pi Dt)^{-3/2} \exp\left[-\frac{(\mathbf{r}-\mathbf{r}_0)^2}{4Dt}\right] F_{r_0}(t). \tag{5.12}$$

The function F(t) depends on time and on the initial distance between the electron and acceptor, and is expressed in terms of the conditional probability P(r, t) by means of the integral relation

$$F_{r_0}(t) = \int d^3\mathbf{r}\, P(\mathbf{r},t).$$

The physical meaning of this function is clear. It determines the probability that the electron that was at distance r_0 from the acceptor at $t = 0$ has escaped capture by the time of observation $t = 0$. Averaging $F_{r_0}(t)$ over the initial distribution,

$$\bar{F}(t) = \int_0^\infty dr_0\, \eta(r_0)\, F_{r_0}(t) = \frac{n(t)}{n_0},$$

gives a quantity that can be experimentally measured.

An equation for $F_{r_0}(t)$ is obtained if the function (5.12) is substituted into (5.7) and the integration over the directions of the vector $\mathbf{r}_0$ and the variable r is carried out:

$$\frac{dF_{r_0}(t)}{dt} = -\Gamma t^{-1/2} \exp\left(-\frac{r_0^2}{4Dt}\right)\left[t^{-1} + \frac{S(r_0,t)}{\alpha r_0}\right] F_{r_0}(t), \tag{5.13}$$

where

$$\Gamma = \alpha(4\pi D)^{-1/2},$$

$$S(r_0,t) = 2\int_0^\infty dr\, r\, W(r) \exp\left(-\frac{r^2}{Dt}\right) \sinh\left(\frac{rr_0}{2Dt}\right).$$

The initial condition is taken as F(t) = 1 at t = 0. Then (5.13) can be solved to give

$$F_{r_0}(t) = \exp\left\{-\frac{\alpha}{r_0}\,\mathrm{erfc}\left(\frac{r_0}{4Dt^{1/2}}\right) - \Psi(r_0,t)\right\}, \qquad (5.14)$$

where

$$\Psi(r_0,t) = \int_0^t \Gamma\,\frac{S(r_0,x)}{x^{1/2}\,\alpha r_0}\,\exp\left(-\frac{r_0^2}{Dx}\right)dx$$

and erfc(x) is the error function.

In the general case, the function $F_{r_0}(t)$ can only be calculated numerically. However, in many interesting limiting cases an analytical solution can be found; for example the stationary limit as $t \to \infty$ has the form

$$\bar{F}_\infty = \int_0^\infty dr_0\,\eta(r_0)\,\exp\left(-\frac{\alpha + R_t}{r_0}\right)dr_0, \qquad (5.15)$$

where

$$R_t = \frac{a^3\nu}{4D}\left\{1 - \left(\frac{r_0}{a} + 1\right)\exp\left(-\frac{2r_0}{a}\right)\right\}.$$

If tunneling is weak, $R_t \ll \alpha$, i.e. $4D\alpha/\nu a^3 \gg 1$, then the effective reaction radius is equal to the size of the Coulomb well α. In the opposite limiting case, the "tunneling" radius R_t plays the leading role. It determines a spherical region around the acceptor so that an electron that is initially there cannot leave and tunnels to the acceptor.

An idea of how $\bar{F}(t)$ varies with time can be gained from the following considerations. At small times electron tunneling transfer will prevail over diffusion. In fact, during time t the electron can diffusively travel a distance $r_D \sim (Dt)^{1/2}$ or tunnel over the distance $r_t \sim \frac{1}{2}a\,\ell n\,\nu t$. Therefore initially for $t < (a^2/4D)\,\ell n^2\,\nu t$ diffusive approach of the reagents plays no role, $r_D < r_t$. Then

$$\Psi(r_0,t) = \nu t\,\exp[-2r_0/a]$$

and

$$\bar{F}(t) = \int_0^\infty dr_0\,\eta(r_0)\,\exp\{-\nu t\,\exp[-2r_0/a]\}.$$

The exponential integrand in (5.16) is small for $r_0 < \frac{1}{2}a \ell n\, \nu t$ and close to unity for $r_0 > \frac{1}{2}a \ell n\, \nu t$. This means that all the electrons that were not far from the acceptor at $t = 0$ will have tunneled to it, while the electrons removed further will effectively not have moved.

In the large-time limit the diffusive approach of reactants begins to have a significant influence on the kinetics of electron scavenging, which is now described by a pseudofirst-order kinetic law

$$\frac{d\bar{F}}{dt} = -\Gamma \left[1 + \frac{a^3\nu}{4D\alpha}\right] t^{-3/2}\bar{F}. \tag{5.17}$$

A similar law can be obtained directly from (5.1), in which tunneling is not taken account of at all [48,49]. Tunneling results here only in an increase of the constant in (5.17), which is of the same origin as the increase of the effective reaction radius (5.10).

Anomalously large effective radii have been observed experimentally for some time (see e.g. [58,59]), although it was not known that they are connected with tunneling effects. However, tunneling processes are now accepted as playing an important role in the interpretation of electron scavenging in liquids (see e.g. the review [60]). Recently an increase of the effective reaction radius was also observed in the reaction of positronium in water–glycerol mixtures [61] and in nonpolar hydrocarbons [62].

9.6. RANDOM WALKS OF PARTICLES OVER LATTICE WITH SCAVENGERS

This section considers the random walk of a particle over a regular lattice that contains scavengers randomly distributed over its sites. This model may be applied to some problems in the diffusion of vacancies and interstices in crystals with impurity centers. Some concepts that appear when studying it will also be useful later for the development of theoretical interpretations of experiments on electron scavenging. As will be seen below, the analysis is carried out under the assumption that the particles transfer only between neighbouring sites. This condition is realized in tunneling transfer, with its strong dependence of the transition probability on distance.

The problem of particle diffusion over a lattice with scavengers is examined in great detail by Balagurov and Vaks [30]. The principal idea of this paper is that the kinetics of scavenging of a randomly-moving particle in the large-time limit is governed by low-probability fluctuations of the spatial distribution of the scavengers. Supposing the concentration of the moving particles to be small, the authors of [30] consider the motion of a single particle. The theory of random walks [63] can be used to describe its motion.

Thus the equation of motion

$$P_{s+1} = \sum_{r'} p(\mathbf{r}-\mathbf{r}')\,(1-\delta_{r'})\,P_s(\mathbf{r}') + \delta_r P_s(\mathbf{r}) \tag{6.1}$$

is written for the quantity $P_s(\mathbf{r})$ that characterizes the probability of finding the particle at site r after it has made s steps. Here the quantity δ_r is unity if a scavenger is at site r and zero otherwise. Therefore the first term on the right-hand side of (6.1) corresponds to the arrival of the particle at site **r** from site **r**′, and the factor $1 - \delta_{r'}$ excludes such a possibility if a scavenger occupies the site **r**′. If there is a scavenger at site r then if particle arrived there at the sth step it will remain at the (s + 1)th step. This is taken account of by the second term on the right-hand side of (6.1).

As usual, we shall be interested only in the values of $P_s(\mathbf{r})$ at sites without scavengers, i.e. the probability for an electron not to be captured. It is therefore convenient to introduce the quantity

$$\widetilde{P}_s(\mathbf{r}) = (1 - \delta_r)\,P_s(\mathbf{r}),$$

which coincides with $P_s(r)$ at sites without scavengers and is zero at sites with scavengers. Only transitions to nearest neighbours are considered. Then multiplying (6.1) by $1 - \delta_r$ gives the equation

$$\widetilde{P}_{s+1}(\mathbf{r}) = \widetilde{D} \sum_{\rho} \eta(\mathbf{r},\mathbf{r}+\rho)\,\widetilde{P}_s(\mathbf{r}+\rho). \tag{6.2}$$

Here $\widetilde{D} = 1/N_c$, where N_c is the first coordination number, so $\widetilde{D}$ characterizes the probability of transition to one of the neighbours. The quantity $\eta(\mathbf{r},\mathbf{r}+\rho)$ is unity if sites **r** and $\mathbf{r} + \rho$ are free of scavengers and zero if a scavenger occupies at least one of them.

An expression for the quantity $P_s(\mathbf{r})$ can be found by means of a decomposition in terms of the eigenfunctions of the Hermitian integral operator $\widetilde{D}\eta(\mathbf{r},\mathbf{r}')$. This, however, is not sufficient, since the explicit form of these eigenfunctions is not known in the general case. An exception is presented by the problem of a particle walking over a one-dimensional lattice with scavengers, in which an exact solution is possible. This is due to the property peculiar to the one-dimensional problem where a particle between two neighbouring scavengers does not feel the effect of the other scavengers. Therefore the problem is that of a particle diffusing over a finite segment with zero boundary conditions (for details see [30]).

In multidimensional systems the problem cannot be solved in terms of just two scavengers, since the particle may pass around one of them and encounter a third scavenger. Nevertheless, the asymptotic behaviour of $\overline{P}(t)$ (the average probability of finding the uncaptured particle in a lattice site) can be found to within exponential accuracy by the following consideration. At large times t the particles will "survive" only in large enough regions that are free of scavengers. The particle motion inside these regions is governed by the conventional diffusion equation, which results from (6.2) for $t \gg \tau$ and $r \gg \rho_0$ (ρ_0 is the nearest neighbour spacing and, τ is the characteristic hopping time). Beyond this region the scavengers are distributed with the average density, and hence the function $\widetilde{P}(t, \mathbf{r})$ decays here over distances of the order of the average spacing between the scavengers, which is assumed to be much smaller than the region size. Therefore $\widetilde{P}(t, \mathbf{r})$ may be sought, to within the required accuracy, as a solution of the equation

$$\frac{\partial P(t, \mathbf{r})}{\partial t} + D\Delta\widetilde{P}(t, \mathbf{r}) = 0, \tag{6.3}$$

with the conditions $\widetilde{P}(t, \mathbf{r}) = 0$ at the region boundary and $\widetilde{P}(0, \mathbf{r}) = 1/N_d$ inside the region, where N_d is the number of lattice sites inside the region; $D = \rho_0^2\widetilde{D}/\tau$.

The probability of finding an uncaptured particle within a region of volume V (the three-dimensional case is implied) at large times is determined by

$$\widetilde{P}_V(t) \sim \exp[-Dk_{0_V}^2 t], \tag{6.4}$$

which follows from the diffusion equation (6.4). Here k_{0_V} is the minimal eigenvalue for the given region V.

The quantity $\overline{P}(t)$ that characterizes the average density of the uncaptured particles is derived from $\widetilde{P}_V(t)$ by averaging over all fluctuation regions. Now the probability of finding a scavenger-free region of volume V in a lattice with randomly distributed scavengers is determined by the Poisson formula

$$f(V)\,dV = c\exp(-cV)\,dV \tag{6.5}$$

(c is the relative concentration of scavengers), which depends only on the volume and does not depend on the shape of the region. If, for the sake of simplicity, a sphere is considered then

$$k_{0_V} = \pi/R, \tag{6.6}$$

where R is the radius of the sphere. Then averaging (6.4) with weight (6.5) gives

$$\bar{P}(t) \sim \exp[-\alpha_s(c^{2/3} t\rho_0^2/\tau)^{3/5}] \tag{6.7}$$

at

$$t\rho_0^2/\tau \gg c^{-2/3}, \quad \text{where} \quad \alpha_s = 5(\tfrac{1}{3}\pi)^{8/5} 2^{-1/5} \approx 4.7.$$

In deriving (6.7), all the fluctuation regions are assumed to be spheres. Since the sphere has the minimal eigenvalue among all regions with the given volume V, taking account of the contributions of non-spherical regions might change only the preexponential factor, while the principal exponential factor remains unchanged.

Thus the most interesting qualitative result of this section is that the asymptotic behaviour of the number of free random walkers in a lattice with randomly distributed scavengers is determined by the fluctuations of their spatial distribution. The result (6.7) as well as the results for one- and two-dimensional systems were compared in [30] with the corresponding results obtained in the gas approximation [64], i.e. neglecting fluctuations. (In this connection, the paper [27] should also be mentioned.) The gas approximation appears to work well only in the two- and three-dimensional cases, and even then only at relatively small times. On the other hand, we have verified in Chapter 8 that fluctuations of the spatial distribution of traps are of great importance when describing particle diffusion in disordered media. Below we shall proceed to the study of the scavenging of particles diffusing in disordered media with randomly distributed scavengers. Then the role of both types of fluctuations will be analysed.

9.7. DIFFUSION-CONTROLLED ELECTRON SCAVENGING; $N_0 \gg n_0$. HOPPING DIFFUSION.

In the course of diffusion-controlled scavenging, as opposed to the single-step process (see Sections 9.2–9.4), an electron moves for a while over the centers (which we shall call traps below), and only then is it captured by a scavenger (acceptor). A model of electron diffusion over randomly distributed traps (without scavengers) is discussed in detail in Chapter 8, where the principal role played by the fluctuations of the trap spatial distribution in the formation of current flow is outlined. On the other hand, Section 9.6 demonstrates the leading role of fluctuations of the scavenger spatial distribution in the kinetics of the scavenging of particles moving randomly over a regular lattice with randomly distributed scavengers. The situation will now be considered in which both types of fluctuations are present.

This section deals with the case when the number of scavengers is much greater than the number of excited electrons ($N_0 \gg n_0$). Then the decrease in the number of free scavengers with time may be disregarded and the quantity N_0 treated as constant. The approach of Balagurov and Vaks [30], described in detail in the previous section, was applied in [31] to the analysis of the kinetics of electron scavenging in this regime. Processes in ordered and disordered media are generally different. The differences are mainly due to the fact that diffusion carrier transport proceeds only over a small proportion of traps, which form a diffusion cluster (DC) (see Chapter 8). All the other (isolated) centers are treated as sources or sinks. That is why the electron motion can no longer be described by the conventional diffusion equation, and use must be made of the modified equation proposed in [65] (see Section 8.4). Here the dispersion of the electron binding energies in the traps is ignored, and they are all assumed to be equal to ϵ_t.

A sphere of radius R, free of scavengers, is considered (taking account of nonspherical fluctuations changes only the proportionality coefficient in the final result [31]). The electron motion is described by the equation

$$\frac{\partial}{\partial t} n(\mathbf{r},t) + \frac{\partial}{\partial t} \int_0^t \Psi(t-t')\, n(\mathbf{r},t')dt' + D\,\Delta n(\mathbf{r},t) = n_0 \Psi(t), \qquad (7.1)$$

where $n(\mathbf{r},t)$ is the density of free electrons at point $\mathbf{r}$ at time t, and $n(\mathbf{r},0) = n_0$. This requirement means that at the initial time $t = 0$ the electron can be found at any point of the fluctuation volume with equal probability.

Since outside the fluctuation volume there are enough scavengers effectively capturing the electrons, (7.1) should be supplemented by the boundary condition $n(\mathbf{r},t) = 0$ at $|\mathbf{r}| = R$. Equation (7.1) is then Laplace-transformed with respect to t and the transformed equation solved. The solution is integrated over the fluctuation volume and the inverse Laplace transformation performed in the large-time limit, $t \gg \tau_R$ (where $\tau_R \sim R^2/D$ is the time necessary for an electron to leave the fluctuation volume by means of the usual diffusion). Averaging the result over the Poisson distribution (6.5) gives [31]

$$n(t) = \frac{n_0}{45D}\,\Gamma\left(\frac{5}{3}\right)\left(\frac{3}{4}\pi N_0\right)^{2/3} \Psi(t), \qquad (7.2)$$

where Γ is the gamma function.

Thus it can be seen that the expression for the dependence of the free electron concentration n(t), (7.2), in the case of a disordered medium differs from (6.7). This is due to the fact that some of the electrons occupy ICs that are distant from the DC. For t $\gg \tau_R$ the kinetics of their outflow from the corresponding fluctuation volumes is controlled by the slowest stage, i.e. the stage of electron flow from the ICs to the DC, which is described by the function $\Psi(t)$ and does not depend on the fluctuation volume. The subsequent motion of the electrons outside the latter proceeds by means of diffusion over the DC within the relatively short time τ_R. As a result, the electron density in the DC does not decrease exponentially according to (6.7), but rather according to a slower power-like law determined by the function $\Psi(t)$.

Apart from the DC electrons, whose behaviour has been discussed up to now, there are always some electrons in the ICs. Determination of the density of these electrons is much simpler, and may be reduced to solution of the simple differential equation (see Section 8.4)

$$\frac{\partial}{\partial t} n_{j,IC} = -W_j(n_{j,IC} - \eta). \tag{7.3}$$

The solution is found neglecting the small quantity η on the right-hand side and averaging over the distances between the ICs and the DC. The resulting mean electron density in the ICs is

$$n_{IC} = n_0 \exp\left[-\frac{4\pi}{3} n_t a^3 \ell n^3 \nu t\right], \tag{7.4}$$

where n_t is the concentration of traps over which the electron may diffuse.

Since the function $\Psi(t)$ is, by definition, the time derivative of the function $Q(t) = n_{IC}(t)/n_0$, it can easily be seen that in the large-time limit most of the uncaptured electrons occupy the ICs, and the kinetic curves of electron scavenging are described by (7.4). However, this relates only to the total number of electrons. If the current flow is measured in the system, i.e. the number of moving electrons is measured, than it is equal to the number of DC electrons, i.e. n(t).

The following conclusion can be formulated on the basis of the above results. In a disordered medium for $n_0 \ll N_0$ the principal role is played by fluctuations of the spatial distributions of the traps rather than of the scavengers, as was the case for scavengers randomly distributed over the sites of a regular lattice (see also [66]).

9.8. DIFFUSION-CONTROLLED ELECTRON SCAVENGING; $N_0 = n_0$. HOPPING DIFFUSION

This section considers the behaviour of the free-electron concentration under the assumption that their number is initially equal to the number of scavengers, $n_0 = N_0$. Since on the capture of an electron by a scavenger the numbers of both free electrons and free scavengers decreases by unity, the equality $n(t) = N(t)$ remains valid throughout the process. Similarly to Section 9.7, the electron is assumed to diffuse over traps whose number is much greater than the number of scavengers.

In the large-time limit electron scavenging may proceed only via the DC. An electron occupying an IC leaves it, travels to the DC, moves for a while in the DC and is then captured by a scavenger. Here it is reasonable to extend the definition of the DC and to include in the latter some ICs that have had time to exchange electrons with the centers entering the original DC. If such a definition of the extended DC (EDC) is used then in the large-time limit only those centers may be treated as isolated that are far removed from their nearest neighbours because of a fluctuation of their spatial distribution. So the characteristic time of electron transition to such a center exceeds the time passed after the start of the process. These ICs might be occupied by electrons that have not yet made a single hop. Some electrons occupy the EDC at the same time; the probability η of finding an electron in an EDC center is small ($\eta \ll 1$) and depends weakly on the number of such centres.

The balance equations for electron exchange between ICs and the EDC are

$$\frac{d}{dt} n_i(t) = W(R_i)(\eta - n_i(t)) \approx -W(R_i)\, n_i(t), \tag{8.1}$$

where $n_i(t)$ is the probability of finding an electron in the ith IC at t if initially $n_i(0) = 1$. The balance equations for the free scavengers are

$$\frac{dN_j(t)}{dt} = -W_A(R_j)\,\eta N_j(t), \tag{8.2}$$

where $N_j(t)$ is the probability of finding the jth scavenger free if initially $N_j(0) = 1$.

$$W(R_i) = \nu \exp\left(-\frac{2R_i}{a}\right), \quad W_A(R_j) = \nu_A \exp\left(-\frac{2R_j}{a_A}\right) \tag{8.3}$$

are the tunneling probabilities from an IC or scavenger respectively to another IC.

Thus electron transition from an IC to a scavenger proceeds in three stages: (i) the electron travels from the IC to the EDC; (ii) it moves over the EDC; (iii) it is captured from the EDC by a scavenger. The EDC motion is rapid, so either stage (i) or stage (iii) is the limiting stage.

Hence two regimes are possible, which may be called the donor and the acceptor (scavenger) regimes. Here the role of donors is played by the ICs.

In the acceptor regime the electrons leave the IC rapidly and start moving over the EDC, from which they sink relatively slowly to the acceptors. Then it may be assumed that nearly all free electrons are in the EDC, i.e.

$$\eta(t) = n(t)/n_t.$$

Thus the kinetics of electron scavenging is governed only by (8.2), whose solutions are

$$N_j(t) = \exp\left[-W_A(R_j)\int_0^t \eta(\tau)\,d\tau\right]. \tag{8.4}$$

Equation (8.4) is summed over all j using the equality of the number of free electrons and acceptors:

$$n(t) = \sum_j N_j(t)$$

The result is then averaged over the distance R_j by means of the Poisson distribution, and the equation for the mean concentration of free electrons at time t is thus obtained:

$$n(t) = 4\pi n_0 n_t \int_0^\infty dR\, R^2 \exp\left\{-\left(\frac{R}{R_t}\right)^3 - \frac{\nu_A}{n_t}\,\mathcal{N}(t)\exp\left(-\frac{2R}{a_A}\right)\right\}. \tag{8.5}$$

Here

$$R_t = \left(\frac{4\pi}{3}\,n_t\right)^{-1/3}, \quad \mathcal{N}(t) = \int_0^t n(\tau)\,d\tau.$$

The second term in the exponent of (8.5) is large for all values of R satisfying the condition

$$R < R_1(t) = \frac{a_A}{2}\,\ell n\left(\frac{\nu_A\,\mathcal{N}(t)}{n_t}\right), \tag{8.6}$$

and is small otherwise. This means physically that by time t all relatively accessible acceptors with $R < R_1(t)$ are occupied, while all the other acceptors with $R > R_1(t)$ are free. This allows the integral on the right-hand side of (8.5) to be calculated approximately; and thus (8.5) is reduced to a differential equation with respect to the variable $x = n(t)/n_0$:

$$\frac{dx(t)}{dt} = -\frac{3\nu_A n_0 x^2}{n_t R_t} \ell n^{2/3}\left(\frac{1}{x}\right) \exp\left[-\frac{R_t}{a_A} \ell n^{1/3}\left(\frac{1}{x}\right)\right]. \tag{8.7}$$

An approximate solution of this equation satisfies

$$\ell n x + \frac{R_t}{a_A} \ell n^{1/3} x + \ell n \beta t = 0, \tag{8.8}$$

where

$$\beta = \frac{\nu_A n_0}{n_t} \frac{a_A}{R_t} \left[3\ell n^2 \left(\frac{1}{x}\right) + \frac{R_t}{a_A}\right]$$

is a new preexponential factor, which does not necessarily coincide with the characteristic electronic frequency 10^{15} s^{-1}.

During the initial time period

$$\beta t < \exp[(R_t/a_A)^3], \tag{8.9}$$

the first term in (8.8) may be neglected. Then

$$n(t) = n_0(\beta t)^{-\gamma_A}, \tag{8.10}$$

where $\gamma_A = (a_A/R_t)^3 \ell n^2 \beta t$. At larger times (the opposite inequality to (8.9)) the first term in (8.8) becomes the leading one; then

$$n(t) = \frac{n_0}{\beta t} \exp\left(\frac{R_t}{a_A} \ell n^{1/3} \beta t\right). \tag{8.11}$$

The physical meaning of (8.10) and (8.11) seems to be as follows. There are many free electrons in the EDC from the start, so the actual scavenging of electrons by acceptors is the limiting stage rather than their approach to the acceptor. This is why the form of the function (8.10) is determined mainly by the spatial distribution of traps around the acceptor. In fact, by time t only those acceptors remain free that have no traps around them at a distance

$$R > a_A \ell n(n_0 \nu t/n_t) \tag{8.12}$$

The factor n_0/n_t reflects the small probability of finding an electron near the scavenger. Substituting (8.12) into the Poisson equation, the formula (8.10) is obtained.

A different situation occurs at larger times. Then electron approach to the acceptor is the limiting stage. The capture proceeds rapidly. Since the number of free electrons is equal to the number of free acceptors, the kinetics of their recombination is governed by the equation

$$\frac{dx}{dt} \sim -x^2, \tag{8.13}$$

whose solution is close to (8.11). The quantity β in (8.11) is proportional to both the probability of an electron meeting an acceptor and the rate of reaction between them.

It is assumed above that the electrons leave the ICs rapidly and gather in the EDC. This assumption holds if the rate of electron transfer to the EDC is larger than the rate of their scavenging. In the opposite case it may be assumed that practically all the electrons occupy the ICs and the scavenging kinetics is described by (8.1), whose solution yields (7.4).

Thus there is a crossover from electron scavenging governed by (8.2), which is called the acceptor regime, to the donor regime, in which electron scavenging is governed by (8.1), if the rate of electron transfer from the ICs to the EDC is equal to the rate of acceptor capturing. It is easy to see that these two rates are always equal in the donor regime, since the capture rate determined by the concentration η of the EDC free electrons will "follow" the rate of the electron loss from the ICs. At the same time, the latter is effectively independent of the situation in the EDC.

To find the conditions necessary to realize one or the other regime, the time derivatives of the functions (7.4) and (8.10) or (8.11) should be equated. At times satisfying condition (7.9) the ratio of derivatives is

$$\chi(t) \sim \exp\left[-\left(\frac{a}{R_t}\right)^3 \ell n^3 \nu t + \left(\frac{a_A}{R_t}\right)^3 \ell n^3 \beta t\right]. \tag{8.14}$$

Here the function $\chi(t)$ grows with time and becomes greater than unity if

$$\mu = \frac{a_A}{a}\,\frac{\ell n \nu t - \ell n(\nu/\beta)}{\ell n \nu t}$$

is greater than unity. In the opposite case $\chi(t)$ decreases. If $a > a_A$ then at all times $\mu < 1$, and $\chi(t)$ decreases with time, so that the donor regime occurs.

If $a_A > a$ then the situation is more complicated. At small times

$$\nu t < (\nu/\beta)^{\frac{a}{(a - a_A)}}$$

the conditions of the donor regime hold. For larger times the acceptor regime sets in. Account must be taken of the fact that crossover is possible within the acceptor regime from the kinetic dependence (8.10) to (8.11). Then the function $\chi(t)$ that determines the type of regime becomes

$$\chi(t) \sim \exp\left[-\left(\frac{a}{R_t}\right)^3 \ell n^3 \nu t - \frac{R_t}{a_A} \ell n^{1/3} \beta t\right], \qquad (8.15)$$

and it can easily be seen that it goes through a maximum and tends to zero as $t \to \infty$. The time when the function $\chi(t)$ is equal to unity is approximately determined from the inequality

$$\beta t > \exp\left[-\lambda\left(\frac{R_t}{a}\right)^{3/2}\right], \qquad (8.16)$$

where $\lambda \leqslant 1$. At such times the acceptor regime is replaced by the donor one.

Unfortunately, there have not been enough detailed experimental studies of multistep hopping diffusion in the kinetics of electron scavenging. The problem is aggravated by the fact that it is difficult to distinguish single-step and multistep scavengings, since the kinetic curves in different cases are obtained by an averaging over the Poisson distribution and appear to be rather close. This is why a more detailed examination is necessary, for example to separate the contributions of the moving and "motionless" carriers (see Section 9.7) and to reveal different regimes (Section 9.8) in order to identify multistep electron scavenging.

References

1. De Vault, D. and Chance B. (1966) *Biophys. J.*, **6**, 825.
2. Goldanskii, V.I. (1959) *Dokl. AN SSSR*, **127**, 1037.
3. Grigorov, L.N. and Chernavskii, D.S. (1972) *Biofizika*, **17**, 195.
4. Hopfield, J.J. (1974) *Proc. Natl. Acad. Sci. USA*, **71**, 3640.
5. Jortner, J. (1976) *J. Chem. Phys.*, **64**, 4860.
6. Smith, D.R. and Pieroni, J.J. (1965) *Can. J. Chem.*, **43**, 876.
7. Tsujikawa, H., Fucki, K. and Kuri, Z. (1965) *Bull. Chem. Soc. Jpn.*, **38**, 2210.
8. Ershov, B.G. and Tseitlin, E.L. (1970) *Khim. Vysok. Energii*, **4**, 186.
9. Mikhailov, A.I. (1971) *Dokl. AN SSSR*, **197**, 136.
10. Zamaraev, K.I., Khairutdinov, R.F., Mikhailov, A.I. and Goldanskii, V.I. (1971) *Dokl. AN SSSR*, **199**, 640.
11. Miller, J.R. (1972) *J. Chem. Phys.*, **56**, 5173.
12. Kiefer, F., Meyer, C. and Rigaut, J. (1971) *Chem. Phys. Lett.*, **11**, 359.
13. Zamaraev, K.I., Khairutdinov, R.F. and Zhdanov, V.P. (1985) *Electron Tunneling in Chemistry*. (In Russian.) Nauka, Novosibirsk.
14. Goldanskii, V.I., Zamaraev, K.I., Mikhailov, A.I. and Khairutdinov, R.F. (1973) In: *Problems of Kinetics of Elementary Chemical Reactions*. (In Russian.) Nauka, Moscow.

15. Anikolenko, V.A. and Mikhailov, A.I. (1976) *Dokl. AN SSSR*, **230**, 102.
17. Zamaraev, K.I. and Khairutdinov, R.F. (1974) *Chem. Phys.*, **4**, 181.
18. Aleksandrov, I.V. and Khairutdinov, R.F., Zamaraev, K.I. (1978) *Chem. Phys.*, **32**, 123.
19. Tachiya, M. and Mozumder, A. (1974) *Chem. Phys. Lett.*, **34**, 77.
20. Miller, J.R. (1975) *J. Phys. Chem. Solids*, **79**, 1070.
21. Stradovskii, Cz. and Kroh, J. (1978) *Rad. Phys. Chem.*, **12**, 175.
22. Baxendale, J.H. and Scharfe, P.H.G. (1976) *Chem. Phys. Lett.*, **39**, 411.
23. Plonka, A., Lefik, W. and Kroh, J. (1979) *Chem. Phys. Lett.*, **62**, 271.
24. Hamill, W.H. and Funabashi, W. (1977) *Phys. Rev.*, **B16**, 5523.
25. Hamill, W.H. and Funabashi, W. (1978) *Phys. Lett.*, **A56**, 175.
26. Hamill, W.H. (1978) *J. Phys. Chem.*, **82**, 2073.
27. Klafter, J., Silbey, R. (1981) *J. Chem. Phys.*, **74**, 3510.
28. Blumen, A., Klafter, J. and Zumofen, G. (1983) *Phys. Rev.*, **B27**, 3429.
29. Klafter, J., Blumen, A. and Zumofen, G. (1983) *J. Phys. Chem.*, **87**, 191.
30. Balagurov, B.Ya. and Vaks, V.G. (1973) *Zh. Eksp. Teor. Fiz.*, **65**, 1939.
31. Chekunaev, N.I., Berlin, Yu.A. and Fleurov, V.N. (1983) *J. Phys.*, **C16**, 2569.
32. Landau, L.D. and Lifshitz, E.M. *Statistical Physics*, Vol. 1. Pergamon, Oxford.
33. Parmon, V.N., Khairutdinov, R.F. and Zamaraev, K.I. (1974) *Fiz. Tverd. Tela.*, **16**, 2572.
34. Tachiya, M. and Mozumder, A. (1974) *Chem. Phys. Lett.*, **28**, 87.
35. Dainton, F.S., Pilling, M.J. and Rice, S.A. (1975) *J. Chem. Soc. Faraday Trans. II*, **71**, 1311.
36. Gailitis, A.A. (1975) *Uchen. Zap. Latv. Univ.*, **234**, 44.
37. Galanin, M.D. (1955) *Zh. Eksp. Teor. Fiz.*, **28**, 485.
38. Oppenheimer, J.R. (1928) *Phys. Rev.*, **31**, 66.
39. Ziman, J.M. (1972) *Principles of the Theory of Solids*. Cambridge University Press.
40. Bullot, J. and Albrecht, A.C. (1969) *J. Chem. Phys.*, **51**, 2220.
41. Khairutdinov, R.F., Berlin, Yu.A. and Zamaraev, K.I. (1977) *Izv. AN SSSR, Khim.*, 1977.
42. Doheny, J. and Albrecht, A.C. (1977) *Can. J. Chem.*, **55**, 2065.
43. Kozhushner, M.A. and Ivanov, G.K. (1978) *Fiz. Tverd. Tela*, **20**, 9.
44. Aleksandrov, I.V. (1980) *Chem. Phys.*, **51**, 449.
45. Ivanov, G.K. and Kozhushner, M.A. (1982) *Khim. Fiz.*, **1**, 1039.
46. Onsager, L. (1938) *Phys. Rev.*, **54**, 554.
47. Smoluchowsky, M.V. (1917) *Z. phys. Chem.*, **92**, 129.
48. Mozumder, A. (1966) *Advances in Radiation Chemistry*, Vol. 1. Wiley, London.
49. Mozumder, A. (1968) *J. Chem. Phys.*, **48**, 1659.
50. Berlin, Yu.A. (1975) *Dokl. AN SSSR*, **223**, 625.
51. Berlin, Yu.A. (1975) *Dokl. AN SSSR*, **223**, 1387.
52. Fabrikant, I. and Kotomin, E. (1975) *J. Luminescence*, **9**, 502.
53. Pilling, M.J. and Rice, S.A. (1975) *J. Chem. Soc. Faraday Trans. II*, **71**, 1563.
54. Berlin, Yu.A. (1976) *Dokl. AN SSSR*, **228**, 628.
55. Kotomin, E. and Fabrikant, I. (1977) *J. Phys.*, **C11**, 4931.
56. Buttler, P.R. and Pilling, M.J. (1977) *Discuss. Faraday Soc.*, **63**, 38.
57. Khairutdinov, R.F. (1976) *Dokl. AN SSSR*, **228**, 149; (1976) *Khim. Vysok. Energii*, **10**, 556.
58. Pikaev, A.K. (1969) *Solvated Electrons in Radiation Chemistry*. (In Russian.) Nauka, Moscow.
59. Hart, E.J. and Anbar, M. (1979) *The Hydrated Electron*. Wiley, New York.
60. Yakovlev, B.S. (1979) *Usp. Khim.*, **48**, 1153.
61. Berlin, Yu.A., Kevdina, I.B. and Shantarovich, V.P. (1981) *Chem. Phys. Lett.*, **77**, 489.
62. Byakov, V.M. and Petukhov, V.R. (1984) *J. Rad. Nucl. Chem. Lett.*, **85**, 67.
63. Montroll, E.W. and Weiss, G.H. (1965) *J. Math. Phys.*, **6**, 167.

64. Ryazanov, G.V. (1972) *Teor. Mat. Fiz.*, **10**, 271.
65. Chekunaev, N.I., Berlin, Yu.A. and Fleurov, V.N. (1982) *J. Phys.*, **C15**, 1219.
66. Blumen, J. Klafter, Zumofen, G. (1986) In: *Optical Spectroscopy of Glasses*. (ed. Zschakke, F.), Reidel Pub. Co., pp. 199–265.

CHAPTER 10
Tunneling Effects in Biology

This chapter discusses biological phenomena in which tunneling effects play an important part.

Tunneling electron transfer is studied in the photosynthesis reaction center.

A possible role of the tunneling effect is examined in the transitions between various conformational substates of biopolymers and a glass-like model of proteins is formulated.

A discussion is given of tunneling chemical reactions in interstellar dust, resulting in the formation of complex organic molecules, which may later serve as a basic material for the early stages of biological evolution.

10.1. ELECTRON TUNNELING IN PHOTOSYNTHESIS

Different biochemical processes in which electrons or some heavier particles are transferred proceed permanently in biological systems. Various temperature regimes have been studied and in the low-temperature regime, in particular, when the rate constant of some processes may be determined by tunneling rather than activation. For example, a plateau at temperatures below 25 K [1] was discovered in a study of the temperature dependence of the rate constant of CO rebinding to hemoglobin — which indicates an essential role for tunneling at these temperatures. And although these temperatures are very far from being normal, information obtained in these experiments helps in giving a better understanding of the system under consideration and hence of its natural function.

A more surprising fact is that tunneling may proceed quite rapidly even at room temperature, and moreover may play a key role in the natural function of some biological systems. The most striking example of this is photosynthesis. Although the way in which photosynthetic reaction centers function is not yet understood in complete detail, it is now quite clear that photosynthesis would be impossible without tunneling.

The first observation of electron tunneling transfer in a biological system was made in [2–4], where an electron was shown to be transferred from cytochrome C to an oxidized form of the bacteriochlorophyll in the photosynthetic center of the *Chromatium D* bacterium at very low temperatures. The time of electron transfer $\tau(t)$ increases exponentially from 10^{-6} s to 10^{-3} s as temperature decreased from room temperature to 130 K, and then remains constant down to liquid-helium temperature. The initial interpretation of this result ascribed the plateau of the temperature dependence $\tau(T)$ to a tunneling transition, while the growth of $\tau(T)$ for $T > 130$ K was related to over-barrier transitions. However, the barrier width in this case ought to be extremely large $d \sim 70$ Å, whereas the height should, in contrast, be small $V_b \sim 0.14$ eV. This agrees poorly with our view of the structure of the reaction center, since more realistic parameters would be $d \sim 10$–20 Å and $V_b \sim 1$–2 eV. In this case the rate of electron tunneling transfer even at room temperature is many orders larger than the activation transfer rate. Then the question naturally arises as to the origin of the temperature dependence of the transfer time $\tau(T)$. This is explained in [5–8] by the fact that emission and absorption of phonons as well as variations of the frequency and the equilibrium positions of the oscillators occur during the electron tunneling transfer.

A semiclassical scheme of Hopfield [7] is now considered. Let the electron be transferred from molecule A to molecule B, with the energy of molecule A changing by

$$\Delta E_A = E_A + V'_A(\{R_i\}) - V_A(\{R_i\}), \tag{1.1}$$

where E_A is the electron binding energy in the molecule; $V_A(\{R_i\})$ and $V'_A(\{R_i\})$ are the potential energies of the nuclei in the molecule before and after the electron departure (see Fig. 10.1). The nuclear coordinates are assumed not to change during the transition time. Now the probability that initially the nuclei in molecule A have coordinates $\{R_i\}$ is written as

$$P_A(\{R_i\})\, dR_i = \frac{\exp[-V(\{R_i\})/k_BT]}{\int \exp[-V(\{R_i\})/k_BT]\, dR_i}\, dR_i. \tag{1.2}$$

Molecule B is characterized in the same way. The only difference is that the energy E_B enters the formula with the opposite sign.

Then the probability of electron tunneling from the molecule A to the molecule B takes the form

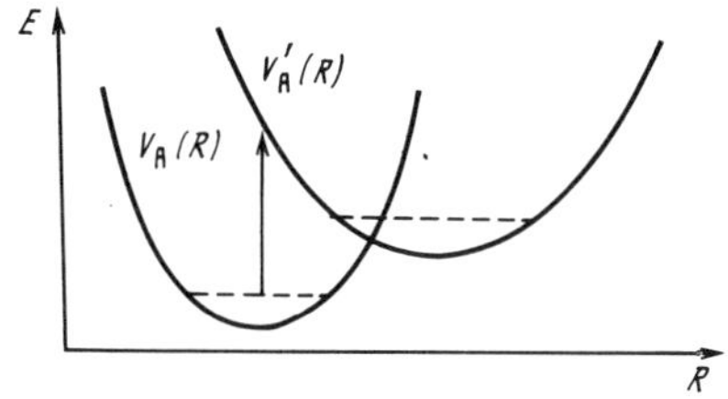

Figure 10.1 A scheme of the nonadiabatic transition in the molecule at the electron tunnel transfer.

$$W_{AB} = \frac{2\pi}{\hbar} |T_{AB}|^2 \int dR_A dR_B P_A(\{R_{iA}\}) P(\{R_{iB}\}) \delta(\Delta E_A + \Delta E_B), \quad (1.3)$$

where $T_{AB} \sim \hbar\nu \exp(-R_{AB}/a)$ is the transition matrix element. A harmonic approximation for the potential energies of the nuclei is assumed:

$$\left.\begin{aligned} V_A(R) &= \frac{1}{2} k_A R^2, \\ V'_A(R) &= \frac{1}{2} k_A (R - R_{0A})^2, \end{aligned}\right\} \quad (1.4)$$

where the elasticity coefficient k_A does not change during the electron transition and the energy minimum is displaced through a distance R_{0A}. Then

$$P_A(\Delta E_A)\, d\Delta E_A = \frac{1}{(4\pi k_B T \Delta_A)^{1/2}} \exp\left[-\frac{(\Delta E_A - E_A - \Delta_A)^2}{4 k_B T \Delta_A}\right] d\Delta E_A, \quad (1.5)$$

where $\Delta_A = \frac{1}{2} k_A R_{0A}^2$. A similar equation is obtained for molecule B (E_A being replaced by $-E_B$). On carrying out thc integration in (1.3), the probability of the tunneling transition is

$$W_{AB} = \frac{2\pi}{\hbar} |T_{AB}|^2 \frac{1}{(4 k_B T \Delta)^{1/2}} \exp\left[-\frac{(E_A - E_B - \Delta)^2}{4 k_B T \Delta}\right], \quad (1.6)$$

where

$$\Delta = \Delta_A + \Delta_B.$$

This formule, derived by Hopfield [7], has formed the basis for the interpretation of many experimental results [9–11] that have been obtained near the room temperature. At lower temperatures, when such an approach using semiclassical concepts is not valid, the results of a more rigorous quantum-mechanical analysis due to Jortner [8] should be applied. Here these rather involved calculations will not be dwelt upon in detail, especially since in many cases the above simplified semi-classical approach is enough to yield qualitative and sometimes quantitative understanding.

Many years of studies on the structure of bacterial photosynthetic reaction centers have produced a detailed knowledge of their chemical structure and scheme of energy processes (see e.g. [12–15]). For example, the total molecular weight of the reaction center of the *Rhodopseudomonas sphaeroides* bacterium is about 70000; it contains four bacteriochlorophylls (BChl) and two bacteriopheophytins (BPh), in which in contrast with BChl the central Mg atom is replaced by two protons. There are also cytochrome C, a divalent Fe atom and two quinone (Q) molecules in the reaction center. Two BChls absorb light with characteristic wavelength 800 nm. Two BPhs absorb light in the region > 500 nm, their absorption spectra being somewhat different from each other. One of the quinones (primary Q) seems to be more closely connected with the Fe atom than the other (secondary Q). As a rule, the reaction center contains two photochemical reaction systems acting in sequence and absorbing light with different wavelengths. The purpose of such a two-step system is as yet not quite clear, although there are some considerations that suggest that some advantages are gained from the point of view of natural selection (see e.g. [15]).

Figure 10.2 depicts an energy scheme for the processes in the reaction center. Only one photochemical system is now considered, for the sake

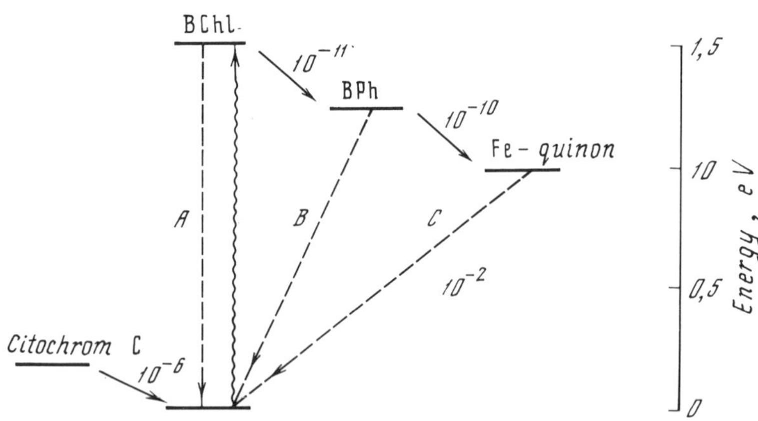

Figure 10.2 Energy scheme of the processes in the reaction center. Solid lines show the main transitions, dashed lines show the parasite ones, wavy line is for the absorbed phonon.

of simplicity. A quantum of light (~ 1.5 eV) excites BChl of the *Rhodopseudomonas sphaeroides* bacterium. After this, the electron transfers from the BChl to the BPh during a time of order 10^{-11} s, it then passes from the BPh to the Fe–quinone complex during 10^{-10} s and from cytochrome C to the reduced BChl during 10^{-6} s. This process results first in a separation of electric charges, so that a double layer forms on the cell membrane to which the reaction center is attached; secondly, the primary energy of the electronic excitation is lowered in the course of the transfer and transforms into chemical energy due to variations of the conformational states of the molecules comprising the reaction center. (The polaron effect in physics and the reorganization of the medium in solid-state chemistry are analogues of such a variation). Thus the energy of a light quantum is transformed into an energy form that can be assimilated by the ATP (adenosine triphosphate) molecules, which are the main energy carriers in the cells.

We do not here consider the photosynthetic processes in detail, especially since not all of them are understood at present (see e.g. [12–14]). Therefore only one question will be discussed: "Why does the tunneling effect play such an important role in photosynthesis?" This question has to be seen in the context of a living organism, for which natural selection acts as an important factor, since only those systems survive that are functionally adapted in the best way. Thus the above question can be reformulated as follows: "What kind of advantages does the tunneling effect give in photosynthesis?"

One of the advantages follows directly from the formula (1.6) for the probability of tunneling transition between two molecules. Such an electronic-nuclear transition is accompanied by a variation of the equilibrium positions of the nuclei. This is accounted for by the exponential factor (Franck Condon factor) in (1.6). Its magnitude depends very strongly on the relation between the energies $E_A - E_B$ and Δ, where the latter is the energy of conformation rearrangement (polaron effect). The probability is at maximum at $E_A - E_B = \Delta$ and decreases rapidly if these energies diverge. Owing to this, the electron and conformational energies may be fitted in such a way that the transformation of light energy into "biochemical" energy proceeds with minimal losses. For this, the Franck–Condon factors of the "parasitic" processes, shown by the dashed lines in Fig. 10.2, must be small.

Attention was first drawn to this fact in [5,6] (see also the book [12]). Another idea was proposed and elaborated in detail in [16]. It is quite possible that a nonlinear polaron corresponds to the conformational transformations to which the molecule is subjected to after the electron is transferred to it. Its electric field causes displacements of individual segments of the macromolecule, so that it goes to a metastable state that does not change for a relatively long time, even after the electron leaves and the electric field causing the conformational reconstruction disappears. This also gives significant advantages, since the possibility arises not only of transforming light energy to an energy form more convenient for biosystems, but also of keeping it during a time long enough for the relatively slow processes of biosynthesis.

Therefore the protein globule may be compared with a simple mechanical machine [16,17], which has one degree of freedom and two fixed equilibrium positions. A further development of this point of view on biomolecules is found in the [18], where a correlation of segmental mobility and tunneling electron transfer is studied in the photosynthetic reaction centers in the *Rhodospirillum rubrum* and *Rhodopseudomonas sphaeroides* bacteria.

10.2. TUNNELING TRANSITIONS BETWEEN QUASIDEGENERATE CONFORMATIONAL SUBSTATES OF BIOPOLYMERS

Not until Chance and de Vault [3,4] discovered a low-temperature plateau of the oxidation rate of cytochrome C by chlorophyll did it become clear that the tunneling effect plays an essential role not only in purely chemical reactions but also in some biological processes. Then only electron displacements were considered; nuclear shifts were studied only with regard to the electronic–nuclear transitions contributing to the Franck–Condon factors (see Section 10.1). However, after the discovery of the quantum low-temperature limit of the rate of a chemical reaction in (formaldehyde polymerization [19,20]), a plateau was reported in heavy-particle tunneling in biologically important reactions, i.e. in the Fe–ligand (CO, O_2) rebinding in heme-containing proteins [1,21].

This section examines heavy-particle tunneling, not in the chemical reactions of biopolymers, but rather in their dynamical properties, i.e. in the transitions between conformational substates or (more accurately) in the quasidegenerate conformational states (QDCS) of biopolymers. In considering this, the tunneling effects in the processes taking place in glasses should be recalled. (These are discussed in Chapter 6.)

As was emphasized in Chapter 6, the most important characteristic of glass is the existence of double-well potentials, giving rise to two-level systems as a direct consequence of not being in equilibrium. The macroscopic state of the glass is characterized by a nonzero entropy S, and may be implemented at zero temperature by means of a large number ($\sim e^S$) of microscopic states with different spatial atomic configurations, between which transitions are possible that proceed at low enough temperatures by means of tunneling.

The principal difference between the crystalline and amorphous states is as follows. A crystal is in equilibrium and is ordered, and its atoms at zero temperature (when $S = 0$ according to the Nernst theorem) are localized geometrically unambiguously,† i.e. the position of each atom is determined to within the zero-point amplitude ($\langle x^2 \rangle^{1/2} \sim 0.1$ Å, where $\langle x^2 \rangle$ is the mean-square atomic displacement). The glass is nonequilibrium and disordered ($S \neq 0$ at $T = 0$, since the Nernst theorem is inapplicable to nonequilibrium systems). So even at low temperatures at least some of its atoms are localized geometrically ambiguously and may occupy different spatial positions. That is why the root-mean-square atomic displacement $\langle x^2 \rangle^{1/2}$ may greatly exceed the zero-point amplitude (especially at high temperatures), since it is determined by the well-to-well spacing in the double-well potential.

Therefore the substance may be specified by the presence (+) or absence (−) of ordering, periodicity and geometrical unambiguity. Thus in this notation the ideal crystal (+ + +) and glass (− − −) are complete opposites. In contrast with simple materials, for which all three pluses (minuses) follow from their being at equilibrium (nonequilibrium), the situation in biopolymers is much more complicated, since a peculiar combination of crystal-like and glass-like properties is inherent to each individual biomolecule.

Over forty years ago, Schrödinger [22] characterized proteins as aperiodic crystals, treating their ordering as the principal factor leading to their functional activity. The aperiodic crystal may be viewed at as a quite definite rigid structure in which the spatial positions of atoms with respect to each other are fixed, this property being inherent to individual macromolecules rather than to the multimolecular system of a crystal. Only the absence of periodicity distinguishes it from an ordinary crystal, i.e. (+ − +). In the middle of the nineteen fifties Linström-Lang (see e.g. [23]) was the first to propose the idea that the dynamical properties of proteins are no less important for their functional activity than their ordering. In the course of time this assertion, which at first seemed dubious, became widely acknowledged and

† This assertion does not hold for *quantum* crystals (see Chapter 7), in which the zero-point amplitude is comparable to the interatomic spacing.

supported by a large number of experimental data obtained by such techniques as isotope (H–D) exchange, NMR and spin, luminescence and Mössbauer labeling [24].

There is as yet no complete understanding of the conformational mobility of proteins. As many authors have emphasized, one of the principal peculiarities of these motions (often hypothetical), corresponding to different conformational states of the protein globule, is their cooperative character, i.e. the mutual participation of a large number of atoms or of large segments of the macromolecule. In line with this, the assumption is made when describing the conformational mobility that the globule is in the ground state corresponding to the deepest absolute minimum in the multidimensional energy surface. All the other conformational states are related to excitations into higher ancillary minima (Fig. 10.3). So they are weakly populated short-lived fluctuations of the ground structure. Because of this, the conformational mobility does not lead to any revision of the notation (+ − +) for the biopolymer macromolecule.

A fundamentally new approach to the dynamics of proteins arose in connection with X-ray dynamical analysis (XRDA) [24,25], which allows determination of the mean-square displacements of individual nonhydrogen atoms in the protein globule. As a result, various structures are found within the same globule, for example "quasisolid" parts where the mean-square displacements do not exceed the ordinary vibrational amplitudes ($\langle x^2 \rangle \sim 0.01$–$0.04$ Å^2) and "liquid-like" parts where the atomic displacements are significantly larger. This allows consideration of conformational heterogeneity of biopolymers and quasidegenerate conformational states (QDCS).

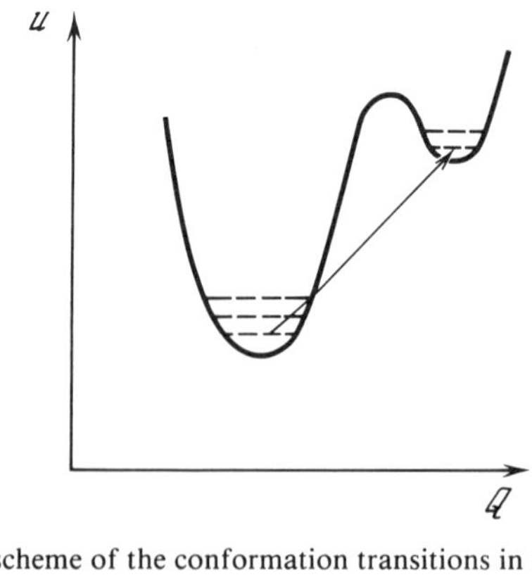

Figure 10.3 Energy scheme of the conformation transitions in the biomolecule according the aperiodic crystal concept.

QDCSs should be understood as different configurations of the same gross structure of a biomolecule, with equivalent or nearly equivalent energies. Their atomic configurations differ only slightly (for example owing to reversible rotations around σ-bonds, shifts of hydrogen bridges, fluctuational ruptures and restorations of the hydrogen bonds, which result in small displacements of large molecular groups with respect to the whole globule). The concept of QDCSs and transitions between them explains the cusp in the temperature dependence of the Mössbauer-effect probability and its rapid decrease at $T \geqslant 200$ K, which was first observed several years ago [26]. This concept was confirmed quantitatively in experiments on Mössbauer spectroscopy [27–29] and Releigh scattering of Mössbauer radiation [30,31] (see also the reviews [32,33]).

Atoms or groups like CH_2, NH are specific for different QDCSs. They are capable of occupying two or more positions with nearly equivalent energies. Hence in the presence of QDCSs the atomic configuration of the biopolymer macromolecule is geometrically ambiguous, and it should therefore be denoted as (+ − −), which is closer to a glass than to an ideal crystal. Such a combination of glass-like and crystal-like properties inherent to an individual molecule is connected with the fact that such a molecule must perform a certain biological function. Each biomolecule is an informational system, which is why it must be ordered. This point was emphasized by Schrödinger [22] in defining a biomolecule as an aperiodic crystal. On the other hand, it must be nonequilibrium, i.e. it must have nonzero entropy, this property remaining even at zero temperature. Thus it must have glass-like properties. In other words, the basic biomolecule does not occupy the absolute energy minimum but is in an ancillary minimum, i.e. its state is metastable, and the above large-scale conformational transitions correspond to excitation of the molecule to even higher ancillary minima (Fig. 10.4). According to quantum-mechanical principles, the ground state of the system is necessarily non-degenerate, while the excited states are as a rule degenerate. The degeneracy of an ancillary minimum means that it is possible to find several such minima with very close energies (i.e. several QDCSs) in the multidimensional energy surface.

A consistent analysis of the QDCS pattern leads to the inevitable conclusion that not only are the structural properties of biopolymers and glasses similar, but there are also many similarities in their dynamical properties. In particular, the transitions between different QDCSs may proceed by means of tunneling. This is why it should be expected that various effects specific for glasses will be observed in biopolymers.

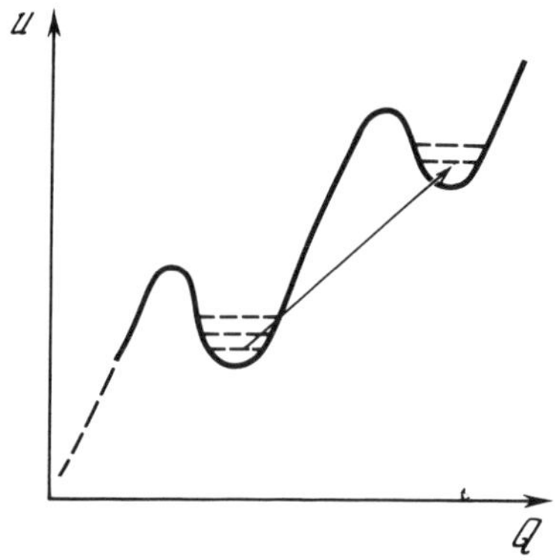

Figure 10.4 Energy scheme of the conformation transition in the biomolecule from the point of view of the glass-like model.

This idea was first put forward in [34], where an analysis of experimental data on the low-temperature specific heat for a number of polymers and biopolymers [35–39] was also carried out. These data appear to obey the linear temperature law characteristic of glasses (see Section 6.3). It is interesting to note that the density of QDCSs (two-level systems) n(0) increases with increasing complexity of the molecules. Thus for collagen [35] and DNA [39] it is two orders of magnitude greater than for simple glasses. Measurements of the specific heat and dielectric constant in metmyoglobin down to 0.1 K show that these quantities behave as in ordinary glasses [40].

Therefore, according to the glass-like model [34], biopolymers and proteins should be treated as a peculiar and interesting class of dielectric glasses where each individual macromolecule is a heterogeneous glass with regions of higher and lower QDCS densities, and hence with lower and higher vitrification temperatures. It is clear that these regions correspond to the "liquid-like" and "quasisolid" regions observed by the XRDA technique at room temperature. However it should not be thought that at such temperatures the transitions are necessarily over-barrier. The results of Chapter 6 show that the tunneling effect in ordinary glasses and possibly in biopolymers plays an essential role even at temperatures where the basic function of the protein is performed.

10.3. COLD PREBIOTIC EVOLUTION

According to the modern concepts of the appearance of life on the

Earth (see, e.g. [15]), the stage of prebiotic evolution comprises the following steps:

(i) the creation of compounds of various complexity up to protein monomers and nucleic acids;
(ii) polymerization (polycondensation) of the monomers in protenoid or polynucleotide chains;
(iii) compartmentalization (i.e. formation of closed multimolecular systems in solution) of the coacervates (according to Oparin [41]) or protenoid microspheres (according to Fox [42,43]), which are the ancestors of the cells;
(iv) catalytic interaction of the polynucleotide and polypeptide systems, the appearance of the genetic code and passover to biological evolution.

At each step, the internal organization of the systems is more and more complicated, i.e. an "antithermodynamical" evolution occurs that is accompanied by a decrease in the entropy of these systems. Such processes are possible since the systems considered are open in the sense that they permanently exchange energy and matter with their environment. That is why the self-organization does not contradict the second law of thermodynamics.

Therefore, we can believe that prebiotic evolution is allowed by the laws of physics, and we may try to develop a theory of the appearance of life on the Earth without appealing to the Creator. However, we still face many unresolved problems when trying to develop a consistent theory of prebiotic and biological evolution, especially the starting steps. In particular, the specific mechanisms of self-organization processes must be studied, and the mechanisms of self-organization processes, and we must be sure that they really do have time to be completed during the ages of the Earth and Universe. For example, if it assumed that the first two steps took place under the conditions of the Earth's atmosphere at biological temperatures (~ 410 K) then a number of difficulties arise that are mainly relevant to the second step, polymerization. Under such conditions it may require a time that exceeds the age of the Earth for its completion.

This is the reason why the possibility of cold prebiotic evolution in interstellar space, where any exothermic reaction with the formation of very complicated products is allowed, is of special interest. This kind of process becomes thermodynamically profitable, since any entropy restrictions are lifted as $T \rightarrow 0$. As a result, complicated compounds storing "negative entropy" are formed. These might have significantly accelerated prebiotic and biotic evolution on the Earth.

However, the idea of the cold prebiotic evolution would have encountered insuperable difficulties if Arrhenius' law had been valid down to the lowest temperatures. The processes of "negative-entropy storage" would have proceeded extremely slowly under the conditions in deep space cold, and a sufficient amount of the complicated molecules could have been accumulated only during a period exceeding the age of the Earth. This problem is solved by the existence of low-temperature tunneling exothermic reactions (e.g. formaldehyde polymerization [19,20]), which allows consideration of the possibility of tunneling cold prebiotic evolution. The finite rate of the tunneling reactions as $T \to 0$ encourages belief that the time needed for "negative-entropy storage" is relatively small [44], and cold prebiotic evolution proceeds sufficiently fast.

According to present knowledge of interstellar chemistry (see e.g. [45–47]), the most widely spread element in the Universe is hydrogen, the abundance of helium relative to hydrogen is 0.09, that of oxygen 7×10^{-4}, carbon 3×10^{-4}, nitrogen 9×10^{-5}, magnesium, silicon, iron (totaling $\sim 10^{-4}$). Among nearly 50 interstellar molecules discovered by radio and optical spectrometry, the most abundant (apart from H_2) are CO ($\sim 10^{-4}$), NH_3, HCN and HNC ($\sim 10^{-6}$). There is no doubt that interstellar space is well populated by water and methane. Formaldehyde is also to be found ($\sim 10^{-8}$); its low-temperature polymerization has been directly observed under laboratory conditions.

The main locations for interstellar chemistry are the interstellar clouds — gigantic (10^{17}–5×10^{19} cm) regions of dust grains and gas. The total mass of the interstellar dust is ~ 0.01 of the gas mass; it is assumed that the grains of dust have central "nuclei" ($r \sim 10^{-6}$ cm) made of silica or graphite surrounded by "dirty ice" ($r \sim 10^{-5}$ cm) — a mixture of the frozen substances mentioned above. There are 10^{12} hydrogen atoms per grain in the cloud.

Diffuse ($n_H \sim 10$–10^3 cm^{-3}) and dense or dark ($n_H \sim 10^3$–10^7 cm^{-3}) clouds should be distinguished. The diffuse clouds are transparant to ultraviolet radiation, which causes ionization of atoms and dissotiation of molecules. The dense clouds are not transparant in the ultraviolet. Photodissotiation processes are therefore absent in the depths of the dense clouds; here an appreciably wider assortment of complicated multiatomic molecules is present (molecules with up to $m = 11$ atoms are found, cyanotatetrayne, $H(C{\equiv}C)_4({\equiv}N)$. This is why the dense clouds are sometimes called "molecular clouds". The temperature of the interstellar dust grains in these clouds is 10–20 K.

The ultraviolet irradiation that penetrates the outer layers of the dark clouds (the current flow $I_{UV} \sim 3 \times 10^8$ cm^{-2} s^{-1} and the interaction cross-section for complicated molecules $\sigma_{UV} \sim 10^{-18}$ cm^2) interacts with each molecule once per $\tau_{UV} \sim 100$ years on the average. The only agents that are capable of inducing ionization or excitation of the molecules in the depths of the dark clouds are cosmic protons with a long free path. The characteristic time of their interaction with an individual molecule in the dust grain is $\sim 3 \times 10^9$ years. This time exceeds by far not only τ_{UV} but also the lifetime of the clouds $\tau_{cl} \sim 10^5$–10^7 years, which is determined by their collisions with each other or by the gravitational collapse.

There are three types of chemical reactions studied in interstellar chemistry: in the gaseous phase; on the surface of the dust grains; and within the bulk of the "dirty ice". Tunneling polymerization is most likely to proceed directly in the interstellar dust grains at T = 10–20 K within the bulk of the "dirty ice" or on its surface. The long-free-path cosmic protons may initiate the polymerization chains even in the depths of the dense clouds. Polymerization reactions have special advantages, since the radiation-chemical yield of these reactions is large. For example, in the case of formaldehyde it is nearly 10^4 reacting molecules per 100 eV of irradiation energy [19,20]. According to the estimates carried out in [48], during a time of 3×10^5 years the cosmic protons deposit enough energy in a grain to polymerize it completely were it made of formaldehyde.

Mixed content of the dirty ice can hinder the polymerization, for example chain growth can stop if the active end group does not find a monomer molecule to react with. On the surface the formation of a new link of the growing chain can be preceded by several ineffectual contacts of the active end group with other molecules attacking the surface from the gaseous phase or diffusing over the surface.

In spite of all the hindering factors, tunneling polymerization of formaldehyde in interstellar dust grains appears to be possible (see the estimates in [48]). There is also direct experimental evidence of existence of complicated polymers in space. Thus Wickramsinghe and Hoyle, on the basis of a comparison of galactic and laboratory infrared spectra, came to the conclusion that the formation of formaldehyde–polyoxymethylene polymers [48,49] and even polysaccharides [50–52] is possible in interstellar space.

The hypothesis of cold prebiotic evolution that is made reasonable by the tunneling effect helps to overcome some difficulties that are encountered in the problem of the appearance of life on the Earth. The dictum can sometimes be heard that life is the struggle against entropy. It is quite possible that in early prebiotic stages, the tunneling effect was the main weapon in this struggle.

References

1. Alberding, N., Austin, R.H., Beeson, K.W., Chan, S.S., Eisenstein, L., Frauenfelder, H. and Nordlung, T.M. (1976) *Science*, **192**, 1002.
2. Chance, B. and Nishimura, M. (1960) *Proc. Natl. Acad. Sci. USA*, **46**, 19.
3. De Vault, D. and Chance, B. (1966) *Biophys. J.*, **6**, 825.
4. De Vault, D., Parks, I.H. and Chance, B. (1967) *Nature*, **216**, 642.
5. Grigorov, L.N. and Chernavskii, D.S. (1972) *Biofizika*, **17**, 195.
6. Blumenfeld, L.A. and Chernavskii, D.S. (1973) *J. Theor. Biol.*, **39**, 1.
7. Hopfield, J.J. (1974) *Proc. Natl. Acad. Sci. USA*, **71**, 3640.
8. Jortner, J. (1976) *J. Chem. Phys.*, **64**, 4860.
9. Potasek, M.J. and Hopfield, J.J. (1977) *Proc. Natl. Acad. Sci. USA*, **74**, 229.
10. Austin, R.H., Hopfield, J.J. (1979) In: *Electron Transport and Oxygen Utilization* (ed. Chien Ho), p. 324. Elsevier North-Holland, New York.
11. Redi, M. and Hopfield, J.J. (1982) *J. Chem. Phys.*, **11**, 5272.
12. Chernavskaya, N.M. and Chernavskii, D.S. (1977) *Tunnel Transport of Electrons in Photosynthesis*. (In Russian.) Izdatelstvo MGU, Moscow.
13. (1979) Tunneling in Biological Systems (ed. B. Chance *et al.*), p. 215. Academic Press, New York.
14. De Vault, D. (1980) *Q. Rev. Biophys.*, **13**, 387.
15. Volkenshtein, M.V. (1978) *General Biophysics*. (In Russian.) Nauka, Moscow.
16. Khurgin, Yu.I., Chernavskii, D.S. and Shnol, S.E. (1967) In: *Vibrational Processes in Biological and Chemical Systems*, p. 42. (In Russian.) Nauka, Moscow.
17. Blumenfeld, L.A. (1978) *Q. Rev. Biophys.*, **11**, 251.
18. Chernavskii, D.S., Frolov, E.N., Goldanskii, V.I. and Rubin, A.B. (1980) *Proc. Natl. Acad. Sci. USA*, **70**, 7218.
19. Goldanskii, V.I., Frank-Kamenetskii, M.D. and Barkalov, I.M. *Dokl. AN SSSR*, **211**, 133.
20. Goldanskii, V.I., Frank-Kamenetskii, M.D. and Barkalov, I.M. (1973) *Science*, **182**, 1344.
21. Peters, K., Applebury, M.L. and Rentzepis, P.M. (1977) *Proc. Natl. Acad. Sci. USA*, **74**, 3119.
22. Schrödinger, E. (1942) *What is Life?* Dublin University.
23. (1977) Equilibrium Dynamics of the Protein Basic Structure, p. 156. (In Russian.) *Nauch. tsentr. biol. issled.*, Pushchino.
24. Frauenfelder, H., Petsko, G.A. and Tsernoglou, D. (1980) *Nature*, **280**, 563.
25. Artymiuk, P.J., Blake, C.C., Grace, D.E.P., Oatley, S.J., Phillips, D.C. and Sternberg, M.J.E. (1980) *Nature*, **280**, 563.
26. Frolov, E.N., Mokrushin, A.D., Likhtenshtein, G.N., Trukhtanov, V.A. and Goldanskii, V.I. (1973) *Dokl. AN SSSR*, **215**, 165.
27. Parak, F., Frolov, E.N., Mössbauer, R.L. and Goldanskii, V.I. (1981) *J. Mol. Biol.*, **145**, 825.
28. Bauminger, E.R., Cohen, S.C., Nowik, X., Ofer, S., Yariv, J. (1983) *Proc. Natl. Acad. Sci. USA*, **80**, 736.
29. Keller, H. and Debrunner, P.G. (1980) *Phys. Rev. Lett.*, **45**, 68.
30. Krupyanskii, Yu.F., Parak, F., Gaubman, E.E., Wagner, F.M., Goldanskii, V.I., Mössbauer, R.L., Suzdalev, I.P., Litterst, F.L. and Vogel, H. (1980) *J. de Physique*, **41**, C1, 489.

31. Krupyanskii, Yu.F., Parak, F., Goldanskii, V.I., Mössbauer, R.L., Gaubman, E.E., Engelman, H. and Suzdalev, I.P. (1982) *Z. Naturforsch.*, **37C**, 57.
32. Debrunner, P.G. and Frauenfelder, H. (1982) *Annu. Rev. Phys. Chem.*, **33**, 283.
33. Goldanskii, V.I., Krupyanskii, Yu.F. and Frolov, E.N. (1973) *Molekular. Biologiya*, **17**, 532.
34. Goldanskii, V.I., Krupyanskii, Yu.F. and Fleurov, V.N. (1983) *Dokl. AN SSSR*, **272**, 978.
35. Mrevlishvili, G.M. (1979) *Usp. Fiz. Nauk*, **128**, 273.
36. Misutani, U. and Massalskii, T.B. (1976) *Nature*, **259**, 505.
37. Finegold, L. and Cude, J.L. (1972) *Nature*, **238**, 33.
38. Fanconi, B. and Finegold, L. (1975) *Science*, **190**, 458.
39. Verkin, B.I., Sukharevskii, B.Ya., Telezhenko, Yu.V., Alapina, A.V., Vorob'eva, N. Yu, (1977) Fiz. Nizk. Temp. **3**, 252.
40. Singh, G.P., Schink, H.J., von Löhneysen, H., Parak, F. and Hunklinger, S. (1984) *Z. Phys.*, **B55**, 23.
41. Oparin, A.I. (1957) The Origin of Life on Earth. *Izdatelstvo AN SSSR*, Moscow.
42. (1965) The Origins of Prebiological Systems and of their Molecular Matrices (ed. S.W. Fox). Academic Press, New York.
43. Fox, S. (1969) *Naturwissenschaften*, **56**, 1.
44. Goldanskii, V.I. (1979) *Nature*, **279**, 109.
45. Gammon, R.H. (1978) *Chem. Engng. News*, **56**, 21.
46. Watson, W.D. (1977) *Acc. Chem. Res.*, **10**, 221.
47. Watson, W.D. (1976) *Rev. Mod. Phys.*, **48**, 513.
48. Wickramasinghe, N.C. (1974) *Nature*, **252**, 462.
49. Wickramasinghe, N.C. (1975) *Mon. Not. R. Astron. Soc.*, **170**, 11.
50. Hoyle, F. and Wickramasinghe, N.C. (1977) *Nature*, **268**, 610.
51. Hoyle, F. and Wickramasinghe, N.C. (1975) *Mon. Not. R. Astron. Soc.*, **181**, 51.
52. Wickramasinghe, N.C., Hoyle, F., Brooks, J. and Shaw, G. (1977) *Nature*, **269**, 674.

Index